T0200986

The PDMA ToolBook 3 for New Product Development

The PDMA ToolBook 3 for New Product Development

Edited by

Abbie Griffin
University of Utah

Stephen Somermeyer
Somermeyer & Associates

1807
WILEY
2007

John Wiley & Sons, Inc.

For general information on our other products and services please contact our Customer Care Department within the U.S. at 877-762-2974, outside the U.S. at 317-572-3993 or fax 317-572-4002.

Wiley also publishes its books in a variety of electronic formats. Some content that appears in print, however, may not be available in electronic books. For more information about Wiley products, visit our Web site at www.wiley.com.

Library of Congress Cataloging-in-Publication Data:

The PDMA toolbook 3 for new product development / edited by Abbie Griffin, Stephen M. Somermeyer.
    p. cm.
  Includes index.
  ISBN-13: 978-0-470-08923-1 (cloth)
  1. New products—Management 2. Product management I. Griffin, Abbie. II. Somermeyer, Stephen M.
  HF5415.153.P3549 2007
  658.5′75–dc22
                                                                2007010261

Printed in the United States of America.

10 9 8 7 6 5 4 3 2 1

This book is dedicated to Paul Belliveau: friend, colleague and former ToolBook co-editor. We hope you will come back on board in the future.

Contents

Appendices

I

The PDMA's Body of Knowledge

Gerry Katz

II

The PDMA Glossary for New Product Development

Contributors

EDITORS:

Abbie Griffin

Abbie Griffin holds the Royal L. Garff Presidential Chair in Marketing at the David Eccles School of Business at the University of Utah. She is a member of the Board of Directors of Navistar International, a Fortune 500 manufacturer of diesel engines and trucks, and was the editor of the *Journal of Product Innovation Management*, the leading academic journal in the areas of product and technology development, from 1998–2003.

Stephen Somermeyer

Stephen Somermeyer is Principal of Somermeyer & Associates, Assistant Wine-Maker at Chateau Thomas Winery in Indiana, and Head of Safety for the garages and pits at the Indianapolis Motor Speedway. He recently helped start-up YourEncore, a temporary resource agency utilitizing retired, technical talent. He spent over 30 years in the pharmaceutical industry at Eli Lilly and Company where he focused on large-scale change projects in R&D. He co-founded an industry R&D benchmarking group and spent significant time identifying best-practices at leading companies and then implementing them at Lilly. He led or was a member of teams that implemented structures and processes such as portfolio management, heavyweight teams and TQM. He consults in the pharma industry and frequently delivers presentations on best practices in benchmarking, metrics, organizational structures and NPD strategy.

Steve is active in the Product Development & Management Association (PDMA) having served as Board member and on numerous committees. He is Co-editor of *The PDMA ToolBooks 1 and 2 for New Product Development* (Wiley) and Technical Editor for *Product Development for Dummies* (Wiley). He has a degree in chemical engineering and a MBA and can be reached at *steve@somermeyer.com.*

CHAPTER AUTHORS

Gunter R. Ladewig (Chapter 1)

Gunter R. Ladewig is president of PRIMA Performance Ltd., a consulting company that specializes in product and process renaissance using TRIZ, Six Sigma, and Lean manufacturing techniques. PRIMA provides fast-track customer teaching tools with *One Look TRIZ* and tough quality solving techniques like *KISS*, Keeping It Statistically Simple. Gunter is an expert in operational efficiency improvement and world-class product design. In 1992, he was winner of IBM's Innovation Invitational, and in addition, was twice on IBM's winning team of the *Government of Canada Award for Business Excellence:* The Gold Award for Productivity, and the Gold Award for Quality.

Gerry Katz (Chapter 2)

Gerry Katz is a recognized authority in the areas of new product development, design of new services, and market research, with nearly 35 years of consulting experience. At Applied Marketing Science, Inc., he has led more than 100 major client engagements employing The Voice of the Customer, quality function deployment (QFD), and a large number of other marketing science applications. He serves on the board of directors of the Product Development & Management Association (PDMA), where he is certified as a new product development professional (NPDP) and is a contributing editor to *Visions* magazine. He is the author of several award-winning papers and received the William O'Dell Prize from the American Marketing Association in 1988. His articles have appeared in *The Journal of Product Innovation Management*, *The European Management Journal*, *The Journal of Marketing Research*, and *Interfaces*. He has lectured frequently at the business schools of MIT (Sloan), the University of Pennsylvania (Wharton), Dartmouth College (Tuck), Carnegie-Mellon University, and Harvard University. He has appeared twice on the *NBC Today Show* and in *The Wall Street Journal*. Mr. Katz received his bachelor of science degree in Management Science from the University of Rochester in 1970 and his master of science degree in management from the MIT Sloan School of Management in 1972.

Nelson Whipple (Chapter 3)

Nelson Whipple is director of RSG Market-Driven Solutions. Nelson holds a BA in Behavioral Sciences from the University of Chicago and has more than two decades of market and product research experience in a variety of consumer and B2B markets. He has served clients in such diverse industries as transportation, chemicals, financial services, professional services, food products, computer hardware, software services, and consumer durables. Nelson was nominated by ESOMAR for the 2005 International Research Paper of the Year for his analysis of the value of brand in consumer mobile handset markets across nine countries, a paper that was published in *Excellence in International Research 2005*. He also co-authored "Tapping the Power of User Trials—The Story of Motorola's Web-Based TRYMEMOTO User Feedback System," presented at the 2005 PDMA International Conference.

Thomas Adler (Chapter 3)

Thomas Adler is president of Resource Systems Group. Tom holds a PhD in Transportation and Management from MIT and has over 30 years' experience directing major research projects. He has authored more than 50 published papers on methods and applications, including the first U.S. government manual on conjoint methods. He has pioneered the application of discrete choice methods to practical product marketing problems, such as the influence of customer inertia in product switching, product line cannibalization, and pricing. Prior to founding RSG, Tom taught market research and statistics at Dartmouth College. Among his current activities, Tom serves as an advisor to the National Academy on panel survey research and spends about 70 percent his time directly involved in client engagements. Tom's product development experience includes work with mobile phones, airline in-flight offerings, and apparel.

Stephan McCurdy (Chapter 3)
Stephan McCurdy is director of RSG Market-Driven Solutions. Steve holds an MBA from Kellogg School of Management, with concentrations in marketing, analytical consulting, and management and strategy. He has 15 years of experience developing business solutions from consumer and B2B research, including market-based strategies for product-line upgrades, service bundling, new technology adoption and migration, and customer retention. Steve led a major initiative to define a corporate value proposition for a division of a Fortune 500 company that would guide all product, marketing, and service activities. He also designed and manages a panel to track the division's value proposition within the highest volume segment of the market. Product development experience ranges from industrial equipment to financial services.

Anne Orban (Chapter 4)
Anne Orban is director of discovery and innovation at Innovation Focus. Anne combines her experience in process design and delivery with a commitment to helping clients grow by releasing the creative potential of their people. Anne has a master's degree in adult and organizational development and is a certified new product development professional. She served as co-chair of the 2006 PDMA Annual Conference and also as Annual Conference workshop chair, 2004 to the present.

Christopher W. Miller (Chapter 4)
Christopher W. Miller is founder and chief executive officer of Innovation Focus. Chris started Innovation Focus in 1987 to help enterprises grow through meaningful innovation. He is a past-president of the Product Development & Management Association and has been acknowledged for his work on the Focused Innovation Technique™ and Hunting for Hunting Grounds™.

Larry Marine (Chapter 5)
Larry Marine is a highly regarded user experience and product design professional. He consults to companies of all sizes, from start-ups to Fortune 100 clients, aligning their business objectives to their users' needs. Larry is the founder of Intuitive Design, a consulting group specializing in creating product designs that synthesize business needs with the high-value tasks users must complete. He started Intuitive Design nearly 20 years ago after earning a degree in cognitive science at UC San Diego. Prior to this, he was a computer systems

specialist for 15 years. Larry conducts user observations in the design process to help clients understand what products must do. The resulting designs are innovative and highly successful, often resulting in products that dominate their market. His repeat clients include Cardinal Health, Ericsson, FedEx Kinko's, American Airlines, GTE, Vanguard Mutual Funds, Fannie Mae, and many more. Larry and his team at Intuitive Design also provide seminars to improve organizations' design processes. After spending several years in the hustle and bustle of San Diego, Larry now lives with his wife, their two girls, and one great dog in the clean air and quiet surroundings of Colorado Springs, Colorado. You can reach him at *Larry@IntuitiveDesign.com*.

Chad A. McAllister (Chapter 5)

Chad McAllister seeks to bridge the gap between technology solutions, users' expectations, and business objectives. He is a senior solutions consultant with LexisNexis, a seminar presenter for Intuitive Design, and an adjunct professor teaching information systems-related courses. Chad has nearly 20 years of experience working directly with customers to develop, integrate, and deploy software systems that meet business objectives and exceed customers' expectations. His roles have encompassed software development, business development, sales, marketing, and professional services. He is a member of the Product Development & Management Association and the Academy of Management. He holds degrees in electrical engineering and a PhD in organization and management. He consults with organizations on the topics of requirements engineering, knowledge management, and product development. Further, he enjoys providing training to organizations that want to improve their success in developing technology products. Chad, his wife, and two children live in the beautiful hills of Monument, Colorado—a venue that receives more than its fair share of sunshine, snow, and lightning—sometimes all in the same day. He can be reached at *chad@ckmcallister.com*.

Peter A. Koen (Chapter 6)

Peter Koen is an associate professor at Stevens Institute of Technology and director of the Consortium for Corporate Entrepreneurship, whose mission is to stimulate highly profitable activities in the Front End of Innovation (*www.frontendinnovation.com*). Peter's research is directed at determining how large companies organize around breakthroughs, in knowledge creation and best practices in the discovery part of the innovation process and how corporate entrepreneurs obtain start-up funds for ideas that are out of sequence with the normal funding cycle. He has 19 years of industrial experience. His academic background includes a BS and MS in mechanical engineering from New York University and a PhD in biomedical engineering from Drexel University.

Thomas C. Holcombe (Chapter 6)

Thomas C. Holcombe is founder and president of THolcombe LLC, a business development consulting company specializing in catalyst and chemical businesses. Over a 35-year career, Tom has run large and small global chemical businesses and founded two companies. He has developed and commercialized many new technologies in the chemical, petrochemical, petroleum refining, and environmental industries. He is an inventor in 15 U.S. patents and is an author of numerous articles in the field. As a new ventures manager at Engelhard (now BASF), he introduced the *market attack team* methodology to the company and led the personal care market attack team cited in this chapter. He holds a BS in chemical engineering from Virginia Polytechnic Institute and State University, an MBA and DPS (doctor of professional studies) in international business from Pace University (New York City), and a professional engineer's license from New York and New Jersey.

Christine A. Gehres (Chapter 6)

Christine A. Gehres leads BASF's global Effect Pigments business in the Cosmetic and Personal Care markets. Christine has almost 20 years of international experience in Speciality Chemicals, mostly in business and technology management and global marketing. She has sponsored and led the commercialization of many new technologies in the plastics, coatings, agriculture and cosmetics industries. As VP Marketing at Engelhard (now BASF), she facilitated the introduction of market attack team methodology to the company, and her division. Christine co-led the "Personal care" market attack team cited in this chapter. She holds a B.A. (Hons) in European Business Administration from Middlesex University, a Diplom Betriebswirt (FH) from Reutlingen and an MBA from the University of Rhode Island.

Brian D. Ottum (Chapter 7)

Brian D. Ottum is president of Ottum Research & Consulting, in Saline, Michigan. His practice is focused on the market research tools used during the early stages of new product development. Ottum Research & Consulting serves clients in the packaged goods, durables, and high-technology industries. Prior to starting his own firm, Brian worked in new product development and international market research for Procter and Gamble. He holds a PhD from the University of Utah, an MBA from Xavier University and a BS in chemical engineering from the University of Wisconsin. Over the past fifteen years, Brian has developed state-of-the-art research methods in consumer needs identification (using KJ analysis), concept screening & optimization (using instantaneous keypad studies), and price-setting (using discrete choice modeling). Brian is co-founder of the Great Lakes Chapter of the Product Development & Management Association (PDMA). He teaches "Top 10 Tools" and "Conjoint Analysis" workshops at PDMA conferences. He wrote the market research questions for the PDMA's New Product Development Professional Certification test. Brian is frequent speaker on innovation topics at national conferences and at company sites. He was selected for Marquis' Who's Who in America, 54th edition, 2000. Contact information: *ottum@comcast.net* (734) 429-8215 *http://mywebpages.comcast.net/ottum/*

Leland D. Shaeffer (Chapter 8)

Leland D. Shaeffer is managing director of PLM Associates, a consulting firm that helps companies improve product lifecycle management processes, together with executing specific aspects of the process such as customer research and product/market strategy. He has held senior positions in engineering and product management at companies ranging from venture capital–backed start-ups to Apple Computer, Unisys, and Imagery/Eastman Kodak. He started his career at McKinsey & Company. Mr. Shaeffer is founding president of the PDMA-Los Angeles Chapter, and he is active in the Technology Council of Southern California and the MIT/Caltech Enterprise Forum. He has instructed executive courses at the California Institute of Technology Industrial Relations Center and the USC Marshall School of Business, and he is a frequent speaker on topics product development and management topics. He holds a BS in electrical engineering from the Massachusetts Institute of Technology and a MBA from Stanford University.

James S. Twerdahl (Chapter 8)

James S. Twerdahl is managing director of James S. Twerdahl & Associates, a marketing and general management consulting firm. Jim gained much of his experience in consumer electronics, having been the president or CEO of Jensen Sound Laboratories (car audio and home loudspeakers), the JBL Incorporated division of Harman International (home and professional audio products), and Marantz Company (audio and video components). He also was chairman of Mayco Colors, Inc., a manufacturer of products for the craft, art, school, and hobby markets. He taught brand and product management in the graduate business school at Loyola Marymount University. He is a graduate of Trinity College in Hartford, Connecticut, and has an MBA in marketing from the Kellogg School of Management at Northwestern University. Mr. Twerdahl has held leadership positions in a number of industry, civic, and charitable organizations, including six years as chairman of the board of the House Ear Institute.

Kenneth B. Kahn (Chapter 9)

Kenneth B. Kahn is an associate professor of marketing and a Stokely Scholar in the Department of Marketing and Logistics at the University of Tennessee. Ken has a PhD from BIE, Georgia Institute of Technology; MSIE, Virginia Polytechnic Institute and State University; and PhD in marketing, Virginia Polytechnic Institute and State University. His teaching and research interests concern product development, product management, new product forecasting, and interdepartmental integration. He has published in a variety of journals, including the *Journal of Product Innovation Management*, *Journal of Business Research*, *Journal of Forecasting*, *Journal of Business Forecasting*, *Marketing Management*, and *R&D Management*. He is the author of the books *Product Planning Essentials* (Sage Publications, 2000) and *New Product Forecasting: An Applied Approach* (M. E. Sharpe, 2006), and is editor of the *PDMA Handbook on New Product Development*, 2nd ed. (Wiley & Sons, 2004).

Sharad Rastogi (Chapter 10)

Sharad Rastogi is a principal based in the Waltham, Massachusetts, office of PRTM Management Consultants. He has advised several companies on IP strategies and the implementation of structured business processes for managing intellectual property. Additionally, he has significant experience in market and product strategic planning; innovation and product development management; portfolio, project and resource management; and codevelopment and alliance management. He currently focuses on the life sciences industry, but also has experience in other industries including computers and peripherals, software, communications, and consumer electronics. Sharad has a bachelor of technology degree from the Indian Institute of Technology, Kanpur, an MS from the Ohio State University at Columbus, Ohio, and an MBA from the Wharton School of the University of Pennsylvania. He can be reached at 781-434-1273 and *srastogi@prtm.com*.

Aritomo Shinozaki (Chapter 10)

Aritomo Shinozaki is a principal based in the Mountain View, CA office of PRTM Management Consultants. He has significant experience in core practice areas, including business strategy, innovation strategy, technology and IP management, product management and development, portfolio management, product strategy, service strategy, and IT operations. His industry experience includes software, computer hardware, networking solutions, and life sciences. His technical background includes business application development, high-performance computing and networking technologies, and object-oriented software development. Ari is a graduate of Princeton University with a BA in physics. He attended the University of Illinois at Urbana-Champaign and graduated with a PhD in physics. He also received an MBA with Distinction from the Johnson Graduate School of Management at Cornell University. He can be reached at 650-864-3582 and *ashinozaki@prtm.com*.

Matthew Kaness (Chapter 10)

Matthew Kaness is a manager based in the New York office of PRTM Management Consultants. Matt has significant experience in working with technology-based and consumer products companies in various industries including food and beverage, sporting goods, footwear and apparel, consumer packaged goods, retail and specialty chemicals. His areas of expertise include new product development and introduction, business and product line planning, intellectual capital strategy and asset management, overhead optimization, and strategic sourcing and offshore contract-manufacturer negotiations. Matt holds a BS in mechanical engineering from the Catholic University of America and an MBA from the Darden Graduate School of Business at the University of Virginia. Matt can be reached at 203-905-5616 and *mkaness@prtm.com*.

Douglas Neff (Chapter 11)

Douglas Neff is a teacher, development coach, and writer. He earned his MA in consciousness studies from JFK University and his BA in music and religious studies from Ithaca College. He is currently the information design manager for Nano-Tex, Inc., an advanced materials company in northern California. He is also a certified transformational leadership trainer, an adjunct professor at JFK University, and an educator for the Eden Alternative, a nonprofit organization that seeks to bring lasting change to the long-term care environment. He is the founder and president of Toucan Learning Systems, an educational consultancy firm that specializes in breakthrough training and development work. Contact Doug at *dougneff@mac.com.*

Kimberly Houchens (Chapter 11)

Kimberly Houchens is an innovator and leader of innovative teams. The teams she has had the privilege to manage have commercially launched a diverse range of products, ranging from food and beverage packages, nano-coatings, commercial wallcoverings, surgical gowns, and military and space equipment. She is currently the vice president of product development—North America for Amcor, a global packaging company. Formerly, she was the CTO for Nano-Tex, Inc., and director of R&D for OMNOVA Solutions. She holds an MS and PhD in Textile Technology and Management from NCSU, a BS from MSU, and an associate's degree in design from FIT. She served the PDMA as a co-founder and president for the Cleveland Chapter. Contact Kim at *Kim.Houchens@AmcorPET.com*

Kevin Schwartz (Chapter 12)

Kevin Schwartz is a Partner with PRTM, a leading global management consulting firm specializing in operational strategy and innovation. As a consultant, he works with start-ups through Fortune 100 companies to optimize their innovation performance. Kevin leads PRTM's Open Innovation initiative and has helped to define leading practices for collaborative development across industries, including biotechnology, electronics, software, and consumer products. Prior to joining PRTM, Kevin worked in new product development in the aerospace industry in various positions with General Electric and Lockheed Martin. He holds a BS in Mechanical and Aerospace Engineering from Cornell University, an MS in Systems Engineering from the University of Pennsylvania and an MBA from the Haas School of Business at UC Berkeley. He is based in PRTM's Mountain View, California, office.

Jennifer Abell (Chapter 12)

Jennifer Abell, who most recently served as a Manager with PRTM management consultants, has experience implementing collaborative development strategies and practices within the biotechnology industry. As a consultant, Jennifer has worked with a range of companies making strategies operational by developing and implementing the processes that support them. She works across several industries—life sciences, hardware/software, consumer electronics, telecommunications, and electrical distribution and controls. Prior to joining PRTM, Jennifer worked for a life sciences start-up in various marketing and technical capacities, for Oracle Corporation managing ERP implementations, and for Eaton Electric in new product development and sales. She holds a BSE in Mechanical Engineering from the University of Michigan, Ann Arbor, and an MBA from The University of California, Los Angeles. She is currently based in the San Francisco Bay Area.

Douglas A. Peters (Chapter 13)

Douglas A. Peters is president of DS Performance Group and has over 20 years consulting experience in training, teambuilding, and organizational development. The DS Performance Group is focused on achieving dramatic and sustainable increases in individual, team and organizational performance; his clients range from start-ups to Fortune 500 companies across a wide variety of industries. Doug is a subject matter expert in personal, interpersonal, team, and organizational effectiveness. He has designed and delivered hundreds of training programs and curriculums and has conducted more than 400 team-building events. Much of his learning on teams comes from a 20-year relationship with 3 M Company, where he worked in over 75 divisions and staff groups to implement new product development, business development and executive teams. Doug's organizational development work includes new product cycle time reduction, culture assessment and change, and organizational effectiveness. One intervention with 3 M resulted in a 50 percent reduction in product development cycle time, according to a Harvard Business School study. Doug is also a theorist and model builder. His latest work, *The Performance Equation—The Formula for Dramatic and Sustainable Performance*, provides a unifying theory of human performance was published in the *Performance Improvement Journal*. This innovative work provides a unifying theory of human performance. Doug can be reached at Douglas A. Peters, DS Performance Group, 545 Main Street N. Hutchinson, MN 55350: *dougpeters@dsperformancegroup.com*; *www.dsperformancegroup.com*; 320-587-0372

Gregory D. Githens (Chapter 14)

Gregory D. Githens has over 25 years of new product development experience in industries that include consumer products, software, pharmaceuticals, medical devices, scientific instrumentation, biotechnology, agribusiness, engineering services, and information technology. Clients have achieved improved time-to-market, better metrics, better strategic alignment and improved risk management, among other benefits. Greg is the author of the "Team Based Risk Management" chapter of *PDMA Toolbook 1 for New Product Development* (John Wiley, 2002). He is the co-author of *Successful Project Management* (4th ed., John Wiley, September 2005). Greg is a contributing senior editor to the Product Development & Management Association's *Visions* magazine, where he has authored more than 25 articles on new product development performance. He has also written and spoken on NPD and innovation topics for Management Roundtable, the Project Management Institute, the International Institute of Research and other cutting-edge organizations. He has contributed chapters to Program & Portfolio Management chapter in *Managing Multiple*

Projects (Marcel Dekker, 2002), and the chapter on Handling Unpleasant Project Tasks, in *People in Projects* (PMI, 2001). Greg holds a BS from The Ohio State University, an MEn. from Miami University, and an MBA from Bowling Green State University. He is a certified *Project Management Professional* and a certified *New Product Development Professional*. Contact him at 419-424-1164 or *GDG@CatalystPM.com*

Ken Bruss (Chapter 15)

Ken Bruss EdD is a global learning and development consultant with PAREXEL, Int'l. He has over 25 years of experience in the areas of change management, organizational development, process improvement, teambuilding, training, and TQM. Prior to joining PAREXEL, Ken was a principal with HDA Consulting, a firm specializing in helping companies bring new products from concept to release. As both an internal and external consultant, he has facilitated process-improvement initiatives resulting in faster cycle time, lower development costs, and increased market share. Ken has conducted training programs in the United States, Europe, and Asia for such diverse companies as Analog Devices, Bank of New England and Polaroid. Ken has a doctorate in adult education from Boston University and holds PDMA certification as a New Product Development Professional. Ken has served as an AQP judge, and is a frequent conference speaker. Ken received the Best Industry Paper award at the 9th Annual Conference of Work Teams. His most recent publication is "Harnessing Employees' Knowledge to Enhance NPD Results," appearing in the SCPD journal *Concurrency*, *15*, no.1 (Summer 2006).

Wayne Mackey (Chapter 16)

Wayne Mackey's expertise is grounded in over 20 years of hands-on leadership of large engineering, manufacturing, and procurement organizations. His management consulting is focused on product / service development, and he is especially effective in collaborative design, metrics, portfolio management and business strategy implementation. He is co-author of the book *Value Innovation Portfolio Management: Achieving Double-Digit Growth through Customer Value*. Mr. Mackey has been a principal with Product Development Consulting, Inc. since 1997. Prior to joining PDC, he worked in industry for 20 years in high tech, aerospace, and automotive fields. He is a natural change agent and leader, having counseled Fortune 500 companies, major universities (Stanford, MIT, Carnegie-Mellon), and government agencies in product development, supply chain management, and rapidly implementing enterprise-wide change. Mr. Mackey also has worked as a senior scientist, program manager, engineering manager, and systems engineering manager. Mr. Mackey is an internationally acknowledged expert in metrics and has been a keynote speaker on achieving rapid organizational change, partnering and product development. He earned a BS in electrical engineering and economics from Carnegie-Mellon University in Pittsburgh, Pennsylvania, and a master of science in engineering from Loyola Marymount University in Los Angeles, California.

Introduction

Welcome to *The PDMA ToolBook 3 for New Product Development*.

As with *ToolBooks 1* and 2, this book has been written and edited by *PDMA* volunteers (royalties from the *ToolBook* series go to the *PDMA*) who are new product development (NPD) experts with a passion for NPD and the desire to contribute to the improvement of the NPD profession. They are NPD professionals (practitioners, service providers and academics) who have committed their effective practice learnings to these pages. Four ToolAuthors are return contributors: Gregory D. Githens, Gerald M. Katz, Peter Koen, and Christopher W. Miller. Peter Koen and Chris Miller have contributed to all three *ToolBooks*. Our hats are off to these dedicated professionals who have taken the time to once again share their expertise with the world.

These chapters provide you with in-depth, how-to knowledge that you can use to improve your organization's operation. *ToolBook 3* is a collection of best-practice tools presented such that you can put down this book and use them immediately.

ToolBook 1 emphasized tools to manage NPD processes and improve them. It presented 16 tools divided into those most suitable for project leaders (four to use before starting the NPD process, and four to use any time during the process), NPD process owners and portfolio managers.

The competitive situation and our global orientation changed materially between *ToolBook 1* (written pre-9/11, published in 2002) and *ToolBook 2* (published in 2004). "Hard" process improvements had become less important than effectively managing the more "soft" organizational issues. *ToolBook 2* thus focused on organizations and culture, the fuzzy front-end (FFE) and learning, both in terms of managing the NPD process and the portfolio and project pipeline. The corporate environment has again changed, and thus *ToolBook 3* differs materially from either of the previous books in the series, evolving in several interesting ways including providing material useful to new audiences, emphasizing tools to improve the information brought into and used by NPD teams, and providing a number of tools that are more strategically oriented than tactically oriented.

For the first time, the *ToolBook* series presents a number of tools that are most appropriate for use in the engineering design and development phases. Part I of this book, with chapters on TRIZ and quality function deployment (QFD), will be explicitly useful to engineers and technical development people, in addition to our traditional audience of project leaders, marketers and market researchers, product managers, NPD process owners, and NPD executives. We are particularly pleased to be able to include a chapter on TRIZ, the theory of inventive problem solving, as previous efforts to include a chapter on this complex set of techniques and tools were unsuccessful.

A significant portion of *ToolBook 3*, as presented in Part II, focuses on tools for improving the market and customer information and research used in NPD. As both the Marketing Science Institute and the Institute for the Study of Business Markets have found that marketing resources in firms have become more decentralized and distributed (read *too-downsized*), there is a critical need to bring the ability to generate and use high-quality market information to the NPD teams themselves. Teams can no longer depend on in-house market research groups, or even marketers in their divisions, to provide these information-developing services. This section provides seven tools that teams can use to develop this information on their own.

Parts III and IV provide sets of strategic tools—some aimed at improving performance across the firm (Part III) and others that target improving NPD project performance (Part IV). *ToolBook 2* moved in content from the more in-the-trenches tools that helped manage the process itself to tools that helped with the softer side of development; *ToolBook 3* takes another step upward into presenting a number of more strategic ones. Especially important, in this era of global competition and codevelopment, is the chapter on managing intellectual property, which leads off Part III.

Most of the material in previous *ToolBooks* was directed at larger, more mature firms. Another difference in *ToolBook 3* is that a number of the tools will be highly useful for smaller firms as well as for medium and larger firms. Many of the market-information tools will allow smaller firms, with lower research budgets, to incorporate higher-quality market information into their NPD projects. Chapter 8 on naming is a prime example of a chapter that present tools applicable to firms of varying scales. Both of the strategic tools in Part III will be very useful for smaller firms, as will most of the project-level strategic tools of Part IV.

HOW TO USE THIS BOOK

Rather than reading the entire book from cover to cover, we suggest that you use this book chapter by chapter, as the needs of your product development program and project dictate. You may find it helpful to read each of the four-part introductions to get a high-level perspective of each part's contents. Then, as you consider a weakness in your NPD process or a problem situation that you'd like to fix, you can go to the particular chapters that apply to the situation and try putting one or more of these tools to immediate use.

Alternatively, you may just be looking to improve some aspect of your NPD on a proactive basis. In this case, we recommend that you look at the chapter(s) that most closely fit the NPD area you are looking to improve. The chapters are full of best-practice tools that can improve the effectiveness of any NPD organization!

Abbie Griffin
Steve Somermeyer

Part 1

Tools For Engineering and Design

The tools of Part I will be most useful during the concept-generation and design stages of NPD. Both of the tools in this short but extremely powerful section of the ToolBook can be defined as creative problem solving techniques.

Chapter 1 presents the seven inventive techniques of TRIZ, the theory of inventive problem solving developed by Genrich Altshuller and his associates. TRIZ is a flexible methodology for engineered creativity that focuses on trying to eliminate performance contradictions, such as increasing fuel economy, while simultaneously increasing engine power. Each of the techniques uses a slightly different approach to overcoming contradictory performance goals. If one of the techniques does not provide a means to overcome a contradiction, others can be applied until the problem is solved. Although TRIZ can be very complex to implement, this chapter provides clear explanations of and examples for each of the seven inventive techniques, allowing teams to explore their application to help make useful performance trade-offs in their projects.

Quality function deployment (QFD), is presented in Chapter 2 as another creative problem-solving technique for the concept-generation and design phases of NPD, and as a companion chapter to "The Voice of the Customer," *Chapter 7 in* ToolBook 2. *Originally popularized in the global automotive industry (Toyota was an early adopter), QFD has proven its worth across many industries. QFD starts from statements of customer needs (the voice of the customer) and explicitly translates each one into performance measures that map to solving those customer problems. The interactions between needs and measures is captured in a matrix that helps visualize the overall situation and summarizes importance and performance data as well. Then the performance measures are translated into features and product solutions in another matrix, the features and solutions translated into manufacturing processes in a third matrix, and then into parts specifications in a final matrix. This methodology provides a mechanism for the team to set priorities on which areas they will focus on providing better solutions to customers' problems.*

1

TRIZ: The Theory of Inventive Problem Solving

Gunter Ladewig
President, PRIMA Performance Ltd.

BREAKTHROUGH PRODUCTS AND PROCESSES WITH SEVEN INVENTIVE TECHNIQUES

Genrich Saulovich Altshuller (1926–1998), wondered, "Could inventions be the result of systematic inventive thinking?" Over half a century, Altshuller and his associates investigated some 50,000 patents. Their work resulted in the breakthrough discovery that in excess of 95 percent of all patents used only seven inventive tools. Less than 5 percent come from breakthroughs in science and brand new ideas. They also found that exceptional patents improved performance by resolving contradictory requirements, like increasing speed without higher fuel consumption. Another revelation was the frequent occurrence of a windfall of benefits that arose from resolving a system's fundamental contradiction. Not only were many costly add-ons and expensive tolerances no longer required, but many systems had inherited valuable, new, product-differentiating capabilities and features. Altshuller also found that if patents were categorized by what they did functionally, rather than by industry, the same problem had been solved over and over again with just a handful of inventive techniques. The result, TRIZ, a Russian acronym for The Theory of Inventive Problem Solving, provides us with a methodology for engineered creativity.

TRIZ is a methodology that provides product and process designers with inventive problem-solving tools that not only accelerate the design process but also help them achieve world-class performance improvements beyond the trade-offs most designers consider unavoidable. Because TRIZ uses a functional approach to problem solving, it is equally applicable to solving business dilemmas faced by a giant steel mill as it is for resolving issues about microchips or potato chips. It reaches across many different functional lines, not just product development.

TRIZ can provide the marketing team with inventive techniques for product renaissance, both through product differentiation and competitive analyses. They can jazz up their products with brand-new applications, or they can put wings on their product or process with desirable new features.

Finally, TRIZ is an inventive problem-solving tool that can be used by the continuous improvement team in charge of Value Analysis/Value Engineering (VA/VE), Lean, or Six Sigma initiatives. In a most uncompromising way, TRIZ can be used to cut off (i.e., eliminate) costly and poor-quality components, and then make the pruned system work again by applying several inventive techniques. Saying it another way, TRIZ defines the problem and then walks around it with inventive techniques to find a solution. The following pages illustrate the use of seven inventive TRIZ techniques, first conceptually and then again by using a real product example, a vacuum cleaner.

But first, we need to include a few words about creativity activation. In facing a problem solution space, most of our knowledge is confined to our industry, background, and education. We can start fresh each time by using trial and error, brainstorming, or other creativity-unleashing methods such as Synectics (www.synectics.com) to generate *out-of-the-box* solutions. But there are other choices. Should we try to get as many ideas as possible (brainstorming), or should we try to get quality ideas? What's more important: head count (i.e., many brainstormers in one room), or head content? Should we emulate the traits of great inventors, or should we use their tools? Our choice is to get the best ideas from the best inventors by using their best tools, TRIZ—a methodology tried and proven in the real world of the worldwide patent base.

This chapter first introduces TRIZ as a general methodology, and then sets up the example that will be used to illustrate the inventive techniques. Then each of seven inventive techniques is explained in detail, and applied to the example. Finally, the chapter closes with keys to success and pitfalls to avoid when using these techniques.

TRIZ FLOW CHART

Figure 1-1, a problem flow chart, provides roadmaps for different system problems. Column 1 provides the roadmap for system performance improvement through contradiction or trade-off elimination. Column 2 is used to determine a product's evolutionary maturity (remaining 'head room' for improvement and benchmarking). Column 3 is used for product differentiation by providing the existing product with brand-new applications and/or unique features. In addition, it can be used as a competitive analysis benchmarking tool. Column 4 is a rather specialized tool used for the improvement of measurement systems.

General Problem Statement

The objective is to improve the efficiency and reduce the cost of power brush vacuum cleaners using seven different techniques. Each technique will be

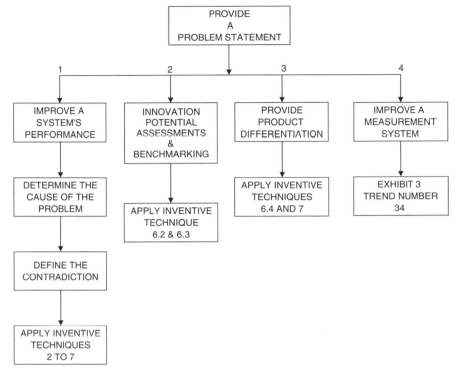

FIGURE 1-1. Problem-solving flow chart.

illustrated, first with generalized conceptual examples, and then by applying them to the vacuum cleaner. Solutions from these techniques will in some cases be the same or have overlapping features. Vacuum cleaners considered include central vacuum cleaners, floor models, and uprights. Vacuum cleaners remove dirt, dust and debris from floors, carpets and other surfaces by supplying both high-suction force to lift or free trapped debris from a carpet and also high airflow to transport the debris away quickly and enable rapid cleaning. Please note that ways to obtain self-cleaning carpets are not being sought, although this could be another goal for TRIZ. In this case, the problem under investigation is to improve the vacuum cleaner's efficiency.

The first step in applying TRIZ is to find the root cause[1] that is constraining the system's main function. An analysis of the vacuum cleaner reveals that its main function, providing suction force, is severely limited by atmospheric pressure, which is an absolute maximum of 14.7 pounds per square inch

---

[1] Use the master detectives for finding the root cause of the problem.

   ◆ Pareto rule: 80% of the defects come from 20% of the issues.
   ◆ R. Kipling: Who, Where, When, What, Why, How?
   ◆ E. Goldratt: Theory Of Constraints for Cause & Effects attribute analysis.
   ◆ D. Shainin: Don't guess, instead, 'listen to the parts, they're smarter than the engineers'

FIGURE 1-2. Electric power brush.

(PSI). In fact, most vacuum cleaners rarely generate three PSI of suction. To overcome this constraint, many vacuum cleaner manufacturers add a costly, electric motor-driven power brush (see Figure 1-2) that provides additional force for separating dirt from the surface being cleaned.

Technique 1: Formulate the Contradiction: Conceptual Example

A contradiction occurs when an improvement in one part of a product or system fundamentally causes deterioration in another part. Our fundamental premise is that all improvements are suboptimal unless the fundamental contradiction, or tug-of-war, causing the problem is unearthed and eliminated without trade-off in performance of any other aspect of the system. The goal is to somehow separate these opposing requirements so they can't exert a detrimental influence on one another.

To formulate the contradiction, follow four steps:

Step 1: State the primary function of the system.
Step 2: Transform the problem into a contradiction statement by defining what it is that is reducing the primary function's effectiveness.
As one example, consider the goals of a manufacturing facility. At a particular time, the primary goal of a manufacturing plant might be to increase throughput. The contradiction then becomes how to provide maximum throughput, while minimizing inventory cost:

> For high throughput($\uparrow$): Increase($\uparrow$) parts supply, but then there are excessive inventory costs.

> For low throughput ($\downarrow$): Reduce ($\downarrow$) parts supply, and then inventory costs are minimized.

The goal, to have both high throughput and low inventory cost, is obtained with a just-in-time (JIT) manufacturing strategy. The contradiction of having parts and not having parts was solved by separating the two opposing requirements in time: provide parts only when needed so that inventory cost is minimized.

Step 3: To help solve a contradiction, intensify it by stating the extremes of the conflict. Using this approach helps get out of the box of current of thinking. For example, what is a picture that is very small, nonexistent, and simultaneously very large, infinite? Answer, a picture in the computer's memory is very small and can't be seen, yet when displayed on the monitor it is very large and can be seen by someone with a notebook computer on the moon.

Step 4: To better visualize the contradiction and energize creative problem solving, draw a picture of the conflict zone.

THE VACUUM CLEANER'S CONTRADICTION   There are three different strategies or levels of solution that can be used to improve the vacuum cleaner:

Strategy 1 would be to eliminate the problem cause—that is, the vacuum cleaner's inherent (without power brush) design contradiction. (See Figure 1-3, the optimized, vacuum motor fan assembly trade-off.)

Strategy 2 would strive to eliminate the contradiction arising from the need of a costly power brush add-on to obtain superior cleaning. This is referred to in TRIZ as a paired object contradiction—that is, the bare-bones vacuum cleaner (poor cleaning) versus one with power brush add-on (costly).

Strategy 3 would eliminate the *root* cause of the problem by providing many multiples of atmospheric pressure cleaning capability (even more than the power brush provides).

Generally, techniques 1 and 4 employ strategy 1, while the remaining techniques generally employ strategy 2.

A different project might have decided to improve the vacuum's performance by eliminating problems associated with clogging of the filter. If the filter were selected, the project would start by defining its primary function and then its contradiction. The filter issue will be revisited at the end of technique 3.

General Problem Statement

This project's goal is to improve the poor cleaning capability of the vacuum cleaner so we can eliminate the expensive electric power brush add-on. The first task is to define the vacuum cleaner's primary function and walk through the strategies:

Step 1: The vacuum's primary function is to clean (remove debris from the carpet).

Step 2: The contradiction is in determining how to provide both maximum suction and flow. (See Figure 1-3, the optimized trade-off performance design chart.)

Step 3: State the intensified extremes of the conflict. If the vacuum is close to the carpet (i.e., stuck to it), we have maximum suction, but we have zero flow, and no debris removal. If the vacuum is too far from the carpet, we have little suction to remove debris, but we have maximum airflow, and again, little or no debris removal. The formulated intensified contradiction can be written as follows:

> For maximum cleaning (↑↑), maximum force (↑↑) is needed, but then there is zero flow (↓↓).

> For minimum cleaning (↓↓), minimum force (↓↓) is needed, but then there is high flow (↑↑).

Step 4: Draw the conflict zone domain (See Figure 1-4).

We need to note several things regarding the conflict, especially the extremes. If the vacuum is close to the carpet (i.e., stuck to it), there is maximum suction, but zero flow, and little or no debris removal. If the vacuum is far from the carpet, there is little suction to remove debris, but there is maximum airflow, and again, little or no debris removal. The contradiction to be solved is this: A vacuum cleaner with both maximum suction and maximum airflow is the answer!

Several points about solving contradictions are in order. First, there are two approaches possible for solving each contradiction: One approach involves improving the first attribute, the suction force of the vacuum cleaner, and the other involves improving the second attribute, the flow of the vacuum cleaner. For most situations the approach that most closely represents the main function, in this case cleaning or maximum suction force, is chosen for the improvement path and then, somehow, maximum flow has to be obtained.

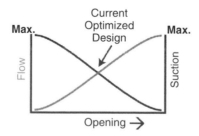

FIGURE 1-3. Trade-off performance design chart.

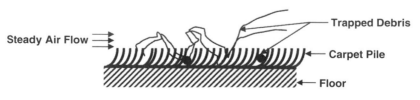

FIGURE 1-4. Conflict zone domain.

However, if the conflict is intensified, two more options arise. If after intensification, it becomes impossible to execute the main function, the nozzle is stuck to the carpet and there is no flow, then the other approach to resolving the conflict maximum flow should be taken as a starting point. This does not mean the overall goal is compromised. The contradiction still must be solved and both of the conflicting attributes maximized. If intensification destroys the product or object the primary function is acting on (in this case, debris), select the conflict that most closely represents the main function, but then try to solve a modified version by slightly backing off from the extreme state of intensification.

Please note that if a satisfactory solution to the problem is not obtained after applying all seven techniques, one can select the opposite version of the conflict and repeat the problem-solving process.

Frequently, just by intensifying the contradiction and by drawing a picture of the conflict zone domain, solutions come to mind and the problem can be solved. The intensified contradiction and the conflict zone picture may suggest oscillating high/low pressure pulses that wiggle and dislodge trapped debris for improved cleaning. If a solution is not obtained at this stage, proceed to the following techniques, which provide different tools for solving contradictions.

Process Quick Check: If the contradiction is experienced by a single object, for example air requiring both high suction force (no flow) and high flow (little suction), skip to technique 4, Physical Separation Techniques. This is an example of contradiction-solving strategy 1. If strategy 2, (paired object contradiction) elimination of the expensive power brush add-on is desired, proceed to the next step.

Technique 2: Formulate the Ideal Final Result and Define the Ideal Machine

The ideal final result (IFR) is a technique that removes us from the psychological inertia of the current way of thinking or doing things. What is important is that IFR frees us from the physical ways of achieving our goal by defining the desired resulting end state. Saying it another way, we work back from the answer, or desired function. An example of an ideal final result would be: Give me the hole, not the drill. This approach provides us with new alternatives to achieve our goal, and sometimes, all on its own, provides us with breakthrough solutions already at this stage of the TRIZ problem-solving process.

The ideal machine is a machine that performs its function but does not exist. We can create the ideal machine simply by transferring the function from machine 1 to another machine (i.e., machine 2). We thereby eliminate machine 1. Machine 1 has become the ideal machine; it performs its function but does not exist as that tool (we get the function for free). One example of an ideal machine would be the carpenter's hammer. It is both a hammer and a crow bar. The crow bar's function has been transferred to the hammer. The crow bar doesn't exist, yet the hammer performs its function. It has become the ideal machine.

The ideal final result and the ideal machine concepts can be used together to eliminate a contradiction. The hubcap on an automobile's wheel introduces the contradiction of providing a pleasing appearance, its primary function, but at a negative function of increased probability that it comes off. The ideal final result desired from the hubcap is improved secure adhesion. Transferring the hubcap's aesthetics function to the wheel's rim—designing and manufacturing a rim with a pleasing appearance (and not using the hubcap)—creates an ideal machine. The primary function of aesthetics is retained, but since the hubcap no longer exists, the negative possibility of losing it has been eliminated. The contradiction is solved.

In the same way, Value Engineering, Lean Manufacturing or Six Sigma aficionados can use the IFR and ideal machine concepts to transfer to another resource the desirable functions and thus prune costly or defect-prone components from the system. Hence, many expensive and time-consuming improvement techniques, such as design of experiments, do not have to be used to solve every problem. The cause of the problem can be identified and dealt with either by using already-existing resources to perform corrective actions or by pruning it away altogether. The cause of the problem was simply identified, and then walked around by using the ideal machine concept. Note that in general terms, TRIZ defines a resource as any substance (component) or field (force and energy) of the problem entity or its environment. Space, time, information and functions are also considered resources.

THE VACUUM CLEANER'S IDEAL FINAL RESULT AND MACHINE The desired ideal final result or function is: "Give me the high force of the power brush" (without the power brush). The question then becomes whether there is there any resource, component, or force in the vacuum cleaner or its environment that can be used to provide this IFR? Is there anything in the existing system or its environment that can provide high force (ideal result) and at the same time eliminate the costly power brush (ideal machine)? One solution is obtained by using the pushing power of the operator to rotate the roller brush and thus eliminate, at a minimum, the power brush motor. Other potential alternatives that come to mind are the use of electrostatic dust-repellent carpets, or the use of electrostatic energy generated by the airflow to assist suction in removing debris. These alternatives will be investigated further as other inventive techniques are applied to the problem.

Technique 3: Solve the Problem using Functional Diagrams and Pruning

TRIZ provides several systematic methodologies that assist in improving the performance of technological systems. The first of these is the use of functional diagrams and pruning.

This TRIZ procedure applies the principles of value analysis, developed by Lawrence Miles, to costly, poor-quality, or inefficient trade-off-causing

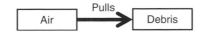

FIGURE 1-5. Substance-field function diagram.

FIGURE 1-6. Functions.

components by simply cutting them off or pruning them. This technique shows how to obtain the ideal final result and ideal machine for free. However, before constructing the vacuum cleaner's functional diagram, a few points about technological systems, functions, and the method for constructing a function diagram are in order.

A technological system consists of a set of objects or subsystems that perform a set of specific functions on a product. Generally speaking, a technological system consists of a working tool, which provides the primary function, an engine that provides energy, a transmission for carrying the energy, control for managing energy flow, and finally, the casing that maintains its structure and provides safety and aesthetics. The vacuum cleaner is one example of a technological system whose primary function, *cleaning*, is performed by a tool, *air*, that exerts a force, *pull*, on a product, *debris*. This air-pulls-debris example represents the most basic unit or model of a technological system and is referred to in TRIZ as a substance-field, or S-F, unit (see Figure 1-5).

The tool and product may also represent two systems, or two substances, or two components, or a system exerting a force on a single component or product. Larger, more complex technological systems like the remaining parts of the vacuum cleaner can be modeled by linking many individual S-F units. Fields may represent any force that's mechanical, thermal, chemical, electrical, magnetic, or electromagnetic. Each field, in turn, can be represented by many subcategories. For example, a mechanical field may be surface tension, friction, centrifugal force, inertia, pressure gravity, and so on. There are three types of fields: the primary field (the one that is in conflict), support fields, and harmful (costly) fields. Each field, in turn, may be scaled for its effectiveness as insufficient or excessive (uncontrolled). Finally, an X indicates a component that's unwanted and should be pruned off (see Figure 1-6 for field or arrow conventions).

To construct a function diagram, follow this procedure:

1. List all system components to obtain many potential choices for finding the ideal machine to which the desirable functions can be transferred while leaving behind undesirable ones to be pruned off.
2. Start the function diagram by connecting the tool with an arrow, which represents the primary field, to the product.

3. Connect other components near the conflict zone to the tool-to-product S-F model, making sure that all components causing the conflict are included.

4. Label all other fields as useful, or harmful.

5. Next scale all fields for their effectiveness. Please refer to the arrow conventions in Figure 1-6.

6. Scaling helps focus our efforts on what has to be improved, controlled, or pruned.

7. Use simple words and phrases for functional field descriptions. Simple phrases instead of restrictive professional jargon provide more choices for obtaining the Ideal final result, or function.

8. Minimize the size of the function diagram by not expanding too far beyond the conflict zone.

9. It is important to make sure that all desirable support functions are included so that they are not unknowingly cut off when the harmful, costly, poor-quality, or low-performance functions are pruned.

10. Prune off components that cause the conflict or other harmful problems, and transfer their desirable functions to other remaining components—that is, to the ideal machine.

If issues arise because a satisfactory component can't be found for transfer (or for intervention to compensate for insufficient or excessive actions), take one of these actions:

◆ Review other components listed in step 1 for possible inclusion in the function diagram.

◆ Review the effect data, technique 7, for possible solutions.

◆ Just transfer desirable functions to various components iteratively and then try to solve issues encountered with inventive techniques 4 to 6.

Here is a general example: A laboratory studies small disks of material specimens by placing them into a crucible, immersing them in aggressive solutions, and subsequently analyzing the solution. Unfortunately, some aggressive solutions not only attack the specimen but also the crucible, which, in turn, causes a contaminated solution and erroneous results. Figure 1-7 shows the function diagram and the solution's harmful action on the crucible. Both the specimen and the aggressive solution must be retained. The crucible indirectly causes contamination. The crucible can be *pruned* and its hold solution function, IFR, transferred to the specimen disk. This can be accomplished by creating a small depression in the specimen disk's top surface to hold a few drops of solution. The specimen disk has become the ideal machine or crucible, one that performs the crucible's holding function, without existing (see Figure 1-8).

USE FUNCTIONAL DIAGRAMS AND PRUNING TO SOLVE FOR THE VACUUM CLEANER   Start by drawing the primary function substance-field diagram by

FIGURE 1-7. **Function diagram with harmful action.**

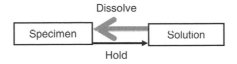

FIGURE 1-8. **`Pruned´ function diagram.**

connecting the tool, air, with the primary field, pulls, to the object or product, the debris to be removed (Figure 1-9). Since the primary function is insufficient to solve the contradiction, it is scaled with a dotted arrow. (See Figure 1-6 for arrow conventions.) Connect with fields (arrows) other nearby components like the vacuum motor, fan, filter, and brush motor to the previously described primary function substance field diagram. Because the brush motor and the brush are not wanted due to their high cost, they are marked with an X for pruning. Since the dust filter resists airflow, especially once it gets clogged with debris, it is added to the diagram and shown as having a harmful, uncontrolled (dashed line) function that resists airflow. In addition, since the dust filter's support functions of holding debris and cleaning air aren't 100 percent effective, they are scaled with dotted arrows as insufficient.

The first objective is to cut off (prune) the brush motor and brush assembly and assign their functions to something else, without adding anything new to the system. Some other component must provide cyclical pushing of debris. But what? Perhaps the fan can provide not only suction flow, but also high-pressure, pulsed jets of low-flow air from its exhaust port (or possibly a second fan impeller).

Next, for even greater performance, the filter may be pruned as well. But with the filter removed, the air is filled with debris, creating another contradiction. Dirty air that removes debris is needed and clean air that doesn't spew debris everywhere also is needed. Can these conflicting requirements now be solved simultaneously? Yes, with cyclonic, self-cleaning air that removes debris particles with centrifugal force. This, in fact, is the principle used by the Dyson Cyclone vacuum cleaner.

To finish, the new pruned system is reconnected and verified for proper operation. The fan inherits a new air-pushing force from its exhaust port. The air gets another primary cleaning function, sonic push, and also a new support function, cyclonic self-cleaning. Since the vacuum cleaner's debris removal performance (pulls), and air's cleaning function (cyclonic self-cleaning) are both much improved, dotted arrows (insufficient) are replaced with solid arrows. Please refer to Figure 1-10 for the pruned and reconnected function diagram of the vacuum cleaner.

EXHIBIT 1

↓ I M P R O V E ↓ ← W O R S E N →

		1	2	3	4	5	6	7	8	9	10	11	12	13
1	Weight of mobile object			15.8. 29.34		29.17 38.34		29.2 40.28		2.8.15 38	8.10. 18.37	10.36 37.40	10.14 35.40	1.35. 19.39
2	Weight of stationary object			10.1. 29.35		35.30 13.2			5.35. 14.2		8.10. 19.35	13.29 10.18	13.10 29.14	26.39 1.40
3	Length of mobile object	8.15. 29.34				15.17.4		7.17.4 35		13.4.8	17.10.4	1.8.35	1.8.10 29	1.8.15 34
4	Length of stationary object		35.28. 40.29				17.7. 10.40		35.8.2 14		28.10	1.14.35	13.14. 15.7	39.37. 35
5	Area of mobile object	2.17. 29.4		14.15. 18.4				7.14. 17.4		29.30. 4.34	19.30. 35.2	10.15. 36.28	5.34. 29.4	11.2. 13.39
6	Area of stationary object		30.2. 14.18		26.7.9 39						1.18. 35.36	10.15. 36.37		2.38
7	Volume of mobile object	2.26. 29.40		1.7.35 4		1.7.4. 17				29.4. 38.34	15.35. 36.37	6.35. 36.37	1.15. 29.4	28.10. 1.39
8	Volume of stationary object		35.10. 19.14	19.14	35.8.2 14						2.18.37	24.35	7.2.35	34.28. 35.40
9	Speed	2.28. 13.38		13.14.8		29.30. 34		7.29. 34			13.28. 15.19	6.18. 38.40	35.15. 18.34	28.33. 1.18
10	Force	8.1.37 18	18.13. 1.28	17.19. 9.36	28.10	19.10. 15	1.18. 36.37	15.9 12.37	2.36. 18.37	13.28. 15.12		18.21 11	10.35. 40.34	35.10. 21
11	Stress or pressure	10.36. 37.40	13.29. 10.18	35.10. 36	35.1. 14.16	10.15. 36.28	10.15. 36.37	6.35. 10	35.24	6.35.36	36.35. 21		35.4. 15.10	35.33. 2.40
12	Shape	8.10. 29.40	15.10. 26.3	29.34. 5.4	13.14. 10.7	5.34.4 10		14.4. 15.22	7.2.35	35.15. 34.18	35.10. 37.40	34.15. 10.14		33.1. 18.4
13	Stability of the object's composition	21.35. 2.39	26.39. 1.40	13.15. 1.28	37	2.11. 13	39	28.10. 19.39	34.28. 35.40	33.15. 28.18	10.35. 21.16	2.35.40	22.1. 18.4	
14	Strength	1.8.40 15	40.26. 27.1	1.15.8 35	15.14. 28.26	3.34. 40.29	9.40. 28	10.15. 14.7	9.14. 17.15	8.13. 26.14	10.18. 3.14	10.3. 18.40	10.30. 35.40	13.17. 35
15	Duration of action of mobile object	19.5. 34.31		2.19.9		3.17. 19		10.2. 19.30			3.35.5	19.2.16	19.3.27	14.26. 28.25	13.3.35
16	Duration of action by stationary object		6.27. 19.16		1.40. 35				35.34. 38					39.3. 35.23
17	Temperature	36.22. 6.38	22.35. 32	15.19.9	15.19.9	3.35. 39.18	35.38	34.39. 40.18	35.6.4	2.28. 36.30	35.10. 3.21	35.39. 19.2	14.22. 19.32	1.35.32
18	Illumination intensity	19.1. 32	2.35. 32	19.32. 16		19.32. 26		2.13. 10		10.13. 19	26.19.6		32.30	32.3.27
19	Use of energy by mobile object	12.18. 28.31		12.28		15.19. 25		35.13. 18		8.35	16.26. 21.2	23.14. 25	12.2.29	19.13. 17.24
20	Use of energy by stationary object		19.9.6 27									36.37		27.4. 29.18
21	Power	8.36. 38.31	19.26. 17.27	1.10. 35.37		19.38	17.32 13.38	35.6.38	30.6.25	15.35.2	26.2 36.35	22.10. 35	29.14. 2.40	35.32 15.31
22	Loss of Energy	15.6. 19.28	19.6. 18.9	7.2.6 13	6.38.7	15.26. 17.30	17.7. 30.18	7.18.23	7	16.35. 38	36.38			14.2. 39.6
23	Loss of substance	35.6. 23.40	35.6. 22.32	14.29. 10.39	10.28. 24	35.2. 10.31	10.18. 39.31	1.29. 30.36	3.39. 18.31	10.13. 28.38	14.15. 18.40	3.36. 37.10	29.35. 3.5	2.14. 30.40
24	Loss of Information	10.24. 35	10.35.5	1.26	26	30.26	30.16		2.22	26.32				
25	Loss of Time	10.20. 37.35	10.20. 26.5	15.2. 29	30.24. 14.5	26.4. 16	10.35. 17.4	2.5.34. 10	35.16. 32.18		10.37 36.5	37.36.4	4.10. 34.17	35.3. 22.5
26	Quantity of substance/matter	35.6. 18.31	27.26. 18.35	29.14. 35.18		15.14. 29	2.18. 40.4	15.20. 29		35.29. 34.28	35.14. 3	10.36 14.3		15.2. 17.40
27	Reliability	3.8.10 40	3.10.8 28	15.9. 14.4	15.29. 28.11	17.10. 14.16	32.35. 40.4	3.10. 14.24		2.35.2	21.35. 11.28	8.28. 10.3	10.24. 35.19	35.1. 16.11
28	Measurement accuracy	32.35. 26.28	28.35. 25.26	28.26. 5.16	32.28. 3.16	26.28. 32.3	26.28. 32.3	32.13.6		28.13. 32.24		6.28.32	6.28.32	32.35. 13
29	Manufacturing precision	28.32. 13.18	28.35. 27.9	10.28. 29.37	2.32. 10	28.33. 29.32	2.29. 18.36	32.28.2	25.10	10.28. 32	28.19. 34.36	3.35	32.30. 40	30.18
30	Object-affected harmful factors	22.21. 27.39	2.22. 13.24	17.1. 39.4	1.18	22.1. 33.28	27.2. 39.35	22.23. 37.35	34.39. 19.27	21.22. 35.28	13.35. 39.18	22.2.37	22.1. 3.35	35.24. 30.18
31	Object-generated harmful factors	19.22. 15.39	35.22. 1.39	17.15. 16.22		17.2. 18.39	22.1.40	17.2. 40	30.18. 35.4	35.28. 3.23	35.28. 1.40	2.33. 27.18	35.1	35.40. 27.39
32	Ease of manufacture	28.29. 15.16	1.27. 36.13	1.29. 13.17	27	15.17. 26.12		16.40	13.29. 1.40	35	35.13. 8.1	35.12	35.19 1.37	11.13. 27
33	Ease of operation	25.2. 13.15	6.13.1 25	1.17. 13.12		1.17. 13.16	1.16. 15.39	18.16. 15.31	4.18. 39.31	28.13. 35		2.32.12	15.34. 29.28	32.35. 30
34	Ease of repair	2.27. 35.11	2.27. 35.11	1.28. 10.25	3.18. 31	15.13. 32	16.25	25.2. 35.11	1	34.9	1.11.10	13	1.13.2 4	2.35
35	Adaptability or versatility	1.6. 15.8	19.15. 29.16	35.1. 29.2	1.35.16	35.30. 29.7	15.16	15.35. 29		35.10. 14	15.17. 20	35.16	15.37. 1.8	35.30. 14
36	Device complexity	26.30. 34.36	2.26. 35.39	1.19. 26.24		14.1. 13.16	6.36	34.26.6	1.16	34.10. 28	26.16	19.1.35	29.13. 28.15	2.22. 17.19
37	Difficulty of detecting and measuring	27.26. 18.35	6.13. 28.1	16.17. 26.24	26	2.13. 18.17	2.39. 30.16	29.1.4	2.18. 26.31	3.4.16	36.28. 40.19	35.36. 37.32	27.13. 1.39	11.22. 39.30
38	Extent of automation	28.26. 18.35	28.26. 35.10	14.13. 17.28	23	17.14. 13		35.13. 16		28.10	2.35	13.35	15.32. 1.13	18.1
39	Productivity/Capacity	35.26. 24.37	28.27. 15.3	18.4. 28.38	30.7	10.26. 34.31	10.35. 17.7	2.6.34. 10	35.37. 10.2		28.15. 10.36	10.37 14	14.10. 34.40	35.3. 22.39
		1	2	3	4	5	6	7	8	9	10	11	12	13

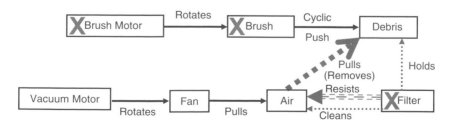

FIGURE 1-9. Vacuum cleaner function diagram.

EXHIBIT 1 (CONTINUED)

↓ IMPROVE ↓ ← ——— W O R S E N ———→

		14	15	16	17	18	19	20	21	22	23	24	25	26	
1	Weight of mobile object	28.27 18.40	5.34.31 35		6.29.4	19.1.32	35.12 34.31		12.36 18.31	6.2.34 19	5.35.3 31	10.24 35	10.35 20.28	3.26.18 31	1
2	Weight of stationary object	28.2.10 27		2.27.19 6	28.19 32.22	19.32 35		18.19 28.1	15.19 18.22	18.19 28.15	5.8.13 30	10.15 35	10.20 35.26	19.6.18 26	2
3	Length of mobile object	8.35.29 34	19		10.15 19	32	8.35.24		1.35	7.2.35 39	4.29.23 10	1.24	15.2.29	29.35	3
4	Length of stationary object	15.14 28.26		1.40.35	3.35.38 18	3.25			12.8	6.28	10.28 24.35	24.26	30.29 14		4
5	Area of mobile object	3.15.40 14	6.3		2.15.16	15.32 19.13	19.32		14.30 32.18	14.15 30.26	10.35.2 39	30.26	26.4	29.30.6 13	5
6	Area of stationary object	40		2.10.19 30	34.39 38				17.32	17.7.30	10.14 18.39	30.16	10.35 18	2.18.40 4	6
7	Volume of mobile object	9.14.15 7	6.35.4		34.39 10.18	2.13.10	35		35.6.13 18	7.15.13 16	36.39 34.10	2.22	2.6.34 10	35.3	7
8	Volume of stationary object	9.14.17 15		35.34 38	35.6.4				30.6		10.39 35.34		35.16 32.18	35.3	8
9	Speed	8.3.26 14	3.19.35 5		28.30 36.2	10.13.19	8.15.35 38		19.35 38.2	14.20 19.35	10.13 28.38	13.26		10.19 29.38	9
10	Force	35.10 14.27	19.2		35.10 21		19.17 10	1.16.36 37	19.35 18.37	14.15	8.35.40 5		10.37 36	14.29 18.36	10
11	Stress or pressure	9.18.3 40	19.3.27		35.39 19.2		14.24 10.37		10.35 14	2.36.25	10.36.3 37		37.36 4	10.14 36	11
12	Shape	30.14 10.40	14.26.9 25		22.14 19.32	13.15.32	2.6.34 14		4.6.2	14	35.29.3 5		14.10 34.17	36.22	12
13	Stability of the object's composition	17.9.15	13.27 10.35	39.3.35 23	35.1.32 15		13.19	27.4.29 18	32.35 27.31	14.2.39 6	2.14.30 40		35.27	35	13
14	Strength		27.3.26		30.10 40	35.19	19.35 10	35	10.26 35.28	35	35.31 40		29.3.28 10	29.10 27	14
15	Duration of action of mobile object	27.3.10			19.35 39	2.19.4 35	6.35 18		19.10 35.38		28.27.3 18	10	20.10.16 28.18	3.35.10 40	15
16	Duration of action by stationary object				19.18 36.40				16		27.16 18.38	10	28.20 10.16	3.35.31	16
17	Temperature	10.30 22.40	19.13 39	19.18 36.40		32.30 21.16	19.15.3 17		2.14.17 25	21.17 35.38	21.36 29.31		35.28 21.18	3.17.30 39	17
18	Illumination intensity	35.19	2.19.6		32.35 19		32.1.19	32.35.1 15	32	19.16.1	13.1	1.6	19.1.26 17	1.19	18
19	Use of energy by mobile object	5.19.9 35	28.35.6 18		19.24.3 14	2.15.19			6.19.37 18	12.22 15.24	35.24 18.5		35.38 19.18	34.23 16.18	19
20	Use of energy by stationary object	35					19.2.35 32				28.27 18.31			3.35.31	20
21	Power	26.10 28	19.35 10.38	16	2.14.17 25	16.6.19	16.6.19 37			10.35 38	28.27 18.38	10.19	35.20 10.6	4.34.19	21
22	Loss of Energy	26			19.38 7	1.13.32 15		3.38			35.27.2 37	19.10	10.18 32.7	7.18.25	22
23	Loss of substance	35.28 31.40	28.27.3 18	27.16 18.38	21.36 39.31	1.6.13	35.18 24.5	28.27 12.31	28.27 18.38	35.27 31			15.18 35.10	6.3.10 24	23
24	Loss of Information	10	10			19			10.19	19.10			24.26 28.32	24.28 35	24
25	Loss of Time	29.3.28 18	20.10 28.18	28.20 10.16	35.29 21.18	1.19.26 17	35.38 19.18	1	35.20 10.6	10.5.18 32	35.18 10.39	24.26 28.32		35.38 18.16	25
26	Quantity of substance/matter	14.35 34.10	3.35.10 40	3.35.31	3.17.39		34.29 16.18	3.35.31	35	7.18.25	6.3.10 24	24.28 35	18.3 16		26
27	Reliability	11.28	2.35.3 25	34.27.6 40	3.35.10	11.32 13	21.11 27.19	36.23	21.11 26.31	10.11 35	10.35 29.39	10.28	10.30	21.28 40.3	27
28	Measurement accuracy	28.6.32	28.6.32	10.26 24	6.28 32	6.1.32	3.6.32		3.6.32	26.32 27	10.16 31.28		24.34 28.32	2.6.32	28
29	Manufacturing precision	3.27	3.27.40		19.26	3.32	32.2		32.2	13.32 2	35.31 10.24		32.26 18.36	13.18 28.18	29
30	Object-affected harmful factors	18.35 37.1	22.15 33.28	17.1.40 33	22.33 35.2	1.19.32 13	1.24.6 27	10.2.22 37	19.22 31.2	21.22 35.2	33.22 19.40	22.10.2	35.18 34	35.33 29.31	30
31	Object-generated harmful factors	15.35 22.2	15.22 33.31	21.39 16.22	22.35 2.24	19.24 39.32	2.35.6	19.22 18	2.35.18	21.35 22.2	10.1.34	10.21 29	1.22	3.24.39 10	31
32	Ease of manufacture	1.3.10 32	27.1.4	35.16	27.26 18	28.24 27.1	28.26 27.1	1.4	27.1.12 24	19.35	15.34 33	32.24 18.16	35.28 34.4	35.23 24	32
33	Ease of operation	32.40 3.28	29.3.8 25	1.16.25	26.27 13	13.17.1 24	1.13.24		35.34 2	2.19.13	28.32 24	4.10.27 22	4.28.10 34	12.35	33
34	Ease of repair	11.1.2 9	11.29 28.27	1	4.10	15.1.13	15.1.28 16		15.10 32.2	15.1 32.19	2.35.34 27		32.1.10 25	2.28.10 25	34
35	Adaptability or versatility	35.3.32 6	13.1.35	2.16	27.2.3 35	6.22.26 1	19.35 29.13		19.1.29	18.15.1	15.10.2 13		35.28	3.35.15	35
36	Device complexity	2.13.28	10.4 28.15		2.17.13	24.17 13	27.2.29 28		20.19 30.34	10.35 13.2	35.10 28.29		6.29	13.3.27 10	36
37	Difficulty of detecting and measuring	27.3.15 28	19.29 39.25	25.34 6.35	3.27.35 16		2.24.26	35.38	19.35 16	19.1.16 10	35.3.15 19	1.18.10 24	35.33 27.22	18.28 32.9	37
38	Extent of automation	25.13	6.9		26.2.19	8.32.19	2.32.13		28.2.27 23.28	23.28	35.10 18.5	35.33	24.28 35.30	35.13	38
39	Productivity/Capacity	29.28 10.18	35.10.2 18	20.10 16.38	35.21 28.10	26.17 19.1	35.10 38.19	1	35.20 10	28.10 29.35	28.10 35.23	13.15 23	35.38		39
		14	15	16	17	18	19	20	21	22	23	24	25	26	

Here is one final comment regarding the air filter problem. The Invention Matrix (Exhibit 1), to be discussed in technique 5, could have been used as well to provide the same answer. If debris removal is improved using a filter, row attribute 26: Quantity of Substance, then column attribute 22 also applies: Energy loss in the filter reduces airflow. At the column/row intersection resides inventive principle 25, Self-service (Exhibit 2)—that is, self-cleaning air. (Note:

EXHIBIT 1 (CONTINUED)

↓ I M P R O V E ↓ ← ———— W O R S E N ————→

		27	28	29	30	31	32	33	34	35	36	37	38	39	
1	Weight of mobile object	3, 11, 1 27	28, 27 35, 26	28, 35, 26, 18	22, 21 18, 27	22, 35 31, 39	27, 28 1, 36	35, 3, 2 24	2, 27 28, 11	29, 5 15, 8	26, 30 36, 34	28, 29 26, 32	26, 35 18, 19	35, 3 24, 37	1
2	Weight of stationary object	10, 28 8, 3	18, 26 28	10, 1 35, 17	2, 19 22, 37	35, 22 1, 39	28, 1, 9	6, 13, 1 32	2, 27 28, 11	19, 15 29	1, 10 26, 39	25, 28 17, 15	2, 26, 35	1, 28 15, 35	2
3	Length of mobile object	10, 14 29, 40	28, 32, 4	10, 28 29, 37	1, 15, 17, 24	17, 15	1, 29, 17	15, 29 35, 4	1, 28, 10	14, 15 1, 16	1, 19 26, 24	35, 1 26, 24	17, 24 26, 16	14, 4 28, 29	3
4	Length of stationary object	15, 29 28	32, 28, 3	2, 32, 10	1, 18		15, 17 27		2, 25	3	1, 35	1, 26	26	30, 14 7, 26	4
5	Area of mobile object	29, 9	26, 28 32, 3	2, 32	22, 33 28, 1	17, 2 18, 39	13, 1 26, 24	15, 17 13, 16	15, 13 10, 1	15, 30	14, 1, 13	2, 36 26, 18	14, 30 28, 23	10, 26 34, 2	5
6	Area of stationary object	32, 35 40, 4	26, 28 32, 3	2, 29 18, 36	27, 2 39, 35	22, 1, 40	40, 16	16, 4	16	15, 16	1, 18, 36	2, 35 30, 18	23	10, 15,6 17, 7	6
7	Volume of mobile object	14, 1 40, 11	25, 26 28	25, 28 2, 16	22, 21 27, 35	17, 2 40, 1	29, 1, 40	15, 13 30, 12	10	15, 29	26, 1	29, 26, 4	35, 34 16, 24	10, 6, 2 34	7
8	Volume of stationary object	2, 35, 16		35, 10 25	34, 39 19, 27	30, 18 35, 4	35			1	1, 31	2, 17, 26		35, 37 10, 2	8
9	Speed	11, 35 27, 28	28, 32 1, 24	10, 28 32, 25	1, 28 35, 23	2, 24 35, 21	35, 13 8, 1	32, 28 13, 12	34, 2 28, 27	15, 10 26	10, 28 4, 34	3, 34 27, 16	10, 18	9
10	Force	3, 35 13, 21	35, 10 23, 24	28, 29 37, 36	1, 35 40, 18	13, 3 36, 24	15, 37 18, 1	1, 28, 3 25	15, 1, 11	15, 17 18, 20	26, 35 10, 18	36, 37 10, 19	2, 35	3, 28 35, 37	10
11	Stress or pressure	10, 13 19, 35	6, 28, 25	3, 35	22, 2, 37	2, 33 27, 18	1, 35, 16	11	2	35	19, 1, 35	2, 36, 37	35, 24	10, 14 35, 37	11
12	Shape	10, 40 16	28, 32	32, 30 40	22, 1, 2 35	35, 1	1, 32 17, 28	32, 15 26	2, 13, 1	1, 15, 29	16, 29 1, 28	15, 13 39	15, 1, 32	17, 26 34, 10	12
13	Stability of the object's composition		13	18	35, 24 30, 18	35, 40 27, 39	35, 19	32, 35 30	2, 35, 10 16	35, 30 34, 2	2, 35 22, 26	35, 22 39, 23	1, 8, 35	23, 35 40, 3	13
14	Strength	11, 3	3, 27, 16	3, 27	18, 35 37, 1	15, 35 22, 2	11, 3 10, 32	32, 40 28, 2	27, 11, 3	15, 3 32	2, 13, 28	27, 3 15, 40	15	29, 35 10, 14	14
15	Duration of action of mobile object	11, 2, 13	3	3, 27 16, 40	22, 15 33, 28	21, 39 16, 22	27, 1, 4	12, 27	29, 10 27	1, 35, 13	10, 4 29, 15	19, 29 39, 35	6, 10	35, 17 14, 19	15
16	Duration of action by stationary object	34, 27 6, 40	10, 26 24		17, 1 40, 33	22	35, 10	1	1	2		25, 34 6, 35		20, 10 16, 38	16
17	Temperature	19, 35 3, 10	32, 19 24	24	22, 33 35, 2	22, 35 2, 24	26, 27	26, 27	4, 10, 16	2, 18, 27	2, 17, 16	3, 27 35, 31	26, 2 19, 16	15, 28 35	17
18	Illumination intensity		11, 15 32	3, 32	15, 19	35, 19 32, 39	19, 35 28, 26	28, 26 19	15, 17 13, 16	15, 1, 19	6, 32, 13	32, 15	2, 26, 10	2, 25, 16	18
19	Use of energy by mobile object	19, 21 11, 27	3, 1, 32		1, 35, 6 27	2, 35, 6	28, 26 30	19, 35	1, 15 17, 28	15, 17 13, 16	2, 29 27, 28	35, 38	32, 2	12, 28 35	19
20	Use of energy by stationary object	10, 36 23			10, 2 22, 37	19, 22 18	1, 4					19, 35 16, 25		1, 6	20
21	Power	19, 24 26, 31	32, 15, 2	32, 2	19, 22 31, 2	2, 35, 18	26, 10 34	26, 35 10	35, 2 10, 34	19, 17 34	20, 19 30, 34	19, 35 16	28, 2, 17	28, 35 34	21
22	Loss of Energy	11, 10 35	32		21, 22 35, 2	21, 35 2, 22		35, 32	2, 19		7, 23	35, 3 15, 23	2	28, 10 29, 35	22
23	Loss of substance	10, 29 39, 35	16, 34 31, 28	35, 10 24, 31	33, 22 30, 40	10, 1 34, 29	15, 34 33	32, 28 2, 24	2, 35 34, 27	15, 10, 2	35, 10 28, 24	35, 18 10, 13	35, 10 18	28, 35 10, 23	23
24	Loss of Information	10, 28 23			22, 10	10, 21 22			27, 22			35, 33	35, 33	13, 23 15	24
25	Loss of Time	10, 30, 4	24, 34 28, 32	24, 26 28, 18	35, 18 34	35, 22 18, 39	35, 28 34, 4	4, 28 10, 34	32, 1, 10	35, 28	6, 29	18, 28 32, 10	24, 28 35, 30	25
26	Quantity of substance/matter	18, 3 28, 40	13, 2 28, 38	33, 30	35, 33 29, 31	3, 35 40, 39	29, 1 35, 27	35, 29 25, 10	2, 32	15, 3, 29	3, 13 27, 10	3, 27 29, 18	8, 35	13, 29 3, 27	26
27	Reliability	5, 11, 1 23		32, 3 11, 23	11, 32	27, 35 2, 40	35, 2 40, 26		27, 17 40	1, 11	13, 35 8, 24	13, 35 1	27, 40 28, 8	11, 13 27	1, 35 29, 38	27
28	Measurement accuracy	5, 11, 1 23			28, 24 22, 26	3, 33 39, 10	6, 35 25, 18	1, 13 17, 34	1, 32 13, 11	13, 35	27, 35 10, 34	26, 24 32, 28	28, 2 10, 34	10, 34 28, 32	28
29	Manufacturing precision	11, 32, 1			26, 28 10, 36	4, 17 34, 26		1, 32 35, 23	25, 10		26, 2, 18		26, 28 18, 23	10, 18 32, 39	29
30	Object-affected harmful factors	27, 24 2, 40	28, 33 23, 26	26, 28 10, 18			24, 35, 2	2, 25 28, 39	35, 10, 2	35, 11 22, 31	22, 19 29, 40	22, 19 29, 40	33, 3, 4	22, 35 13, 24	30
31	Object-generated harmful factors	24, 2 40, 39	3, 33, 26	4, 17 34, 26							19, 1, 31	2, 21 27, 1	2	22, 35 18, 39	31
32	Ease of manufacture		1, 35 12, 18		24, 2			2, 5, 13 16	35, 1 11, 9	2, 13, 15	27, 26 1	6, 28, 11, 1	8, 28, 1	35, 1 10, 28	32
33	Ease of operation	17, 27 8, 40	25, 13 2, 34	1, 32 35, 23	2, 25 28, 39		2, 5, 12		12, 26 1, 32	15, 34 1, 16	32, 26 12, 17			15, 1, 28	33
34	Ease of repair	11, 10 1, 16	10, 2, 13	25, 10	35, 10 2, 16		1, 35 11, 10	1, 12 26, 15	7, 1, 4 16	35, 1 13, 11		34, 35 7, 13	1, 32, 10	34
35	Adaptability or versatility	35, 13 8, 24	35, 5, 1 10		35, 11 32, 31		1, 13, 31	15, 34 1, 16	1, 16, 7 4	15, 29 37, 28	1	27, 34 35	35, 28 6, 37	35
36	Device complexity	13, 35	2, 26 10, 34	26, 24 32	22, 19 29, 40	19, 1	27, 26 1, 13	27, 9 26, 24	1, 13	29, 15 28, 37		15, 10 37, 28	15, 1, 24	12, 17 28, 24	36
37	Difficulty of detecting and measuring	27, 40 28, 8	26, 24 32, 28		22, 19 29, 28	2, 21	5, 28 11, 29	2, 5	12, 26	1, 15	15, 10 37, 28		34, 21	35, 18	37
38	Extent of automation	11, 27 32	28, 26 10, 34	28, 26 18, 23	2, 33	2	1, 26, 13	1, 12 34, 3	1, 35, 13	27, 4, 1 35	15, 24 10	34, 27 25		5, 12 35, 26	38
39	Productivity/Capacity	1, 35 10, 38	1, 10 34, 28	32, 1 18, 10	22, 35 13, 24	35, 22 18, 39	35, 28 2, 24	1, 28, 7 10, 25	1, 32 10, 25	1, 35 28, 37	12, 17 28, 24	35, 18 27, 2	5, 12 35, 26		39
		27	28	29	30	31	32	33	34	35	36	37	38	39	

If we were really stuck for a solution, use of the Physical Effects software[2]
database, discussed in technique 7, might

[2] Invention Machine Corporation, Goldfire Innovator™ TRIZ Software:

♦ Inventive Principles and examples from industry for solving contradictions,
♦ The Trends of System Evolution for optimizing and differentiating systems,
♦ Physical effects database with examples from industry, and
♦ Function diagram mapping software.
   The Web site is: *www.invention-machine.com*

EXHIBIT 2

40 Inventive Principles

1. SEGMENTATION

a. Divide an object into independent parts: bicycle chain, braided wire.

b. Make an object modular: LEGO set, telescopic pointer, computer components.

c. Increase the degree of fragmentation: escalator, roller conveyor.

2. EXTRACTION/REMOVAL

a. Extract (remove or separate) a disturbing part or property from an object: I-beam Vs solid beam, use a glass fiber to separate the hot laser source from where the light is needed.

b. Extract only the necessary part or property: Polaroid sunglasses, a strainer.

3. LOCAL QUALITY

a. Transition from a homogeneous structure of an object to a heterogeneous structure: concrete, plywood, anisotropic materials.

b. Make each part of an object function in conditions most suitable for its operation: toolbox with different-sized compartments, the nail apron.

c. Make each part of an object fulfill different useful functions: Swiss army knife, manicure set.

4. ASYMMETRY

a. Replace a symmetrical form with an asymmetrical form: asymmetrical shapes for foolproof assembly, contoured handles for better gripping.

b. If an object is already asymmetrical, increase its degree of asymmetry: increase the curvature of hockey stick's blade in order to increase its puck-shooting velocity.

5. COMBINING

a. Combine in space identical or similar objects, assemble identical or similar parts to perform parallel operations: honeycomb, transistors.

b. Combine in time homogeneous or contiguous operations: synchronize manufacturing operations, parallel manufacturing operations, fan with multiple vanes instead of a single vane.

6. UNIVERSALITY, MULTIFUNCTIONALITY   Have the object perform multiple functions, thereby eliminating the need for other object(s): laser for cutting, fusing, cleaning.

7. NESTING

a. Place the object inside another, which, in turn is placed inside a third object: paper cups, Russian Matrioshka dolls.

b. Pass an object through a cavity of another object: telescopic pointer, mechanical pencil, retractable seat belt.

8. COUNTERWEIGHT, LEVITATION

a. Compensate for the object's weight by joining with another object that has a lifting force: float for fishing, lifejacket, balloon.

b. Compensate for the weight of an object by interaction with an environment providing aerodynamic or hydrodynamic forces: sail, wings, tides for surfing.

9. PRIOR COUNTERACTION, PRELIMINARY ANTI-ACTION

a. If an action has useful and harmful effects, replace it with an action that controls the harmful effect: heat-treat material by annealing it to minimize the harmful effects of stress.

b. Create actions in an object that will later oppose harmful actions: Provide anti-tension in advance for concrete by prestressing the concrete, use masking tape to prevent overspray.

10. PRELIMINARY ACTION

a. Carry out all or part of the required action in advance: pre-tinned electronic components, self-adhesive bandages.

b. Prearrange objects so they can be used without losing time: nail apron, toolbox.

11. CUSHION IN ADVANCE, COMPENSATE BEFORE   Compensate for the relatively low reliability of an object with counter measures taken in advance: plating, provide redundancy, tolerances, and lifeboats.

12. EQUIPOTENTIALITY   Change the working conditions so that an object need not be modified, raised, or lowered: pit for oil changes, flexible coupling, a conveyor at a height so that parts can slide unto it rather than having to be lifted.

13. INVERSION, THE OTHER WAY AROUND

a. Instead of an action dictated by the specifications of the problem, implement an opposite action: to remove a part from a shrink assembly, cool the inner part instead of heating the outer part, heat shrink tubing.

b. Make a moving part of the object or the outside environment immovable and the non-moving part movable: escalator, treadmill.

c. Turn the object or process upside down: hourglass, cook from above rather than below using infrared radiation.

14. SPHEROIDALITY, CURVILINEARITY

a. Replace linear parts or flat surfaces with curved ones; replace cubical shapes with spherical shapes: replace typewriter keys with one IBM print ball, use circular endless subway tracks rather than straight end-to-end tracks.

b. Use rollers, balls, spirals: ball bearings instead of a sliding mechanism, use a wheel-type pizza cutter instead of a knife.

c. Replace linear motion with rotational motion, utilize centrifugal force: use a router instead of plane to remove wood.

15. DYNAMICITY, OPTIMIZATION

a. Make an object or its environment automatically adjust for optimal performance at each stage of operation: self-adjusting tinted glasses, a belt that continuously changes from straight to curved.

b. If an object is immobile, make it mobile or adaptive. Make it interchangeable: use a ballpoint pen instead of the fountain pen, interchangeable screw driver heads/bits.

c. Divide an object into elements that can change position relative to each other: chain, adjustable wrench.

16. PARTIAL OR EXCESSIVE ACTION   If it is difficult to obtain 100 percent of a desired effect, use somewhat more or less to greatly simplify the problem: dip and then skim or spin off the excess, approach the target quickly, but slow down just before it's reached.

17. TRANSITION INTO A NEW DIMENSION

a. Transition one-dimensional movement, or placement, of objects into two-dimensional; two-dimensional to three-dimensional, etc.: robots with six degrees of freedom.

b. Use multilevel objects instead of single level: multilayer printed circuit boards, stacks of paper.

c. Tilt the object or place it unto its side: laptop screen, adjustable mirror.

d. Use another side: Mobius strip, double sided tape, knife with two sharp edges (sword).

e. Project optical lines unto neighboring areas, or unto the opposite side of an object: reflecting telescope.

18. MECHANICAL VIBRATION/OSCILLATION

a. Set an object into oscillation: hammer drill, mixing by using shaking.

b. If oscillation exists, increase its frequency, even as far as ultrasonic: vibration plus ultrasonic cleaning, ultrasonic and thermo-sonic bonding.

c. Use an object's resonance frequency: destruction of kidney stones with ultrasonic resonance.

d. Use piezoelectric vibration instead of mechanical vibration: quartz crystal clocks.

e. Combine electromagnetic energy with ultrasonic vibration: use microwaves to melt materials and ultrasonic vibration to mix the liquids.

19. PERIODIC ACTION

a. Replace a continuous action with a periodic or pulsed action: DC versus AC, parts sampling.

b. If an action is already periodic, change its frequency: microprocessor frequency (ever-increasing frequency of Pentium chips), water sprinkler.

20. CONTINUITY OF A USEFUL ACTION

a. Carry out an action continuously (i.e., without pauses) so all parts operate at full capacity: the synchronized assembly line.

b. Remove idle and intermediate motions: rotary cutters for cutting in any direction, during thermo-cycling of computers, perform diagnostic testing.

21. RUSHING THROUGH   Perform harmful or hazardous operations at very high speed: inoculation gun instead of syringe to reduce pain, pass through high temperature quickly to prevent melting.

22. CONVERT HARM INTO BENEFIT, ``BLESSING IN DISGUISE´´

a. Utilize harmful factors or environmental effects to obtain a positive effect: gas from manure, waste recycling.

b. Remove a harmful factor by combining it with another harmful factor: use explosives to put out oil well fires, use an acid to neutralize a base.

c. Amplify a harmful factor to such a degree that it is no longer harmful: fight fire with fire by eliminating the main fire's fuel (wood/trees).

23. FEEDBACK

a. Introduce feedback: cursor on computer screen, inspection, Statistical Process Control (SPC).

b. If feedback already exists, reverse it or change its magnitude: part inspection: sampling Vs 100 percent inspection, increase the frequency of feedback for critical situations or parameters.

24. MEDIATOR, INTERMEDIARY

a. Use an intermediary object to transfer or carry out an action: chisel plus hammer instead of just using the hammer, primers for paint adhesion.

b. Temporarily connect an object to another one that is easy to remove: magnet to hold photo onto fridge, air in air mattress, dry ice for cooling ice cream.

25. SELF-SERVICE, SELF-ORGANIZATION

a. Make the object service itself and carry out supplementary and/or repair operations: use a cyclone and have air clean itself, boomerang returns on its own, knife holder that sharpens.

b. The object should service or repair itself: a tire with an internal fluid that plugs a puncture.

c. Make use of waste material and energy: use a flywheel to store excess rotational energy; during low demand periods for electricity, have the generators pump water into elevated reservoirs for later use to rotate the generator turbines.

26. COPYING

a. Use a simple and inexpensive copy instead of an object that is complex, expensive, fragile or inconvenient to operate: mock-ups, CAD drawings.

b. Replace an object by its optical copy or image: digital computer image, computer animation/simulation, optical inspection, projection lithography.

c. If optical copies are already used, replace them with infrared or ultraviolet copies: use infrared detection for seeing enemy soldiers in the dark.

27. INEXPENSIVE, SHORT-LIVED OBJECT INSTEAD OF EXPENSIVE, DURABLE ONE   Replace an expensive object by a collection of inexpensive ones, forgoing certain properties like longevity, or cost: paper/plastic bags, plastic eye lenses, paper towels.

28. REPLACEMENT OF A MECHANICAL SYSTEM

a. Replace a mechanical system with an optical, acoustic, thermal, or olfactory system: use a laser pointer instead of a mechanical pointer, use optical or acoustic measurement instead of mechanical measurement.

b. Use an electrical, magnetic, or electromagnetic field for interaction with the object: magnets to hold things, eddy currents.

c. Change from static to movable fields, from unstructured fields to structured ones: alternating current Instead of direct current.

d. Use fields in conjunction with field-influenced materials (e.g., magnetic materials): solenoids, liquid crystal displays (LCDs).

29. PNEUMATICS AND HYDRAULICS

a. Use gaseous or fluidic objects instead of solid objects. These parts can now use air or water for inflation, or use pneumatic or hydrostatic cushions: air bearings, shock absorbers, vacuum pick-and-place.

b. Use Archimedes force to reduce the weight of an object: a bridge or dock with pontoons.

c. Use negative or atmospheric pressure: to reduce the size of down-filled pillows, put them in a plastic bag and remove the air.

d. Use foam to provide both liquid and gaseous properties plus light weight: injection molding to obtain foamed plastics.

30. FLEXIBLE MEMBRANES OR THIN FILM

a. Replace traditional constructions with those made from flexible membranes or thin film: beer can, plastic shrink-wrap.

b. Isolate an object from its environment using flexible membranes or thin film: paint, surfactants on ponds to minimize evaporation.

31. USE OF POROUS MATERIAL

a. Make an object porous or add porous elements: sintered metal, bricks, air or liquid filters, strainers.

b. If an object is already porous, use pores to induce a useful substance or function: capillaries for suction, heat pipes, carbon for filtering or odor removal.

32. CHANGING COLOR OR OPTICAL PROPERTIES

a. Change the color of an object or its surroundings: RGB color mixing, bug lights.

b. Change the transparency of an object or its environment: transparent tape, glasses with dynamic transparency adjustment, optical lens coatings, polarized glasses.

c. Use color additives to observe an object, or process that is difficult to see: add pigment to water entering a septic system for leak detection.

d. If color additives are already used, employ luminescent tracers: use UV paint to enhance readability.

33. HOMOGENEITY   Make those objects that interact with a primary object out of the same material or a material that is close to the primary object's properties: tooth fillings, to prevent contamination, try to use like materials.

34. REJECTION AND REGENERATION

a. Make portions of an object that have fulfilled their functions disappear (be discarded, dissolved, or evaporated) or modified during the work process: digestible medicine capsules, biodegradable bottles.

b. Restore used up parts of an object during its operation: toilette reservoir, ice cube dispenser.

35. TRANSFORMATION OF PROPERTIES

a. Change an object's physical state (to solid, gas, or liquid): use icebergs to transport water, use dry ice to cool and then disappear, Popsicle instead of liquid.

b. Change the concentration, consistency, rheology: magnetorheoligal materials, thicksotropic materials like ketchup, super-saturated solutions.

c. Change the degree of flexibility: change the air pressure in shock absorbers.

d. Change the temperature or volume: balloon.

36. PHASE TRANSFORMATION   Exploit changes in properties that occur during phase transitions of a substance: use boiling water to maintain a constant temperature of 100° C., freeze water and change it from liquid to solid, Curie point where materials change from magnetic to nonmagnetic.

37. THERMAL EXPANSION/CONTRACTION

a. Use a material that expands or contracts with heat: Use heat-shrink tubing to hold separate items, use thermal compression for assembly of parts.

b. Use materials with different coefficients of expansion: bimetallic springs.

38. USE STRONG OXIDIZERS, ENRICHED ATMOSPHERES, ACCELERATED OXIDATION

a. Replace normal air with enriched air: breathing apparatus.

b. Replace enriched air with pure oxygen: oxy-acetylene torch.

c. Change oxygen to ionized oxygen.

d. Use ionized oxygen.

e. Replace ionized oxygen with ozone: to promote complete combustion.

39. INERT ENVIRONMENT OR ATMOSPHERE

a. Replace the normal environment with an inert one: Use nitrogen during soldering to minimize oxidation of solder joints, hermetic enclosures.

b. Add neutral or inert additives to an object: To prevent oxidation, shield an object with argon.

c. Carry out the process in a vacuum: vapor deposition of metals.

40. COMPOSITE MATERIALS   Replace a homogeneous material with a composite (multiple) material: alloys, fertilizer, plastics with fillers, carbon fiber composites.

provide answers. By entering search words like *filter*, the database would suggest numerous alternatives, including the use of centrifugal force.)

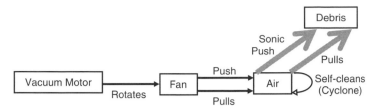

FIGURE 1-10. `Pruned´ and reconnected vacuum cleaner function diagram.

Technique 4: Solve using Physical Separation Techniques

These TRIZ techniques are used when the contradiction is experienced by a single object, as opposed to a pair of objects or subsystems fighting one another. For example, tongs need to be hot at the end that is moving food around a grill and yet remain cold at the other end where they are held. As another example, glasses are desired that transmit and retard light transmission, subject to the light's intensity. In low light, they transmit, while at higher light intensities, they partially block transmission. There are four ways to separate mutually exclusive requirements acting on a single object:

1. *Space:* Separate conflicting requirements acting on a single object by putting distance between the requirements. As already mentioned, tongs must be hot for holding hot objects, and they must be cold so that the hand holding them will not burn. This contradiction may be solved by allowing sufficient distance between the hand and the end of the tong holding the hot object. Another example of separation in space would be the bifocal lens in glasses, where the top part of the lens helps wearers focus on objects in the distance, while the bottom part of the lens provides focus for close-up situations such as reading.

2. *Time:* Separate conflicting requirements with time. The airplane wing has to be large for maximum lift at takeoff and it must be small for minimum drag when at high speed. These mutually exclusive requirements of large versus small can be separated by adjusting the wing's size from initially large to small later on by having retractable flaps or extensions. Another example would be JIT inventory management. By managing the delivery time of inventory, there is no inventory when it is not needed and simultaneously, inventory shows up right when it is needed.

3. *Condition:* Separate based on the situation encountered. An example might be the requirement for an object to attract and not to attract iron. One way to provide both holding and release capability is with the use of an electromagnet. When turned on, it attracts iron. When off, however, there is no attraction. The self-adjusting glasses that change their opaqueness subject to the light intensity are another example.

4. *Whole versus portion:* The whole entity has one characteristic, and portions of the whole have other characteristics. One example is the

strainer; the whole strainer retains the spaghetti, yet individual holes allow for separation of the water from the pasta. Another example is the bicycle chain. Individual links are rigid, yet the whole chain is flexible.

USING PHYSICAL SEPARATION TO SOLVE THE VACUUM CLEANER PROBLEM
The vacuum cleaner requires high suction (no flow), and high flow (no suction). Since a single entity, the air, experiences both requirements, physical separation techniques can be used to solve this problem.

The process is to try to apply each of the four physical separation techniques—space, time, condition, or whole versus portion—to the air in turn, to see if any of them can be used to satisfy both requirements of the conflict. The separation technique of condition does not seem to apply to solving this problem. In considering the separation principle of space, one way to try to solve the flow-versus-suction conflict is by providing areas of high flow, large openings, and areas of low flow (high suction) in the vacuum cleaner's end effecter. In fact, many current designs use cutouts in portions of the bristle brush of the end effecter to provide areas of high flow and other areas of high suction. The whole versus portion separation technique also is already used in vacuum cleaner designs. On the one hand, the bristle brush around the vacuum head already tries to solve this contradiction by providing low flow (high pressure) with the small spaces between individual bristles and the floor. On the other hand, high flow (low pressure) aggregated flow is provided by the whole brush-to-floor interface.

The separation in time principle suggests another potential solution not yet incorporated into standard vacuum design. Introducing air pulsations satisfies both seemingly exclusive requirements. Air pulsation oscillates between maximum suction (no flow) and maximum flow (no suction), using the dynamics between the two maxima to potentially clear debris more effectively.

Technique 5: Solve the Contradiction by using the Invention Matrix

Altshuller refers to the distinguishing properties of objects, or of the subsystems that make up a technological system, as their attributes. He discovered that, despite the immense variety of technological systems, any technological system could be completely defined with only 39 attributes, such as strength, weight, reliability, and complexity, to name a few. Frequently, improving performance in one attribute inherently comes at the detriment of performance on another attribute, creating a contradiction. Since great inventions resolve system contradictions, any contradiction could now be defined in an *invention matrix*, (see Figure 1-11), consisting of improved attributes (*Y* axis) versus deteriorating attributes (*X* axis).

Altshuller's actual invention matrix is shown in Exhibit 1. Improving attributes are listed in the left column. The same 39, but worsening, attributes also are listed across the top of the matrix. After defining inherently occurring conflicts such as Improving Strength (Y axis, attribute 14) versus Increased, Worsening Weight (X axis, attributes 1 or 2), or Improving Productivity (Y axis, attribute 39) versus Deteriorating Precision (X axis, attribute 29), Altshuller researched the worldwide patent base for the very best solutions to these conflicting requirements across all available systems. He and his associates discovered that only 40 *inventive principles* were used over and over to resolve conflicts between these 39 attributes. Exhibit 2 provides definitions and brief descriptions of examples for all 40 inventive principles.

The result is the invention matrix of Exhibit 1, which allows any system conflict to be defined and suggests a number of potential solutions based on the inventive principles. These potential solutions can be found in the intersection of the conflicting row and column attributes. For example, an increase in Productivity/Capacity, system attribute 39, located at the bottom of the left column in the matrix, might lead to worsening Manufacturing precision, number 29, located in the top row of the matrix. At the intersection of these two X-Y contradictory attributes are numbers for the inventive principles that have been found, from a review of the patent literature, to have resolved this conflict in previous inventions. At the crossroads of attributes 39 and 29 are four numbers: 1, 10, 18, and 32. These four numbers represent four high-potential, analogous solutions, based on the inventive principles, to the productivity versus precision conflict. The four inventive principle potential solutions suggest using:

- ◆ 1: Segmentation
- ◆ 10: Preliminary Action
- ◆ 18: Mechanical Vibrations/Oscillations. Suggested solutions under this principle might include any of the following:

  - ◆ Using a hammer drill
  - ◆ Increasing the frequency by going to ultrasonics
  - ◆ Using resonance as used in the destruction of kidney stones
  - ◆ Using piezoelectric vibration as used in quartz crystal watches

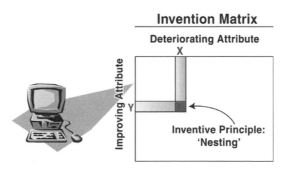

FIGURE 1-11. Invention matrix structure.

◆ Using electromagnetic energy in combination with the above vibrations

◆ 32: Changing Color or Optical Properties,

In addition, combinations of these four principles could be used to solve our conflict.

CASE STUDY:

Although not necessary, commercial software is available that not only helps solve the contradiction by providing inventive principles (solutions), but also provides examples of solutions from many industries. Please see endnote 2 of this chapter for details.

As one example of how to use the inventive matrix, consider improving the performance of a pointer. First, define the contradiction for a pointer: It should be both long (to reach the board) and short (to fit in a pocket). Then:

◆ Locate on the Y axis, the attribute (1 to 39) to be improved: Length. Select the attribute that most closely represents the desired need, in this case (3) Length of Mobile Object

◆ Locate on the X axis the attribute that deteriorates when *conventional* means are used to obtain a long pointer; for example, it doesn't fit into the pocket, its volume increases, in this case (7) Volume of Mobile Object

◆ At the XY intersection, the matrix suggests four Inventive Principles that may apply to how the contradiction can be resolved:

   ◆ 7: Nesting: Place objects inside one another; paper cups, mechanical pencil.
   ◆ 17: Transition into a new dimension: Go in other directions, project optical lines.
   ◆ 4: Asymmetry: Replace symmetrical objects with asymmetrical objects, or vice versa.
   ◆ 35: Transformation of properties: Change object's physical state, solid, gas, liquid, rheology, magnetorheological materials.

Note: Inventive principles are listed in order of highest probability for solving the contradiction.

The inventive principles are only generalized, analogous solutions. The problem solver must interpret these suggestions to find a solution that is appropriate to their specific application. For example, in our pointer problem, inventive principle 7, Nesting, might suggest a telescopic pointer. For more examples of specific ideas for these inventive principles please refer to Exhibit 2.

Of course there are many other ways to formulate this contradiction besides length versus volume. Many contradictions should be formulated to allow problem solvers to find the best solution. This, in itself, is a somewhat creative act, which may require several different attempts. For example, by formulating the conflict as one of Weight of a Mobile Object (attribute 1) versus Volume of a Mobile Object (attribute 7) inventive principles 29, 2, 40, and 28 would have been suggested as potential solutions. Inventive principle 28, Replacement of a Mechanical System with a Field: Magnetic, Electric, or Electromagnetic, might have led to the idea of inventing a laser pointer.

USING THE INVENTION MATRIX TO SOLVE THE VACUUM CLEANER PROBLEM
On the Y axis of the matrix, locate one attribute to be improved: Rate of dust particles removal from the carpet maps to attribute (1), Weight of Mobile Object. Refer to the highlighted row of the matrix. On the X axis, four attributes that deteriorate when *conventional* means are used to carry away heavy particles have been identified by boxing the intersected cells in the matrix:

1. Airflow is reduced (to increase suction force)—that is, Volume of Mobile Object, attribute (7).
2. We have to vacuum longer—that is, Duration of Action of Mobile Object, attribute (15).
3. The amount of removed debris is reduced (reduced flow)—that is, Quantity of Substance, attribute (26).
4. The vacuum cleaner's design becomes more expensive and complex due to costly add-ons, like electric power brushes—that is, Device Complexity, attribute (36).

Next, record the highlighted inventive principles at *each* XY intersection. For the first intersection, row 1 and column 7, inventive principles 2, 28, 29, and 40 are suggested as potential solutions. The intersection of row 1 and column 15 adds inventive principles 5, 31, 34, and 35 as potential solutions. Repeat this exercise for other realistic attribute combinations. For example, the selected objective may be to Increase the Vacuum's Suction, attribute (10), Force, which is the second highlighted row in the matrix. Alternatively, the objective may be to Reduce the Complexity/Cost of Expensive Add-Ons, attribute (36), Complexity, the third highlighted row.

Record all of the inventive principles (highlighted) at the XY intersections and consider using the *most frequently* occurring inventive principles to solve the vacuum cleaner problem. In this case, the most frequently occurring inventive principles, with each arising three times, were:

◆ 18. *Vibrations:* Consider adding pulsating spikes to the air suction to increase the debris separation force and thus possibly eliminate the need for the electric power brush.
◆ 26. *Copying:* Rather than using one large vacuum head with diminishing suction at its extremities (i.e., furthest from the vacuum tube outlet), consider using multiple, miniature vacuum heads. Each multifurcated flow opening is in close proximity with the carpet for maximum suction.

◆ 29. *Pneumatics and Hydraulics:* Consider replacing the electric power brush with low-flow but high-pressure pneumatic, focused-air jets.

Or consider using combinations of these inventive principles to solve the vacuum cleaner problem. For example, use pulsating, low-flow but high-pressure focused air jets (instead of rotating electric brushes) in combination with high suction flow.

Technique 6: Solve with the Trends of System Evolution

During his research of the worldwide patent base, Altshuller also discovered that technological systems tended to evolve along certain prevailing *vectors* (each with discrete phases), which he termed the *trends of system evolution.* Exhibit 3 defines 34 trends of system evolution and provides examples for each. The following six major vector groupings of the 34 trends listed in Exhibit 3 define how most systems tend to evolve:

1. Transition to a higher level, or multiobject system
2. Nonuniform rate of subsystem evolution
3. Shortening of energy path
4. Increasing flexibility, from rigid mechanical to pliable to electrical
5. Transition from macro to micro-level
6. Increasing ideality

EXHIBIT 3

34 Trends of System Evolution

Trend 1, Ideality, serves as a high-level model of what all the trends ultimately try to achieve. For technological process or product improvement use trends 2 to 33. For improving measurement systems use trend 34.

1. IDEALITY

$$\frac{\uparrow \sum F \text{ Useful}}{\downarrow \sum F \text{ Harmful}} \qquad F = \text{function: Increase}\uparrow, \text{Decrease}\downarrow$$
$$\sum = \text{Sum of: Quantity, Magnitude, and Rate}$$

This trend represents an integrated summation of all the trends that follow. Technological systems tend to evolve toward providing greater value (i.e., more useful functions and fewer harmful functions). Useful functions include the primary performance-related functions of the system, support functions, functions for other applications (i.e., a laser pointer that's also a pen light or level), and desirable features of the system.

Harmful functions include those that incur cost or deteriorate useful functions. The objective is to increase the quantity, magnitude, and rate of improvement of, useful functions. For harmful functions, the objective is the exact opposite. One technological system that's probably closer to ideality than any other is the computer. Not only does it perform its basic function, computation, faster and faster, it performs many other functions like those of a phone, a book, or a fax. Furthermore, it has many desirable features like mobility and being lightweight. However, its harmful functions—cost, dollars per computation, and poor quality—have dramatically improved from the early 80 foot-by-30 foot, vacuum tube ENIAC computer.

2. SPACE SEGMENTATION

Monolith → single cavity → multiple cavities → pores → capillaries with active additives

Example: Cooling device: Solid block → single fin → multiple fins → porous fins → capillary heat pipes

3. SURFACE SEGMENTATION

Flat → wavy → dimpled → with active breathing pores

Example: Paper: Flat → corrugated → bubble-wrap → scented wraps

4. SEGMENTATION (CUTTING)

With Solid (axe) → segmented (saw) → liquid → gas or plasma → field

Example: Cutting tool: Knife → grinding wheel → water jet → plasma arc → laser

5. TRIMMING (PRUNING)

Multipart system → Reduced part system → Single component system

Example: Automobile's display: many gauges, miniature LCD's → one LCD monitor

6. COMPLICATION/SIMPLIFICATION (REDUCING)

Few functions per item → many functions per item

Example: Separate phone, fax, printer, calculator, copier → the computer

7. INTRODUCTION OF VOIDS

Two objects → voids into one object → voids external to one object
→ voids around both objects → voids between objects

Example: Bearings: Bushing → sintered → roller bearing . . . → air bearing

8. INCREASING ASYMMETRY   To provide additional function

Example: Mittens → gloves, left and right → gloves with grip surfaces

Mistake-proof assembly designs (e.g., Poke-yoke)

9. FLOW SEGMENTATION

Single stream → bifurcated → several streams → many streams

Example: To provide more controllability, flows are segmented so that flows can be diffused, focused, differentiated, (e.g., water nozzle): single stream → many → mist

10. GEOMETRIC EVOLUTION (LINE)   Geometric structures evolve from a single point toward complex, three-dimensional structures.

Point → line → 2D curve → 3D curves → 3D complex curve

Example: Hydraulic tubing: straight → U-shaped → 3D spiral → curved 3D → spiral

11. GEOMETRIC EVOLUTION (SURFACE)

Flat → cylindrical → spherical → complex

Example: Skylights, mirrors

12. GEOMETRIC EVOLUTION (SPACE)

Cubic → cylindrical → spherical → egg-shaped → spiral

Example: Vases, fuel tanks, loudspeakers

13. INCREASING SYSTEM FLEXIBILITY

rigid → jointed → multi-jointed → elastic → liquid/gas → field → nothing

Example: Measurement: Ruler → folding ruler → tape measure → sonic detector → laser

14. COORDINATION OF ACTIONS

Nomatching → forced → buffered matching → self-matching

Example: Production: Machines working at different rates → rate controlled by slowest process → use of buffer stock → autonomous self-control (own power source & sensory feedback)

15. COORDINATION OF FORCE DYNAMICS

Continuous action → pulsed action → resonance → several actions → traveling wave

Example: Surface cleaning: continuous water jet → pulsed jet → jet tuned to surface's resonance frequency → e.g., combinations of pulsed and continuous → sweeping motion jet

16. MACRO TO MICRO TRENDS   System based on: different components same components same small components substance structure molecular phenomena atomic phenomena fields

Example: Bolts, rivets → thread, zipper → powder, aerosol → crystals, solder → glue → ionized materials, isotope → heat, light, magnetic or electromagnetic fields

17. POLY-FUNCTIONALITY: MONO-BI-POLY, WITH SIMILAR OBJECTS

Mono → bi → tri → poly systems

Example: One lead pencil → multilead automatic pencils

Example: One transistor → multitransistor integrated circuit

18. MONO-BI-POLY, WITH VARIOUS OBJECTS

Mono → bi → tri → poly systems

Example: Knife → Swiss army knife

Screwdriver → Multibit screwdriver

19. ANTI-BI SYSTEMS   Pencil with eraser, heater/cooler, Peltier transistor

20. MONO-BI-POLY SHIFTED SYSTEMS

One color pencil → multicolor automatic pencil

21. MONO-BI COMPETING SYSTEMS   Turbo-prop plane, balloon-propeller plane, and telescope with mirrors and lenses (Maksutov System)

22. MONO-BI COMPATIBLE SYSTEMS   Two-part epoxy, symbiotic systems: Wasted heat used to heat another system, compensating systems: tinted glasses

23. INCREASED DYNAMICITY   To increase responsiveness:

One tolerant state → several tolerant states → dynamically tolerant
→ artificially tolerant (via feedback) → intolerant

Example: Foundation → switch → car F15 Jet fighter → Nitroglycerin

24. INCREASING HUMAN EXPERIENCE

a. Increase use of senses: Taste | smell | vision + touch + hearing
b. Color: Monochrome → binary → visible spectrum → full spectrum (Maxwell's spectrum)
c. Transparency: Opaque → partially transparent → transparent → with active elements (glasses that adjust for brightness)
d. Value: performance → reduced cost → reliability → features → other new uses
e. Product: Commodity → new product → service → experience → transformation

Example: Bread → iPod → concierge → river rafting → spiritual/religious transformation

25. INCREASING CONTROLLABILITY

Uncontrolled system → manual → manual with power assist → self-controlled/feedback
→ smart self-control

Example: Exit opening → door → switch-actuated door → door with motion detector → badge reader actuated (expert systems/artificial intelligence assisted)

26. INCREASING DEGREES OF FREEDOM, DOF

1 → 2 → 3 → 4 → 5 → 6 DOF

Example: X direction $\rightarrow X + Y \rightarrow X + Y + Z \rightarrow X + Y + Z + X/R$ rotation $\rightarrow X + Y + Z + X/R + Y/R \rightarrow X + Y + Z + X/R + Y/R + Z/R$, or any in-between combinations, such as the robot arm

27. INTRODUCTION OF SUBSTANCE   For two interacting objects, A & B:
1—Introduce internal to A or B, 2—External to A/B, 3—To environment around A & B, and 4—Between A & B

28. INTRODUCTION OF MODIFIED SUBSTANCES   Introduce modified versions of A or B, in similar ways as outlined above in trend number 27

29. INTRODUCTION OF FIELDS (FORCE)   For two interacting objects, A & B: Same as for item 27

30. EVOLUTION OF FIELDS (FORCE)

   Mechanical $\rightarrow$ Thermal $\rightarrow$ Chemical $\rightarrow$ Electrical $\rightarrow$ Magnetic $\rightarrow$ Electromagnetic

Examples: Mechanical: gravity, friction, centrifugal force, surface tension, vibration, sound,

Thermal: heating, cooling, evaporation, condensation, sublimation, radiators

Chemical: Explosions, combustion, polymerization, catalysts, taste, smell

Electrical: Electric current, electrostatics, electrolysis, piezoelectric

Magnetic: Magnetizing, demagnetizing, induction, magnetic solids/liquids

Electromagnetic: electrostatics, light (infrared to ultraviolet), radiowaves

31. REDUCED HUMAN INVOLVEMENT

Human $\rightarrow$ human + tool $\rightarrow$ human + semi-automated tool $\rightarrow$ human + fully automated tool $\rightarrow$ autonomous tool

Example: Human $\rightarrow$ human augur drill $\rightarrow$ human with electric drill $\rightarrow$ human with automatic drill press $\rightarrow$ robot

32. CLEVER MATERIALS THAT OVERCOME CONTRADICTORY REQUIREMENTS   Example: Materials with shape memory: straight when pulled, curled when wound

Hard and soft: Ice $\rightarrow$ water

Large and small: Heat-shrink tubing, piezo materials

Hot and cold: Peltier transistor

Magnetic and nonmagnetic: Curie point materials

33. REDUCED ENERGY CONVERSION

            (To increase efficiency) Many conversions $\rightarrow$ Zero?

Example: Propeller plane, Chemical $\rightarrow$ thermal $\rightarrow$ mechanical (rotation) $\rightarrow$ mechanical (pressure drop)

Hand glider: Gravity pressure drop/velocity

---

34. REDUCED NEED FOR MEASUREMENT

Measure attributes → Detect (yes/no) → No direct measurement (self-regulating, or measure: a byproduct or a model)

Example: Measure temperature in degrees → Detect if substance is > or < $X°$ → Self-regulate the temperature of a system at 100°C by using water that can't exceed its boiling temperature of 100°C.

Using a byproduct: Detect disease by analyzing a human being's: blood, temperature, etc.

Using a model: Use a digital (model) picture of a human being's fingerprint or pupil to identify them.

Note: If a measurement system needs to be improved we can do it with trends listed above. Trends 27, 28, and 29, introduction of substances or fields, are particularly useful.

The evolution of the printing industry provides an example of all six trend vector groupings at work. At the start, the printing press had a single, rigid print plate. Transition to a higher level, multiobject system (1) occurred with the invention of the typewriter, which provided the flexibility of printing up different pages, one after another. However, typewriter cost reduction efforts eventually hit a major roadblock. Typewriter keys proved difficult to cost reduce and could not keep up with the rate of cost reduction improvements of other typewriter components—that is, the nonuniform evolution of subsystems (2). This system conflict was removed with IBM's Selectric typewriter ball. It was one typewriter key with all the alphanumeric symbols on it. This, in turn, shortened the energy path (3) via removal of numerous mechanical typewriter linkages. Increasing flexibility (4), in turn, was achieved in the Selectric typewriter through use of electromagnetic instead of mechanical drive mechanisms. Transition from macro to micro-level (5) occurred with the transition of mechanical print mechanisms, typewriter keys, to ink jet, and then laser printers. A higher level of ideality (6) was reached when the printer was totally eliminated, through use of the computer monitor (ideal machine).

However, not all trends, or even vector groups, may apply to any one system. Applicability is subject to the action or entity that needs to be satisfied. The user must scan through the trends to assess their applicability. For product or process improvement, use trends 1 to 33. For measurement improvement use trend 34. In addition, trend 1, Ideality, serves as a high level explanation of what all trends ultimately try to achieve.

The trends of system evolution are very flexible in how they can be applied. Like the invention matrix, the trends can be used to develop the means to improve system performance by solving contradictions. However, they can also be used in three other ways: to assess a product's innovation potential, for competitive analyses or benchmarking, and to help differentiate products already in the market.

PRODUCT PERFORMANCE IMPROVEMENT   To improve system performance by eliminating a contradiction, ask: Can the *action* or *attribute* be improved with trend 1, Ideality, *or* trend 2, Space Segmentation—all the way to the last trend 34, Reduced Need for Measurement? The different trends are reviewed one by one for applicability to the situation to be improved until one is found with the potential to do so. Applicability of the trends is illustrated using the pointer example again.

The pointer's desirable attribute is adaptability, the ability to point at things wherever they are. Scanning through the trends and their examples finds trend 13, Increasing System Flexibility: rigid → jointed → multi-jointed → elastic → liquid/gas → field → nothing. Hence, a one-piece pointer would be in phase 1 of its evolution, and a multijointed (telescopic) pointer in phase 3 of this evolutionary progression.

The question of "Whether the *action* or *attribute* can be improved with/by trend'X'" for eliminating contradictions for this example becomes: Can the pointer's *adaptability* be improved with/by progressing through *rigid → jointed → multi-jointed → elastic → liquid/gas → field → nothing*? Yes of course, using a field, such as a laser pointer, the length of the pointer can be both very small and very long. Better yet, evolve to nothing, the ideal machine. For example a computer monitor's curser produces a pointer that is very short and yet extremely long when viewed by many during a Webcast presentation.

PRODUCT INNOVATION POTENTIAL ASSESSMENT   The second way in which the trends of evolution can be used is to assess a product's innovation potential. By repeating the process of defining applicable trends to one entity at a time for a product (its components, subsystems, the system, super-system), a trends-of-evolution *spider diagram* can be created. This diagram defines the current state of evolution for any one-product entity and also the remaining headroom for improvement (see Figure 1-12).

Figure 1-12 might, for example, represent eight applicable trends of evolution for one entity, the bristles of a toothbrush. Trend vector 1 might represent surface segmentation (trend 3) with phase 1: being a flat surface of bristles; phase 2: a profiled surface more contoured to the tooth/gum profile;

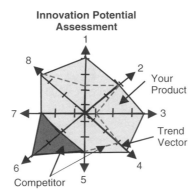

FIGURE 1-12. Toothbrush bristle: innovation potential assessment, and competitive analysis.

phase 3: a dimpled surface for more efficient cleaning; and phase 4: a porous, breathing surface where hollow bristles are filled perhaps with an aromatic and/or disinfecting fluid. Other trends might be increasing human experience (24b) for different colored bristles, or Mono-bi-poly functionality (18), for bristles with different functions such as massaging, abrasion, deep-crevice cleaning or extraction, and so on.

PRODUCT COMPETITIVE ANALYSES OR BENCHMARKING   Competitive analyses also can be performed by repeating this process for a competitor's product and then overlaying the two spider diagrams. By comparing the firm's position on individual trend vectors with those of a competitor, the following can be determined: comparative relative evolutionary positions, trend vectors where the firm lags the competitor, where the firm leads the competitor, and where there is *head room* to improve the products performance (see Figure 1-12). On some occasions, an evolutionary phase may have been skipped, and developing a product based on that phase might provide a more cost-effective alternative, or a new niche application. Although not necessary, once again, commercial software is available to facilitate analyses using the trends of system evolution.

PRODUCT DIFFERENTIATION   The evolutionary trends provide a fourth application, product differentiation through the addition of brand new applications and features to the product. This technique can be applied to a system component, a subsystem of the system, the system itself, or the super-system (i.e., what the system itself, belongs to). For example, the typing keys on a computer's keyboard are defined as components. The keyboard is a subsystem of the computer. The computer is the system. And finally, the Internet is the super-system the computer belongs to. To achieve a differentiated product, simply scan through the trends, and for each phase of a trend ask, "What new applications or features will: $X \rightarrow Y \rightarrow Z \rightarrow \ldots$ provide?" This becomes a disciplined yet very broad approach to achieving lateral thinking.

Different trends in various combinations can be used to increase product differentiation. For example, for Increasing a Chair's Differentiation, phase 4, scented pores of trend 3, Surface Segmentation, could be combined with phase 5, Traveling Wave (Body Massaging) of trend 15, Coordination of Force Dynamics, and with phases for hearing and vision of trend 24, Increasing Human Experience, to provide a virtual reality chair. These options can be thought of as many different combinations within a multi-dimensional space, or for simplicity and ease of visualization, a Product Differentiation Cube (Figure 1-13). The X and Y axes of the cube represent the 34 trends of system evolution for component 1 and component 2, respectively, and the Z axis represents the phases of each trend.

Thus, we can combine different trends and phases to obtain different feature and application combinations for different component, subsystem, and super-system combinations of a system. The Product Differentiating Cube, with multiple axes (more than X, Y, and Z), provides for exploring a multitude of component combinations using a disciplined methodology for uncovering an almost infinite number of potential, lateral thinking choices of new ideas, potentially not yet considered by customers.

In summary, 34 trends of system evolution, each with discrete phases associated with how systems evolve over time, provide the following capabilities:

◆ Product performance improvement through elimination of its contradiction

◆ Product innovation potential assessments

◆ Product competitive analyses or benchmarking

◆ Product differentiation with brand new applications and features

USE TRENDS OF SYSTEM EVOLUTION TO SOLVE FOR THE VACUUM CLEANER
The function is force. Scanning through the trends produces trend 15, Coordination of Force Dynamics, as a potential pathway for considering improvements:

Continuous action → Pulsed action → Resonance → Several Actions
→ Traveling Wave

Trend 30, Evolution of Fields (Force), is another potential improvement pathway:

Mechanical → Thermal → Chemical → Electrical → Magnetic
→ Electromagnetic

The vacuum cleaner is in the first phase of evolution for each trend, with plenty of head room for improvement.

The next step in the performance improvement process is to ask whether the force of the vacuum cleaner can be improved with/by trends 15 or 30. For trend 15, pulsed action suggests using ultrasonic air. Resonance suggests using air as an amplifier that tunes itself to the resonance frequency of the carpet

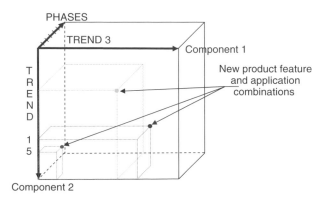

FIGURE 1-13. Virtual reality product differentiation cube.

fibers. Several actions suggest steady-state high flow with an intermittently pulsed high pressure. For trend 30, electromagnetics suggests the possibility of replacing or augmenting mechanically forced air, with electro-static force, perhaps obtained from the airflow's friction.

In using the trends to increase a vacuum cleaner's differentiation, just one subcomponent will be looked at briefly: air suction. Trend 19, the Anti-bi System, suggests a potentially new application for the vacuum cleaner by using it as an anti-system to its current function, and turning into a mini-compressor—perhaps operating as a leaf blower, for example. Alternatively, applying phase three, Resonance, of trend 15, Coordination of Force Dynamics, could lead to the feature of deeper cleaning of carpet fibers, while pulsed on-off action (phase 2 of the same trend) could lead to the added feature or benefit of being able to clean blinds without pulling them off their rack.

Using several of the trends, it is quite common to obtain numerous new, product-differentiating applications and features for a single entity like a simple chair or vacuum cleaner. For additional new features and applications, the Product Differentiation Cube concept should be applied to other vacuum cleaner components, subsystems, the vacuum cleaner itself, or the larger system it's a part of (i.e., the vacuum cleaner/carpet/human super-system).

Technique 7: Solve the Problem using the Physical Effects Database

Thousands of physical effects from science and examples from many industries have been captured in a software knowledge base using written descriptions, animation, technical, and patent references (see endnote 2 for details). By using various functional search words, the software provides numerous product differentiation examples that can be used to create new product applications and features. In addition, these databases are especially helpful when trying to improve a system's performance or when it must be made to work again after costly and poor-quality functions are pruned away. The effects database provides us with answers from industries or areas of expertise outside our knowledge base.

As an example, suppose that NASA was forced to reduce the weight of its space probes. During NASA's pruning process, heavy batteries, a source for electricity, were removed. Is there some in-situ source of electricity in space? Typing into the computer various search words like temperature and electro-motive force (EMF) produces the Seebeck effect as a potential solution (see Figure 1-14). This is an effect whereby voltage is generated across a metallic object that experiences a temperature differential, such as a space probe. The probe's side facing the sun experiences blistering heat, whereas the side facing interstellar space is at subzero temperature.

USE THE PHYSICAL EFFECTS DATABASE TO SOLVE FOR THE VACUUM CLEANER

Using search words such as *separate* or *move substance*, the database repeatedly provides examples recommending high-pressure, pulsating air jets.

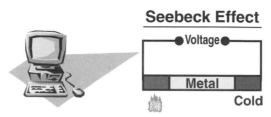

FIGURE 1-14. Seebeck effect.

SUMMARY AND CONCLUSIONS

This chapter introduces seven inventive TRIZ techniques, first by defining them and using general conceptual examples and subsequently by applying them to a mature technological system, the vacuum cleaner. For each inventive technique, our goal was to improve the vacuum cleaner's performance by eliminating the trade-off that deteriorated the system's primary function, suction. In addition, Inventive technique number 6, The Trends of System Evolution, was used to illustrate not only how it can be used for system performance improvement but also for the following:

1. Product renaissance via genesis of brand new product applications and features
2. Innovation potential assessments that define a system's potential for performance improvement
3. Competitive analyses and benchmarking

Using technique 1, the contradiction was defined as producing maximum suction and maximum flow concurrently, and the concept of high-low oscillating pressure pulses was obtained as a possible solution.

Technique 2, the ideal final result and the ideal machine concepts, suggested use of electrostatic force generated by the air's friction, to assist the vacuum cleaner's suction.

Technique 3, Pruning suggested:

- A fan that supplies both: suction (high flow), and high pressure (low flow)
- Air (assisted by pulsating high pressure air jets, operating at much more than the three PSI of current vacuum cleaners)
- Cyclonic self-cleaning air (no filter required) (e.g., the Dyson Cyclone vacuum cleaner)

Technique 4, Physical Separation, suggested separating with time the contradictory requirements of high suction and high flow by using pulsating high/low pressure suction.

Technique 5, The Invention Matrix, suggested inventive principles to improve performance:

◆ 18, Vibration, pulsating air

◆ 26, Copying, a multifurcated nozzle for localized high pressure

◆ 29, Pneumatics, compressed air for high pressure

◆ Combinations of the above

Technique 6, The Trends of Evolution, suggested:

◆ Trend 15, Coordination of Force Dynamics: ultrasonics, resonance, and intermittent steady-state flow with pulsed pressure spikes

◆ Trend 19, Anti-bi system: the combination of suction and pressure

◆ Trend 30, Evolution of Fields (force): use of electrostatic force

Technique 7, a software database search recommended, once again, pulsed, high pressure air, and the use of centrifugal force to separate the debris from the air and thus eliminate the need for the filter.

From these recommendations, three stand out as potentially world-class solutions:

1. The use of the vacuum cleaner's fan to provide both high-flow and high-pressure pulsed airflow. This solution should allow for removal of the expensive electric power brush.

2. The use of resonance as an amplifier to transform low suction into a high-output force through increased carpet fiber oscillation. This phenomenon is the same as using a violin's amplification effect, resonance, to break a wine glass.

3. The elimination of clogging filters through the use of self-cleaning cyclonic air, such as the Dyson Cyclone vacuum cleaner.

And finally, the following list of additional surprise benefits is a secondary outcome:

◆ Improved, deeper in-pile carpet cleaning through the use of carpet fiber resonance

◆ Reduced weight and size resulting from the elimination of the electric power brush and air filter

◆ Intermittent, on-off air pulsations facilitate cleaning of drapes without pulling them of their rack

◆ The additional capability to use the vacuum cleaner as a mini-compressor

◆ Synergistic force amplification resulting from the combined force differentiation of suction with pressure

◆ The many-fold increase of force over the three PSI provided by current vacuum cleaners.

KEYS TO SUCCESS IN APPLYING TRIZ TECHNIQUES

The terminology used to describe the invention matrix (attributes, inventive principles) and the trends of system evolution is, at the start, a bit confusing, but with some practice, becomes very easy to use. Another point to keep in mind: Don't become a prisoner of your words when formulating contradictions, the ideal final result, or function diagrams. Avoid the use of technical jargon, such as *centrifugal motion*, and use simple, all-encompassing phrases or verbs like *rotate* instead.

Preoccupation with software can be disappointing without a fundamental understanding of TRIZ. It's analogous to using spreadsheet software without a basic knowledge of mathematics. Acquiring a thorough understanding of techniques 1 to 6 is recommended before using the TRIZ software packages. Software automates analyses, but most problems can be solved without it.

In order for TRIZ techniques to become part of an organization's culture, it is of key importance that an enduring commitment is made to it, and that individuals who genuinely enjoy problem solving and teaching are selected to become their corporate TRIZ champions.

REFERENCES

Altshuller, G. S. 1984. Creativity as an Exact Science: The Theory of the Solution of Inventive Problems. New York: Gordon and Breach Science Publishing.

Altshuller, G. S., and I. M. Vertkin. 1994. How to Become a Genius. Life Strategy of a Creative Person, Belarus: Minsk (in Russian).

Altshuller, G. S. 1998. 40 Principles: TRIZ Keys to Technical Innovation. Worchester: TIC.

Dettmer, H. W. 1998. Goldratt's Theory of Constraints, Milwaukee, WI: Quality Press.

Roza, V., ed. 1999. TRIZ in Progress, Detroit: Ideation International, Inc.

Sklobovsky, K. A., and R. H. Sharipov. 1995. Theory, Practice and Applications of the Inventive Problems Decision, Protva-Prin: Obninsk (in Russian).

Terninko, J., A. Zusman, and B. Zlotin. 1998. Systematic Innovation: An Introduction to TRIZ (Theory of Inventive Problem Solving). Boca Raton, FL: Saint Lucie Press.

2

Quality Function Deployment and the House of Quality

Gerry Katz
Executive Vice President, Applied Marketing Science, Inc.

Quality Function Deployment, or QFD, is a methodology used to gather and "deploy" customer needs throughout the new product development process. It consists of a series of four matrices—to be filled out by a cross-functional team—which start with a prioritized set of customer wants and needs (the Voice of the Customer) and lead to a prioritized set of detailed design specifications for the new product or service. Although always a bit controversial—loved by some and reviled by others—it remains one of the best formal methodologies available for careful analysis, prioritization, and good decision making in new product development.

The controversy over QFD has little to do with the technique's inherent worth. However, a great many users of QFD have experienced significant frustration and mixed results. On the plus side are comments such as these:

"It forced us to take a fresh look at everything."
"We thought we knew what our customers wanted, but came to realize we didn't know it at all."
"We experienced the classic benefits of cross-functionality."

But the negatives are equally compelling:

"QFD is just unbelievably boring and tedious."
"It took forever and we never really finished it."
"We never really figured out why we were doing it."

This chapter is intended for project leaders and team members who are considering using QFD for their product, service, or process design (or improvement) projects. The purpose of this chapter is to provide a practical how-to guide on the use of QFD—to demystify its intricacies, to provide users with some practical dos and don'ts, and to answer the most important question about QFD that almost never gets asked: Why should anyone bother doing QFD?

This chapter is meant to be a companion piece to Chapter 7 in the *PDMA Toolbook* 2 titled "The Voice of the Customer". The Voice of the

Customer is a necessary first step in the execution of QFD. However, there is so much that has been learned about the process of gathering and organizing Voice of the Customer information that the topic often stands alone, separate from QFD. Thus, this chapter will assume the collection of a complete and rigorously derived Voice of the Customer as a prerequisite for the material that follows here.

THE HISTORY OF QFD

Although there is considerable disagreement as to its exact origins, most writers allege that QFD began at a Japanese shipbuilding firm, Kobe Shipyard, around 1972. Kobe Shipyard builds large fluid-transport vessels, mostly oil tankers. These are extremely complicated devices, requiring literally thousands of design decisions and potential tradeoffs. The story of exactly what happened at Kobe Shipyard will be covered later in this chapter.

From there, the method began to be studied by a few academics within Japanese universities, most important among them Dr. Yoji Akao at Tamagawa University. Akao continued to experiment with the technique at other Japanese corporations throughout the 1970s, and by late in the decade, the technique had made its way into the Japanese auto industry, then emerging as a worldwide quality leader. The first important journal paper about the technique was written by Nakahito Sato, a director at Toyota Auto Body. It originally appeared (in Japanese) in 1983. In it, he describes the formal methodology for QFD, including the four cascading *Houses of Quality*, in great detail (see Figure 2-1).

Two interesting observations emerge. First, the name of the paper is translated as "Quality Function Expansion and Reliability," rather than "Quality Function Deployment." Over the years, there have been a number of different names for the technique, primarily owing to different translations of the three Japanese words: *Hin-shitsu Ki-no Ten-kai*. But no matter which title is used or which translation, one of the greatest problems for QFD is its rather esoteric name: the words *quality function deployment* have almost no intuitive meaning in English! However, that is the name that has stuck, and so, unfortunately, we are saddled with it to this day.

The second issue has to do with the term *House of Quality*. Many QFDers point out (correctly) that, although the terms *House of Quality* and QFD are often used interchangeably, they are not exactly the same thing. In actuality, the term *House of Quality* refers only to the first of the four matrices used in traditional QFD. In the aforementioned Sato paper, there is a diagram of this first matrix with its distinctive *roof line*. According to Don Clausing of MIT, the Japanese, being literal thinkers, observed that this matrix looked like a diagram of a house. Thus, the term *House of Quality* came from a sentence in the paper that reads:

"The mode of this is termed the House of Quality *expansion at our company in view of its shape."*

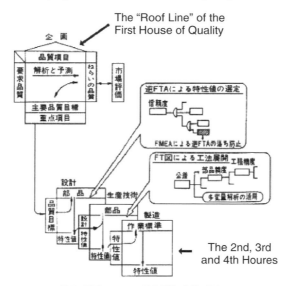

FIGURE 2-1. Diagram of the fair cascading ``Houses of Quality`` from the 1983 article, ``Quality Function Expansion and Reliability`` by Nakahito Sato, appearing in the Japanese journal *Standardization and Quality Control, Volume 36, No 3.*

Although the 1980s represented the height of U.S.-Japanese competition in the auto industry, there was, in fact, a great deal of cooperation going on within engineering circles. And by the mid-1980s, two professional societies were actively talking about and teaching QFD: the American Society for Quality Control (ASQC, now known as the American Society for Quality) and the American Supplier Institute (ASI). Furthermore, all of the (then) big three U.S. automakers were actively using QFD in one form or another by this time. And while there was spirited debate as to who had come first, the Ford Taurus emerged as an early and important market success, which validated the technique's worth.

Although many papers were beginning to appear in engineering journals, the appearance of the paper ``The House of Quality`` by John Hauser and Don Clausing of MIT in the *Harvard Business Review* in 1988 represented the first important English language description of QFD in a popular management-oriented journal. This served to further legitimize this rather technical topic, which was being widely touted among engineers and total quality management (TQM) practitioners. From there, the use of QFD exploded far beyond its origins within the automotive industry. Today it is used for all kinds of manufactured goods, services, and process improvement applications. In addition, as one of the keystones of Design for Six Sigma (DFSS), QFD has continued to achieve widespread use, despite its drawbacks.

THE PROBLEM AT KOBE SHIPYARD

When the product development people at Kobe Shipyard began work on their next generation tanker design project in 1972, there was already a strong orientation toward first studying customer needs, and then trying to design the vessel around those needs. This was already somewhat radical thinking for its time, in that most Western companies approached new product development as an activity that belonged in the R&D laboratory or at the engineer's bench. They would try to innovate around what they *thought* customers needed, hoping that when customers saw the finished product, they would agree. It took more than a decade for Western companies to approach new product development in the way that Kobe Shipyard did.

As each customer need was identified, the design engineers at Kobe came up with potential solutions to these customer needs. They dubbed the needs the *whats* and the solutions were referred to as the *hows* (Figure 2-2). They began the exercise thinking that there was a clean one-to-one correspondence between each *what* and each *how* (Figure 2-3). However, it quickly became apparent that their world was not so tidy. In fact, they came to realize that every *how* seemed to impact more than one *what*, and sometimes in conflicting ways (Figure 2-4). In trying to deal with all of this complexity, one of the engineers at Kobe came up with the idea of turning the *what's* and *how's* end-to-end, thus creating a two-dimensional matrix (Figure 2-5). Then, the

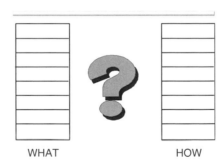

WHAT   HOW

FIGURE 2-2. Linking the WHAT's to the HOW's

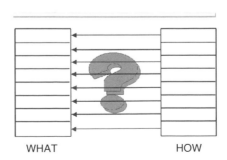

WHAT   HOW

FIGURE 2-3. Causality for Linking the WHAT's to the HOW's

Linking the WHAT's to the HOW's

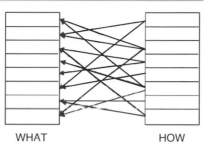

WHAT   HOW

FIGURE 2-4. Multiple interactions in linking the WHAT's to the HOW's

Linking the WHAT's to the HOW's

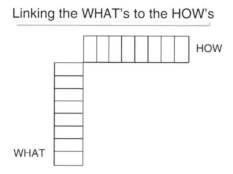

FIGURE 2-5. The origins of QFD: linking the WHAT's to the HOW's

center of the matrix could be used to indicate how strongly each *how* impacted each *what*.

This was the beginning of the QFD matrix structure, and it has today evolved into the traditional House of Quality matrix that will be discussed in detail a bit later.

WHAT YOU NEED TO GET STARTED

QFD is a best done as a cross-functional team process. Although any individual could fill out a QFD House of Quality matrix, there is often a great deal of disparate interpretation coming from people in different functional areas of the company. And many argue that the greatest value from the use of QFD owes to the fact that it forces people from different functions to articulate and work out these differences in interpretation in an attempt to reach consensus. A good cross-functional team for QFD usually consists of about 8 to 10 people with high levels of market, customer, and technical knowledge, historical perspective, responsibility, respect, and clout within the organization. Here are nine criteria for choosing good team members:

1. *The team should be truly cross-functional.* A good QFD team will include individuals from design engineering, manufacturing, research and development, marketing, market research, sales, finance, industrial design, customer service, product maintenance, technical support, and so on. Almost any function that will be involved in the product's design or will have a stake in its outcome should be considered for inclusion on the team.

2. *The team will need an administrator and an "advocate" for the Voice of the Customer.* The role of the administrator is simply meant as a central point of contact for scheduling meetings, arranging for rooms, refreshments, and other logistics. And while all of this same team should have been involved in the process of gathering the Voice of the Customer, it is useful to have at least one member who has been even more highly involved than the rest of the team throughout, and is thus more familiar with all of the details, to act as a representative of customers when the inevitable question comes up as to exactly what customers meant by a particular need.

3. *The team members should be people who have ultimate responsibility to act on the results.* Although there is nothing wrong with including staff members on QFD teams, these people will have the added burden of having to convince their line managers of the correctness of their analysis and conclusions. If key line managers are part of the effort, they are already convinced and merely need to act on the results. This lesson was learned during two highly contrasting applications for two major electric utility companies. In one, the exercise was carried out by an all-staff team. They did a great job, but when they presented their results and recommendations to management, requesting additional resources to carry out their recommendations, they were turned down. In the other application, the team included several key line managers—people who already controlled many of the critical process functions within the company, along with control of their own budgets. They didn't need to convince anybody else—they simply acted!

4. *Team members should have real knowledge of current practice and significant historical perspective.* Throughout a QFD effort, questions will arise as to how things were done in the past and what happened. Having a few people on the team with this type of technical knowledge and long-term historical perspective can often help to better inform the process.

5. *Team members should have the broad respect of their peers.* Many people within a company are inherently skeptical about new product development. For instance, sales people, who often believe that they have superior market knowledge, are often skeptical of any activity going on at the *home office*, especially when it is being carried out by what they might see as "those techies" in engineering who lack their superior level of hands-on market knowledge. Including a

highly respected salesperson on the team can often serve to allay their skepticism. Likewise, including some of the better known, more seasoned engineers might be a better choice than some newly hired recent graduates—even if they had superior technical skills—for much the same reason.

6. *You should include some big names—people with clout.* Following a QFD exercise, there is inevitably going to be a requirement for resources in order to move forward. Including some of these high-level people lends credibility to these budget requests, thus easing the way.

7. *It is widely held that good teams do not include people from extremely different levels of the organization, but ...* The reason for this is a belief that higher-level people will dominate the discussion, making it difficult for lower-level people to express their honest views. However, experience has shown that this is more a function of the personalities of the people involved, rather than their level within the organization. So long as the participants are all good team players, encouraging and listening to opinions expressed by people from all levels of the organization, this should not be a major problem. A good facilitator, whether internal or external to the organization, can also help along these lines.

8. *Look for "constructive contrarians."* Contrarians can be both a force for good as well as ill within a QFD team. The negative kind are those who simply enjoy arguing for its own sake. Whatever the issue, they want to take the opposite point of view, believing that they are "thinking outside the box" and expanding the discussion. More often than not, however, they are merely complicating and slowing the process down. A constructive contrarian, on the other hand, is one who simply thinks of things a little differently from others and does indeed stretch their thinking. However, when their peers are not convinced, they are easily able to let go and move on to the next question. Experience has shown that these constructive contrarians can be an important asset on a team.

9. *Finally, don't try to be too democratic.* While it is obviously good to have a representative from each of the important areas of the company, the previous eight criteria should take precedence. It is far better to leave a certain function out than to include an individual who is likely to contribute little and possibly disrupt things a lot.

THE HOUSE OF QUALITY MATRIX

The traditional House of Quality matrix that has evolved today includes six sections that are euphemistically referred to as the *rooms* in the house. Each of these will be discussed in turn (Figure 2-6).

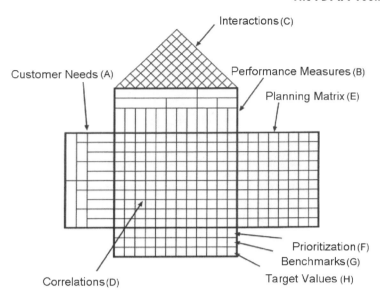

FIGURE 2-6. A traditional House of Quality matrix.

The Customer Needs

The left-hand room contains the customer needs, or the *whats* (Figure 2-6, Room A). This is exactly what the Voice of the Customer (VOC) is intended for, and the need for this type of information resulted in the creation of the many VOC techniques in use today. Most QFDers suggest about 15 to 25 needs as the ideal level of *granularity* for QFD. Since the VOC process often identifies 100 or more such needs, an affinity diagram (see *ToolBook 2*, Chapter 7) is often used to group or cluster the needs into the right level of detail. These needs are listed on the left-hand side of the matrix where they define the rows going forward.

The Performance Measures

Historically, the columns of the matrix have gone by a number of different names: Hauser and Clausing refer to them as *engineering characteristics*, but they are also referred to as *technical quality characteristics, design metrics, performance measures,* and to this day, the *hows*. In Six Sigma lingo, they are called CTQs (critical-to-quality measures). This chapter will use the term *performance measures* as it carries with it the advantage that it works equally well for both manufactured goods as well as services (Figure 2-6, Room B).

Whatever you choose to call them, they carry with them the same definition. They can be thought of as: The performance knobs we can turn such that, if moved in the proper direction, will result in greater satisfaction of customer needs over time.

What exactly is meant by this? The term *knobs* implies that we are looking for things that are directly controllable by the company in the design of the new product or service. (A customer satisfaction measurement would not qualify, as it is not directly controllable by us; rather, it indicates, after the fact, how well we have satisfied a need.) These performance measures should ideally have five key characteristics.

1. *A good performance measure is directly related to one or more customer needs.* If it is not related to an important customer need, there is no point expending any budget or effort on such a performance measure.

2. *A good performance measure is predictive of customer satisfaction.* If moving this performance measure in the desired direction does not have a positive impact on pleasing the customer, then again, there is little point in expending any effort on it.

3. *A good performance measure must be measurable.* Although the team may later choose to measure some things and not others, there must be some objective way of deciding whether design *A* will improve things on this measure more than design *B*, or vice versa.

4. *A good performance measure is directly controllable by the team in the design of our product.* It ought to be a matter of choice in how we design the product or system.

5. *A good performance measure is independent of implementation.* A common mistake among the less experienced is that they choose Boolean criteria as their performance measures. For instance, for an office-seating product, they might define two performance measures such as "foam seat cushion" and "poly-filled seat cushion"—features that are either present or absent. The performance measures should ideally be continuous rather than Boolean variables. A better performance measure would be one that would help the team to evaluate whether foam or poly has the better impact on satisfying the customer's needs. So a performance measure like *amount of displacement on the seat cushion under 175 pounds of pressure*, or *moisture retention of seat cushion measured 10 minutes after spray test* would be superior performance measures. These types of measures can then be used to evaluate whether foam or poly is a better design decision.

The process of choosing performance measures ought not to be taken lightly, for experience has shown it to be the most creative part of a QFD exercise. The process by which the team debates and chooses performance measures usually provokes a good deal of creative thought, which leads to more creative design decisions. For instance, in the process of designing a new office cubicle, one team was grappling with the customer need *gives me the maximum amount of desk space to work on*. The team's initial thinking was to use as its performance measure *the surface area of the desktop*—that is, the length times the width. However, one team member declared that that was too simplistic, because of all of the parts of the desktop that taken up by things such as computers, telephones, in-baskets, staplers, tape dispensers, and so on. So, another team member suggested that perhaps the performance measure ought to be *the surface area of the desktop minus the typical footprint of a computer, telephone, in-basket, and so on.*

The minute the team agreed to this, there was an awkward silence in the room. What were team members thinking about? In that split second, their minds quickly turned to designs that would remove some of these footprints from the desktop. This resulted in ideas such as mounting the phone on the cubicle wall, suspending the monitor from overhead, hanging the in- and out-baskets from the overhead shelf, and so forth.

Debating the best performance measures often leads to this kind of breakthrough creative thinking.

The Interactions

The *interactions* is the section that gave the House of Quality its distinctive name—for it resembles the roofline of a house (Figure 2-6, Room C). Ironically, it is not used all that often, for it has no mathematical impact on the rest of the matrix. Its purpose is simply to highlight any strong positive or strong negative interactions between any two of the performance measures. By this it is meant that many of the performance measures may not be independent of one another. Moving one performance measure in a positive or negative direction may make it harder or easier to move another. When this is believed to be the case, the team goes up to the diagonal block, which represents the intersection of these two performance measures and simply put a plus (+) or a minus (−) sign to indicate where these strong positive or strong negative interactions might occur. Some QFDers actually use a five-point scale ranging from a double plus (++) to a double minus (−−).

Consider the previous example of giving people *the maximum amount of desk space to work on* in a cubicle. This would likely bump up against another important need for the space planners: the ability to *fit as many workers as possible into a given floor plan.* If this were the case, there would clearly be a conflict between the two resulting performance measures: *the area of usable desk space per cubicle* versus *the number of cubicles able to fit per thousand square feet of floor space.* Without some highly creative solution, increasing one will almost certainly make it harder to increase the other—a strong negative interaction.

Although many QFDers skip this section of the matrix entirely, those who use it tend to swear by it. They claim that it definitively points the way to the most important design conflicts that will need to be addressed in order to create a successful new product or service. The interactions section helps to highlight these suspected points of conflict, and while it doesn't actually help the designer solve all the necessary tradeoffs and technical problems, it does point the way toward those places where new technologies or new creative solutions might be sought out to overcome them. For this reason, and given that it can usually be completed in just an hour or two, it is now generally recommended that it be included.

The Correlations

The center of the matrix is referred to as the *correlations* (Figure 2-6, Room D). Unfortunately, this is the most tedious part of the QFD process, the part that often bogs a team down and sinks it. The necessary task is to go through every cell on the matrix and answer the question of how strong is the relationship between each performance measure and each customer need. Typically, QFD has used a four-point scale:

1. Strong relationship
2. Moderate relationship
3. Slight relationship
4. No relationship

Mathematically, these four points are usually scaled as 9, 3, 1, and 0 (actually, the zeros are simply left blank), respectively, although there is no reason why these particular numerical values must be used. Most likely, the originators of QFD wanted more of a logarithmic scale in which the interval between strong and moderate would be an order of magnitude greater than the interval between moderate and slight. Although some prefer to show the actual numerical values in each cell, the Japanese use a set of symbols as shown in Table 2-1.

Why is this process so tedious? It is common for a team to generate two to two-and-a-half times as many performance measures as customer needs. Thus a matrix with 20 rows is likely to have 40 to 50 columns, or as many as 1,000 cells overall. If a typical cell requires anywhere from 3 to 10 minutes of discussion and debate, it is easy to see why this part of the process can grind a team to a halt.

There are a few practical ways to reduce this tedium. The first is to ask each team member to fill out a complete matrix individually, entering their 9s, 3s, 1s, and blanks into a spreadsheet. Then, one person can compile all of the individual spreadsheets into an overall frequency table. What is often the case is that there is a great deal of consensus on anywhere from two-thirds to three-quarters of the cells in the matrix. If almost everyone on the team feels

TABLE 2-1.
Symbols used for Strength of Correlation

Strong relationship	9	◉ or ●
Moderate relationship	3	○
Slight relationship	1	△
No relationship	0	(blank)

that a particular cell points to a strong relationship or no relationship, why bother to waste valuable team time debating these high-consensus cells? Simply fill them in with the agreed-upon value or symbol being used. This will usually leave about 200 to 300 cells where there is a pronounced lack of consensus, and these are the ones that deserve team discussion, debate, and an attempt to reach consensus.

Another common technique to reduce the tedium is to split the team into subgroups around specific areas of expertise such as marketing, manufacturing, or materials science, and then to assign particular rows and/or columns to various subgroups based on their area of expertise. Each sub-group then completes the correlations for their section, with the possibility of bringing a reduced number of the more controversial cells back to the entire group for wider discussion and resolution.

Note that simple averages are never used. Rather, the team needs to spend time trying to understand why some team members believe there is a strong or moderate relationship while others see only a slight or no relationship. These disagreements usually occur not because people are not thinking clearly, but rather, because people from different functions and different parts of the organization often see things quite differently. The process of exposing this incongruent thinking is where much of the learning in QFD occurs, and it is important to get it out on the table. It is often the case that there will be compelling arguments on both sides of these issues, and a healthy debate will ensue. After, say, 5 to 10 minutes of discussion, the team facilitator calls for a revote to see if a consensus can now be reached. If not, the debate can continue for a few minutes more, or the team can simply choose to end the debate by choosing to compromise on a central value.

Another potential time saver requires a bit of additional technology. Experienced QFD facilitators will tell you that the process of counting these revotes adds significantly to the tedium. As people tire, it is often unclear as to whose hand is up or down, and the larger the team, the more complicated this becomes. A number of automated voting devices are now used in live market research (such as focus groups) that can speed this process. At the end of the debate, team members are simply asked to use their voting device to type or dial in their vote. A central PC tabulates the votes and presents the results on a screen for everyone to see. In addition to reducing the tedium, this also adds an element of fun to the process.

Finally, it should be noted that nothing in QFD should ever be thought of as "set in stone." The team can always return to add, delete, or redefine performance measures, or to rethink the correlations. Although fatigue alone is likely to limit the amount of such rethinking, it is quite common for one or two team members to lose a debate on a point that they feel quite strongly about and to want to revisit it later in the process. This possibility should never be denied. Some teams create a formal "appeals process" in which any team member can bring an off-line request to the administrator to re-open discussion on any cell at any time. The administrator then decides which of these to bring back to the entire team.

In the early days of QFD, it was not uncommon for teams to spend anywhere from 4 to 10 days on the correlations alone, leading many teams to simply give up before completing them. But today, using some of the techniques described above, it can usually be completed in two days or less—still significant, but not intolerable!

The Planning Matrix

The right-hand "room" in the house is referred to as the *planning matrix* (Figure 2-6, Room E). It contains the importance and performance (or satisfaction) ratings from the Voice of the Customer process. A more detailed example of a planning matrix is shown in Figure 2-7.

The key output from the planning matrix is the final column on the right, which is known as the *raw weight* (column 7). The raw weight is simply a somewhat "judgmentally enriched" version of the customer-provided importance scores (column 1); it is the key variable that is used in the final calculation of the matrix (to be addressed in the next section on "Calculating the Matrix"). What is meant by the expression *judgmentally enriched*? Although each need is given an average importance score by the customers (column 1), QFD allows for the team members to adjust these scores in a number of ways.

First, it is often the case that, for any of a number of reasons, a team may want to put some extra emphasis on certain customer needs over others. These might include a need for which the company owns an emerging technological advantage, or a need in which the company lags behind a key competitor who is currently *best in class* on that need and thus the project requires extra emphasis in order to catch up or even pull ahead. The team considers the *current performance* (column 2) relative to *competitive* or *best in class performance* (column 3) and judgmentally enters a *targeted performance* value (column 4). Then, an index called an *improvement ratio* is calculated (column 5)—actually, just a ratio of the two—which can later be multiplied against the customer-provided importance score (column 1), thus increasing its value if the targeted performance is greater than the current performance (an improvement ratio greater than 1.0).

For instance, suppose the need *gives me the maximum amount of desk space to work on* received an importance rating of 72 on a 100-point scale (column 1). And suppose that the firm's current cubicle product received a performance / satisfaction score of 8.2 on a 10-point scale (column 2), while the best in class product received a 9.0 (column 3). If the team concluded that it needed to at least match the best in class product, then it would give this need a targeted performance score of 9.0 (column 4). So, the resulting improvement ratio would be $9.0 \div 8.2 = 1.10$ (column 5). And multiplying the importance score of 72 by the improvement ratio of 1.10 (column 1 × column 5) would give us a raw weight for this need of 79.

Another potential adjustment is referred to as a *sales point* (column 6). Many of the customer needs will require solutions that are almost invisible

	Importance	Current Performance - Us	Current Performance - Best in Class	Targeted Performance	Improvement Ratio (Index)	Sales Point (Index)	Raw Weight
	1	2	3	4	5	6	7
Need 1	1						1
Need 2	2						2
Need 3	3						3
Need 4	4						4
Need ...	5						5
	6						6
	7						7
	8						8
	9						9
	10						10
	11						11
Need n	12						12

FIGURE 2-7. The Planning Matrix.

to the customer. But some might create a distinctive market advantage that could and should be used in the product's marketing communications and sales efforts. These can also be expressed as an index, which is multiplied against the original importance score. So to extend this example, if the team felt that responding to the need to provide *more desk space to work on* would be a potent selling point for a new cubicle design, then they might assign a sales point index of 1.20. In this case, the original importance score of 72 would be multiplied by both the improvement ratio index of 1.10 and the sales point index of 1.20, now resulting in a raw weight for this need of 95 ($72 \times 1.10 \times 1.20 = 95$).

Thus, the raw weight (column 7) is simply the initial importance score (column 1) multiplied by these two potential adjustment factors: a targeted improvement ratio index (column 5) and a sales point index (column 6). A caution is recommended here. If the importance scores themselves are assigned judgmentally, there is probably little harm in judgmentally adjusting them. However, ideally, both the importance and performance scores have been obtained from customers in the Voice of the Customer process. If this is the case, these judgmental adjustments should be limited in scope. There is little point in spending all of that time and effort to obtain customer-supplied values for the importance scores, only to judgmentally adjust them to suit the team's thinking. For instance, if the customers indicate that a particular need is in the bottom quartile in importance, it makes little sense to suddenly kick it into the top quartile just because we think it is deserving of additional effort or would make a good sales point. Thus, it is recommended that these adjustments be limited to approximately 10 to 20 percent—that is, adjustment indices of 1.10 to 1.20. In fact, many teams choose not to make any adjustments at all.

The planning matrix should take only one to two hours to complete, and again, the only necessary output is the final column—the raw weight—which represents a final adjusted importance weighting for each of the customer needs (the rows) in the matrix.

Calculating the Matrix

Finally, the bottom "room" on the matrix is where the results of the matrix calculation will be stored (Figure 2-6, Row F). The calculation itself is really just a large "weights times rates" calculation. Mathematically, it is expressed as:

$$\text{Criticality} = \sum (\text{correlations} \times \text{Raw weights})$$

For each column, we simply multiply each correlation (the 9s, 3s, and 1s) by its row's raw weight (the final column at the right of the planning matrix), and add all of the values together. The sum is recorded at the bottom of each column, in the first row, which is labeled *prioritization*. As an example, consider the first column in the House of Quality matrix shown in Figure 2-8 (which will be discussed in the next section). For this performance measure, the team believes there is a strong relationship (scaled as a 9) with needs #1 and #5, a moderate relationship (scaled as a 3) with need #11, a slight relationship (scaled as a 1) with need #6, and no relationship with any of the other needs. So, multiplying each of these scale values (the 9s, 3s, and 1s) by their respective raw weights for each row yields a prioritization score for that column of:

$$(9 \times 101) + (9 \times 99) + (1 \times 97) + (3 \times 107) = 2214$$

These column totals actually have no units to them—they are simply a set of numbers. They are often referred to as *utility points* or *criticality scores*. What they represent is the overall strength of impact—across all of the customer needs—of moving that performance measure in the desired direction by a significant amount.

Once the value has been calculated for all of the columns, it is useful to re-sort the performance measures from highest to lowest prioritization value, with the clear interpretation that those values on the left—those with higher values—should receive more emphasis in the product's design than those on the far right.

Under what conditions would a performance measure obtain a high score under this scheme? In most cases, two things need to occur. First, a column with more 9s and 3s is likely to receive a much higher criticality score than one with only one 9 or one 3. And second, a column where those 9s and 3s occur on needs with higher raw weights is likely to end up with a higher criticality score. What does this imply? Simply that a performance measure that strongly impacts multiple customer needs, and where that impact is on the more important customer needs, is likely to deserve higher prioritization in the product's design than those with lower scores. This is exactly as it should be!

Unfortunately, QFD does not answer every question necessary for the product developers. Although this highly rigorous process does an excellent job of identifying which performance measures deserve emphasis, it stops short of actually helping the team set new target values for those performance measures (Figure 2-6, Row H). This part is most often done through a combination of benchmarking (Figure 2-6, Row G) and judgment. In those cases where the value of the performance measure is either already known or can easily be obtained, the team simply looks at where its current product falls on that performance measure, where the competition falls, and then makes a judgment as to how far they need to "turn the knob."

However, it is often the case that many of these performance measures are new, and thus, their value is unknown for both the firm's own products and their competitors' products. Measurement is never free, and thus the team will have to decide where to put its resources in terms of actually determining current values before being able to set target values for the future. Obviously, those performance measures that turn out to be more critical are the ones that are most deserving of resources for this new measurement process.

AN EXAMPLE OF A COMPLETED HOUSE OF QUALITY

Figure 2-8 shows an example of a completed House of Quality matrix using the *Voice of the Customer* from the office-seating products example in Chapter 7 of the *PDMA ToolBook 2*.

To begin, the customer needs are listed down the left-hand side of the matrix, and the team begins the process of generating performance measures. For instance, in response to the need *easy to scoot around in my workspace*, the team decides that it should measure the amount of force needed to move the chair on its casters under a typical load of 175 pounds (an average body

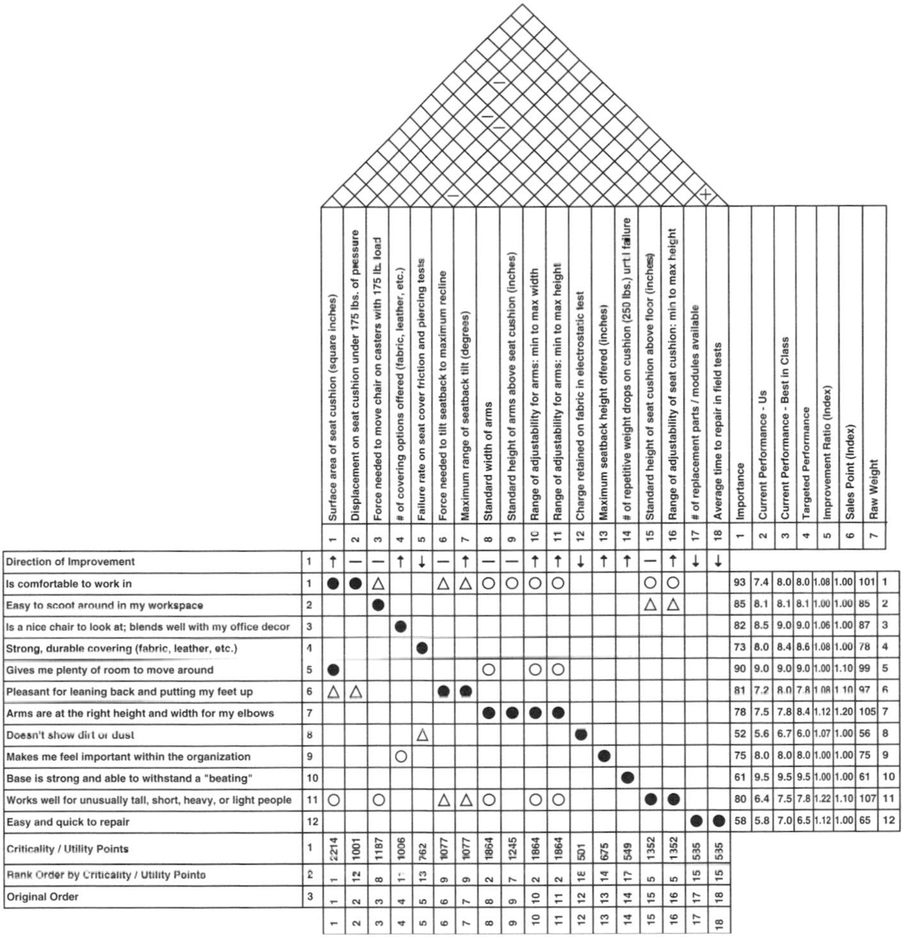

FIGURE 2-8. An example of a completed House of Quality matrix.

The matrix columns (performance measures) are:

1. Surface area of seat cushion (square inches)
2. Displacement on seat cushion under 175 lbs. of pressure
3. Force needed to move chair on casters with 175 lb. load
4. # of covering options offered (fabric, leather, etc.)
5. Failure rate on seat cover friction and piercing tests
6. Force needed to tilt seatback to maximum recline
7. Maximum range of seatback tilt (degrees)
8. Standard width of arms
9. Standard height of arms above seat cushion (inches)
10. Range of adjustability for arms: min to max width
11. Range of adjustability for arms: min to max height
12. Charge retained on fabric in electrostatic test
13. Maximum seatback height offered (inches)
14. # of repetitive weight drops on cushion (250 lbs.) until failure
15. Standard height of seat cushion above floor (inches)
16. Range of adjustability of seat cushion: min to max height
17. # of replacement parts / modules available
18. Average time to repair in field tests

Right-side columns:

1. Importance
2. Current Performance - Us
3. Current Performance - Best in Class
4. Targeted Performance
5. Improvement Ratio (Index)
6. Sales Point (Index)
7. Raw Weight

Customer Need	Importance	Current Perf. Us	Current Perf. Best	Targeted	Improvement	Sales Point	Raw Weight
Is comfortable to work in	93	7.4	8.0	8.0	1.08	1.00	101
Easy to scoot around in my workspace	85	8.1	8.1	8.1	1.00	1.00	85
Is a nice chair to look at; blends well with my office decor	82	8.5	9.0	9.0	1.06	1.00	87
Strong, durable covering (fabric, leather, etc.)	73	8.0	8.4	8.6	1.08	1.00	78
Gives me plenty of room to move around	90	9.0	9.0	9.0	1.00	1.10	99
Pleasant for leaning back and putting my feet up	81	7.2	8.0	7.8	1.08	1.10	97
Arms are at the right height and width for my elbows	78	7.5	7.8	8.4	1.12	1.20	105
Doesn't show dirt or dust	52	5.6	6.7	6.0	1.07	1.00	56
Makes me feel important within the organization	75	8.0	8.0	8.0	1.00	1.00	75
Base is strong and able to withstand a "beating"	61	9.5	9.5	9.5	1.00	1.00	61
Works well for unusually tall, short, heavy, or light people	80	6.4	7.5	7.8	1.22	1.10	107
Easy and quick to repair	58	5.8	7.0	6.5	1.12	1.00	65

Bottom rows (by performance-measure column 1–18):

- Criticality / Utility Points: 2214, 1001, 1187, 1006, 762, 1077, 1077, —, 1245, 1864, 1864, 501, 675, 549, 1352, 1352, 535, 535
- Rank Order by Criticality / Utility Points: 1, 12, 8, 11, 13, 9, 9, —, 7, 2, 2, 18, 14, 17, 5, 5, 15, 15
- Original Order: 1, 2, 3, 4, 5, 6, 7, 8, 9, 10, 11, 12, 13, 14, 15, 16, 17, 18

weight). Likewise, in response to the need *makes me feel important within the organization*, the team learned during the Voice of the Customer process that office workers equate a higher seatback with greater organizational importance. Thus, the team decides that the maximum seatback height offered would be an appropriate performance measure.

Notice that all performance measures have one of three characteristics. With some, more is better, and so the team should strive to increase the measure (↑ in the "Direction of Improvement" row in Figure 2-8). With others, less is better, and so they should strive to reduce the measure (↓). And with a few, there is some central target that would be best, such that any value greater or less than the target value would be worse (−). For the seatback height, higher would always be better (within practical limits, of course). But with regard to the force needed to move the chair on its casters, there is likely to be some target that is ideal. If it requires too much force, it is not "easy to scoot around," while if it requires too little force, the chair becomes unstable and might result in people falling when trying to sit down. The team will likely have to conduct some prototype testing with customers in order to determine the exact target.

Notice, also, that in choosing performance measures having to do with the width and height of the arms, the team began to think about making them adjustable, and so in addition to using standard width and height measures, they have also included measures regarding the *range of adjustability*. This is an example of the creativity that often results from debating about the performance measures referred to earlier in this chapter.

This team identified only a handful of interactions for the roofline of the matrix. For instance, they believe that increasing the number of replacement modules will reduce the average repair time (columns 17 and 18)—a good thing—while increasing the durability of the fabrics offered will increase their ability to hold an electrostatic charge (columns 5 and 12), which, in turn, will attract more dust and dirt—a bad thing!

In the planning matrix for this example, the importance scores were gathered using a 100-point scale, while the performance scores made use of a 10-point scale. Notice that, in choosing targeted improvement values, the team decided that there were some measures where they wanted to match the best in class, others in which they only felt they could play catch-up, and a few in which they believed they should try to set a new best-in-class standard. Likewise, for the sales point indices, there were a few that they believed could have some impact on their marketing messaging, and one that they thought could have a somewhat larger impact. Thus, they gave them indices of 1.10 and 1.20, respectively. Remember that these items are entirely a judgment call.

The resulting raw weights in column 7 are sometimes equal to the initial importance scores in column 1 and other times are somewhat greater—demonstrating the concept of what was earlier referred to as *judgmentally enriched* importance scores.

After completing all of the above rooms in the matrix, the team is now ready to perform the matrix calculations. These scores are shown in the row at the bottom of the matrix in Figure 2-8, labeled "Criticality/Utility Points." A higher score indicates that this would be a more advantageous measure to work on than one with a lower score. Figure 2-9 simply reorders the columns from highest to lowest criticality score, thus aiding the interpretability of the matrix. The indication in this example is that increasing the surface area of the seat cushion would be the most beneficial thing the team could do, and that making the seat height adjustable and the arms adjustable with regard to both their height and width would be close runners up.

Interestingly, improving the durability of the chair frame and the fabrics would likely prove to be far less critical in this example to satisfying customers. This final identification and prioritization of technical specifications is precisely what QFD is intended for, and the development team can now proceed to

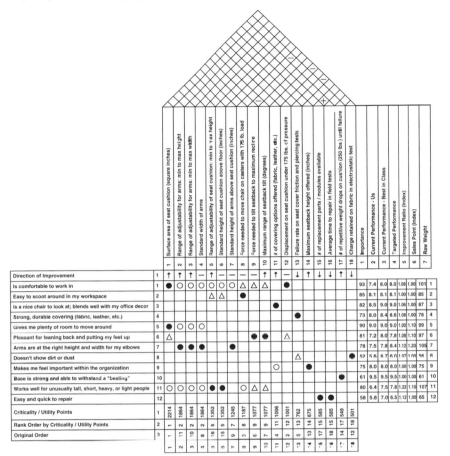

FIGURE 2-9. An example of a completed House of Quality matrix sorted by criticality.

focus its attention on a smaller set of rigorously determined critical design parameters.

AFTER THE HOUSE OF QUALITY, WHAT'S NEXT?

After completing a House of Quality matrix, there are a number of potential next steps. This section looks at those steps.

Build More Houses

In its most rigorous form, QFD includes a series of four matrices (Figure 2-10). What are these subsequent matrices for? Their purpose is to deploy the Voice

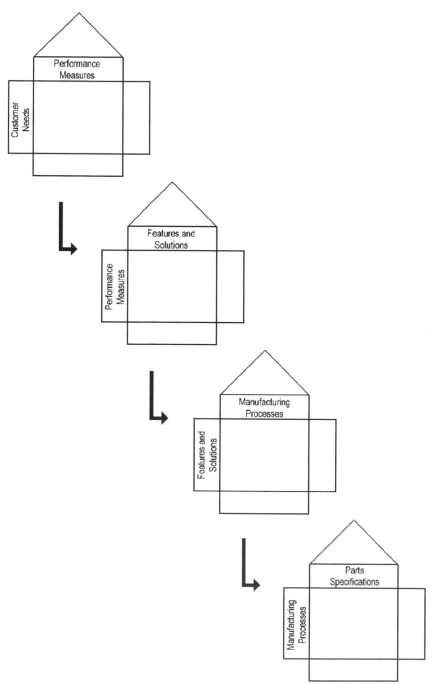

FIGURE 2-10. The four cascading Houses of Quality.

of the Customer throughout the decision-making process for the new product. So for instance, the second matrix usually takes the most critical performance measures as defined in the columns of the first House of Quality matrix and lists them down the side of a new matrix. Then, the team can generate potential *features* or *solutions* that could move these performance measures, and use them as the columns in the new matrix. Thus, this matrix will answer the question of which features or solutions will best move the performance measures—which will, in turn, address the customer needs.

Likewise, a third matrix can take the winning features and solutions, listing them down the left-hand side, and the team can then generate a set of *manufacturing processes* as the new column definitions. In turn, a fourth matrix can take the winning manufacturing processes and test them against, for instance, *parts specifications*.

The logic here is impeccable. These matrices can help a team make decisions as to which parts specifications aid in certain manufacturing processes, which, in turn, address certain features or solutions, which, in turn, address the key performance measures, which, in turn, address the customer needs—the Voice of the Customer. But as a practical matter, some who have been through these multiple matrices refer to them as *matrix hell*. It already requires a great deal of time and effort to get through the first House of Quality matrix. Although the productivity techniques just described have generally reduced the typical amount of team meeting time required for a single complete House of Quality matrix from 8 to 10 days down to about 4 to 5 days, this still represents a substantial investment in people resources. And while some teams utilize other somewhat less desirable shortcuts, such as only doing part of the matrix, it is rare for it to take less than 2 days.

Most QFDers claim that 80 percent or more of the value to be obtained occurs in that very first matrix. Thus, relatively few QFD initiatives go all the way through the four matrices. That first House of Quality matrix is usually highly beneficial by itself and is a good logical stopping point.

There is one reasonable exception. Professor Stuart Pugh of the United Kingdom suggested a slight variation on the second House of Quality, which today is referred to as *Pugh Concept Selection* (Figure 2-11). In this variation, the key performance measures are listed down the left-hand side of the matrix. (Many people simply list the customer needs, although technically, that is not what Pugh suggested.) The columns are then defined as follows: The first column, referred to as the *datum*, represents the existing product, which serves as the initial reference point. Then, a number of conceptual designs are shown in the subsequent columns to the right. It is useful to give these new concepts a name or even a symbol so that the team members can easily recall what they are about. The team then evaluates each concept on each performance measure (or need), asking the question of whether such a design would be better or worse than the datum.

The scale that is usually used is identical to the one used for the *interactions*: a plus (+), a minus (−), or a blank, depending on whether that concept is

		Datum - Current Product	Concept A	Concept B	Concept C	Concept D	Concept E
		1	2	3	4	5	6
Performance Measure 1	1			+	−	−	+
Performance Measure 2	2		+	−		+	
Performance Measure 3	3		+		+		−
Performance Measure 4	4			−	+		+
Performance Measure 5	5			−		+	+
Performance Measure ...	6			−	+	−	−
	7						
	8						
	9						
	10						
	11						
Performance Measure n	12			−	+	+	−
Net Score - Sum of Plusses and Minuses	1	0	1	-2	3	-1	1
		1	2	3	4	5	6

FIGURE 2-11. Example of a Pugh Concept Selection matrix.

believed to be better, worse, or equal to the datum. (Again, some QFDers also allow the use of a double plus ($++$) or a double minus ($--$).) Once each of the concepts has been rated, the net number of plusses and minuses is calculated for each column and the score listed at the bottom. The winning concept now becomes the new datum and additional concepts can be generated and evaluated against it. The goal, of course, is to iteratively reach a final concept that takes advantage of all of the best parts of each of the previous concepts and leaves aside the worst parts.

Pugh concept selection can usually be done quite quickly and is often a good way to summarize and complete the QFD process.

Choose Which Metrics to Focus On

Another post–House of Quality activity is simply to have the team choose exactly which of the performance measures it wishes to focus on going forward. There are several ways to do this. Some simply choose the top *n* (i.e., those performance measures that achieve the highest criticality scores). This decision usually hinges on the magnitude of resources available for continuing R&D and engineering.

Many teams also choose to include some of the *low-hanging fruit*—those items that are further down in the priority list, but are easy, fast, or inexpensive to address. There are always some performance measures that can be moved through something as simple as a policy change in the way the company sells or services a product. And if these things can be implemented quickly and easily, they might as well be included, even though they are not among the highest-priority things to do.

Performance Measure Analysis (PMA)

A somewhat more rigorous process involves an analysis of each of the performance measures from the House of Quality matrix. For each measure, the analysis should include a discussion of the following:

1. *Who owns this measure?* Many performance measures have a clear organizational "owner." Some clearly belong to a single function such as manufacturing, design engineering, marketing, or product maintenance.
2. *What is the current value?* As described earlier, in some cases we already have good information as to where we stand on that performance measure, while in others this is going to require new measurement.
3. *Do we know enough to set a new target value?* In most cases, the answer will be no, but in those where it *is* possible, we ought to get it down on paper right now.
4. *How can this performance measure be moved?* Because of the creative nature of QFD, a team will often have come up with several really good ideas as to how to move that performance measure—for instance, the earlier example about removing items from our cubicle's desktop. Whoever the process owner might be, it is worthwhile to catalog any of these ideas as a good starting point for the person responsible.

Cost-Benefit Analysis

Perhaps the most rigorous post-QFD activity is to perform a formal cost-benefit analysis. This requires the team to estimate in very rough terms the amount of money and time it will take to move that performance measure by the required significant amount. This represents the *cost* part of the equation. The *benefit* part is simply represented by the criticality scores. Then, calculating the ratio of these two produces a formal mathematical way of weighing cost against benefit.

These numbers can also be risk-adjusted by including a "probability of success" factor. Once this has been done, the team may learn, for instance, that while the twelfth most important performance measure only delivers half as many criticality points as the third most important, it may be a better item to address because it will only require one-fifth as great an engineering budget.

Pitfalls To Avoid

There are a number of common pitfalls that inexperienced QFDers often find themselves falling into. So, a few additional cautions and suggestions are in order. Most of these have to do with the *correlations* matrix (Figure 2-6, Room D).

It is important that the team members adopt a uniform and appropriate sense of scaling as to what constitutes a strong, moderate, or slight relationship. Although there are no hard and fast rules as to what this ought to be, it is suggested that the scale be stringent enough such that about 60 percent to 75 percent of the matrix remain blank. Any less than this might imply that the team is using too low a threshold as to what constitutes a relationship. Likewise, on the high end, most performance measures should have no more than about two to four cells having a strong relationship with a need, that is, a 9. Again, anything more than this implies that the team is using too low a threshold. Most teams come to some kind of an implicit agreement on scaling after just a few columns.

Another check on the correlations matrix is that, at completion, every row and every column ought to have at least one 9 (although it is more important that this be true for the rows). A row without a 9 implies that the team has not yet come up with a good performance measure to address that customer need. In this case, the team should try again to generate one or more performance measures that will get a 9 for that need. A column without a 9 implies that the performance measure has little impact on any of the customer needs and thus probably ought to be dropped. However, even if it is not, it is likely to come out extremely low in the final prioritization of the matrix.

Another common mistake is in the logic that teams sometimes use to determine a correlation score. It is important to remember that the thing we are trying to evaluate is the impact of an improvement in the *performance measure* on the *customer need*, not the other way around—that is, the impact of the column on the row. Sometimes teams forget this and reverse the logic, asking themselves: "If we improve customer satisfaction on this customer need (the row), how great an improvement will we experience on this performance measure (the column)?" For example, consider a need such as *easy to learn to operate* and a performance measure such as *number of repetitions required to master a certain task*. It is easy to see how one could use the logic: "If we make it easier to learn to operate our product, then will it take fewer repetitions to master a certain task?" However, the correct logic should be this: "If we can find a way to reduce the number of repetitions needed to master a certain task, then will customers be more satisfied over time on the need *easy to learn to operate?*" As simple as this sounds, even experienced teams sometime fall into this trap.

Most of these pitfalls can be avoided with just a little bit of help and experience. QFD facilitation is a skill that can be easily internalized after doing just one or two real-world cases. But trying to do it "cold" after just a few days of training often leaves teams a bit lost and subject to some of the pitfalls discussed. For this reason, many teams choose to engage an experienced facilitator for their first one or two cases, after which they become quite self-sufficient. There are many good, experienced facilitators available.

WHEN AND WHEN NOT TO USE QFD

A common question is whether it makes sense to use QFD on every development project. Certainly not! QFD requires a fairly large investment in both staff time and out-of-pocket expenses for VOC data collection. Thus, using it on every product, service, or process redesign initiative would be quite prohibitive. Experience has shown that it only makes sense for major projects such as the development of a major new product platform or the reengineering of a major process such as a telephone customer service center or a company's ordering, shipping, and billing functions.

QFD for Services

Most of the early applications of QFD and almost all of the early papers about it dealt with manufactured goods. In fact, most were about automotive products. Because of this, there has always been some question as to its applicability for services. This queasiness has now been shown to be largely unfounded.

The primary difference in service applications is simply that the performance measures are usually "softer" than for manufactured goods. With the latter, most performance measures are physical constructs that can be measured in the laboratory: size, weight, force, time, and so on. For services, measures must often be *invented*, which are harder to determine, but still necessary. For instance, in a QFD on telephone customer service, one of the more important customer needs was *they treat me with courtesy and respect*. How can a company measure this phenomenon without having to rely on an after-the-fact customer satisfaction type of measurement? What the team decided to institute was a form of self-evaluation in which several times a month, some of their most respected peers (telephone reps, themselves) would secretly listen in on calls and rate them on their level of *courtesy and respect*.

In good Deming fashion, the scores were only published for the group as a whole as they were only interested in measuring the system, not the individual person. Each month, the reps would meet to go over the scores and discuss unusual calls and ways to deal with them, and gradually the scores improved. What made this most amazing was that it involved a unionized workforce. Had management tried to impose such a system on them, or had it been the management doing the measurement, such a system would never have flown. The fact that something is difficult to measure does not relieve one from the obligation to try to measure it.

Why Bother?

In his book, *QFD: How to Make It Work for You*, Lou Cohen lists a series of benefits that are summarized here. QFD, he argues, allows a team to do the following:

◆ *Tie design decisions directly to customer needs.* In many companies, the product development decision-making process too often is determined by which sales rep or officer can shout the loudest. They typically base their opinions on largely anecdotal evidence: "Well, my customer says that if we do this, they'll buy a gazillion units!" While this may in fact be true, it does not answer the question of whether that is the best action overall. QFD provides a methodical, "dispassionate" approach to work out all of these difficult decisions. It allows the team to tie each of these decisions directly to customer needs as derived from a broad sample of customers.

◆ *QFD allows the firm to enlist all functional groups in the success of the product.* Since the 1980s, when cross-functionality became widely adopted in response to Japanese competition, QFD has provided a perfect platform in which all of the functional groups can come together and work on these decisions concurrently. In fact, many have observed that QFD provides a comfortable "bridge" in which both technical team members (scientists and engineers) and sales and marketing people can sit down together and work out their sometimes-tenuous relationship. QFD brings together both soft qualitative data—such as the customer needs—which is generally more comfortable for marketing people, and hard quantitative data, which is usually more comfortable for the scientists and engineers. Experience has shown that there is something for everybody, and most teams complete the process far more unified than they began.

◆ *QFD creates a common language that promotes communication across functional barriers.* Sometimes, the same words have different meaning to people from different parts of the corporation. The debating process in QFD usually clarifies these different interpretations and gets team members to agree on a common vocabulary.

◆ *QFD provides a traceable decision process.* Over the course of a typical one- to three-year development horizon, team membership often changes. Some team members leave the company or are transferred to other projects, and new people take their place. A House of Quality matrix often provides a good audit trail, helping new team members to understand why certain decisions were made prior to their entry onto the team.

◆ *QFD reduces mid-stream development changes that destroy time-to-market opportunities.* In his book *Better Products in Half the Time*, Bob King describes an analysis he did at Ford Motor Company. An important question was whether the additional time and expense of using Voice of the Customer and QFD actually helped or hurt time-to-market. The answer was quite striking. By reducing midstream changes, Ford actually reduced its time-to-market quite significantly. Although this conclusion seemed counterintuitive at the time, there is little doubt about it today.

◆ *QFD fosters and promotes creativity.* Clearly, the time used to study customer needs and to analyze how to address them, particularly the process of generating performance measures, forces teams to think about new needs and new features, things that were not on their minds at the beginning of the process. Too many product development efforts concentrate on features that are already known. This rarely results in anything more than a "me-too" product. The goal is to come up with new features and new ways to address customer needs, solutions that create breakthrough products.

◆ *Finally, QFD helps teams to come up with the big "ah-hahs."* Many QFD exercises end up with a prioritization that is highly unexpected to the team members. That is, some performance measures that they thought were critical often turn out not to be, and other performance measures that they thought were only moderately important turn out to be the ones that are among the most critical. The initial reaction to such an outcome is often one of incredulity. Team members might question whether they did it right, and some even want to toss the matrix aside entirely, despite all of the effort expended in creating it. But usually, cooler heads prevail. Someone on the team is likely to conclude, "Hey, maybe our gut feeling going in wasn't right. Maybe we've learned something here, and maybe the most critical items are not the ones we initially centered our thinking around." This kind of "ah-hah" can produce a most satisfying outcome, for the team ends up building a very different product from the one they thought they were going to build before embarking on the QFD process. They leave with a clearer understanding of the customers' needs, a rich set of new performance measures, a set of potential new features or solutions—and a greater likelihood of ultimate product success.

THE OVERALL GOAL OF QFD

Many QFD teams often lose sight of why they are even going through this process. There is actually a very practical goal to be achieved. Most companies require the creation of a formal product requirements or product specifications document. The purpose of this vehicle is to formally lay out exactly what the new product is going to be about: its key features and specifications, along with target values for each of those specifications. This is exactly what QFD sets out to do—and in a highly structured and rigorous way! It asks the product-development team to first identify the key customer needs, to then translate them into performance measures that become the product's specifications, to identify which of these specifications are most critical, and finally, to set target values for them.

This chapter has now come full circle. And instead of deciding these things based on mere whim or anecdote, the team has made these decisions in a completely logical and analytical manner.

REFERENCES

Akao, Yoji. 2004. *Quality Function Deployment: Integrating Customer Requirements into Product Design.* University Park, IL: Productivity Press, Inc. A comprehensive set of some of the best case studies from Japan, edited by (and some written by) Professor Akao.

Akao, Yoji, and Shigeru Mizuno. 1994. *QFD: The Customer-Driven Approach to Quality Planning & Deployment.* Asian Productivity Organization. An English translation of the famous book by the original inventors of QFD.

Cohen, Lou. 1995. *Quality Function Deployment: How to Make QFD Work for You.* Reading, MA: Addison-Wesley Publishing. Probably the best basic "how-to" book about QFD. While it certainly covers all of the underlying theory, its primary thrust is on the many practical implementation issues.

Hauser, John, and Don Clausing. 1988. *The House of Quality.* Harvard Business Review (May–June). The first English-language description of QFD in a major general management publication. Remains one of the most frequently ordered reprints at *HBR* more than 15 years after its publication. A short, excellent starting point.

Katz, Gerald. 2001. "Is QFD Making a Comeback?" *PDMA Visions*, 25 (2) (October). A discussion of why, after several years of apparently falling out of favor, QFD seems to be regaining popularity.

King, Bob. 1987. *Better Designs in Half the Time: Implementing QFD in America*. Methuen, MA: GOAL/QPC. The first complete English-language book about QFD.

Sato, Nakahito. 1983. "Quality Function Expansion and Reliability." *Standardization and Quality Control (in Japanese), Volume 36, No. 3*. First journal article to show a pictorial depiction of the four cascading matrices that define formal QFD and the first to use the term "House of Quality".

Part 2

Tools To Improve Customer And Market Inputs To NPD

Part II presents seven tools that help teams develop better information and use it more effectively in the NPD process. The first four tools will be most helpful in generating interesting concepts and designs. The fifth tool (Chapter 7) will organize market information to help you figure out how to position your new product, while the sixth tool will help you develop the information that will allow you to name your product. The final chapter helps you figure out what information is needed to develop better market forecasts. Many of these tools are cross-functional modifications of tasks that might be considered more traditionally in the purview of the marketing function in the organization. However, given the thinning of the management and marketing ranks in many firms, these tasks are more and more becoming the responsibility of the NPD team. Also useful is that most of these tools do not require a large budget—medium and even smaller, start-up firms can implement them without breaking the bank.

Chapter 3 leads the section off by explaining the maximum difference scaling technique for determining the trade-offs that customers make when considering products with different feature sets. This tool helps teams decide which needs to address, given available resources, allowing them to focus on solving the more important needs—from the customer's perspective. The method as presented in the chapter has several real strengths. First, sophisticated computer programs are not needed, as they are with many trade-off determining methodologies. Simple spreadsheet programs provide all the power needed. Second, a significant amount of additional information is available online to supplement the chapter. For example, while the experimental design for which needs to present in what order is presented in the chapter for 15 needs, the experimental designs for situations ranging from 5 needs to 15 needs are available online. Finally, the chapter not only presents the method, but also demonstrates how to develop the story behind what the specific numbers are showing.

In Chapter 4, another method for developing breakthrough ideas is presented—the Slingshot, which is a multistage group-based process. One of the interesting features of this process is the use of prosumers, individuals from outside of the firm who are simultaneously acting as new product development professionals and as consumers. They participate in both capabilities, professional, nonbiased knowledge to the focus group part of the process and consumer eyes to the idea generation and investigation part of the process. Slingshots can be completed for as little as $10,000 and within three to four weeks.

People sometimes have difficulty articulating their problems. Chapter 5 presents observation – driven product design, a methodology not just for observing users and imputing needs from those observations, but also for first developing actionable business objectives and then integrating those objectives with the user observations to identify higher-potential product opportunities. As the old saying goes, "If you don't know where you are going (don't have explicit objectives), any route will get you there." Starting from a set of explicit project objectives allows the team to more efficiently organize their observation process. Additionally, using two rounds of observations allows a deep exploration of needs.

Another short-term exploratory process for developing breakthrough products is presented in Chapter 6. Market and technology attack teams are chartered for three months to seek large-potential new business opportunities and build a strategy for developing them for the firm. On the one hand, because this four-phase process goes after large and inherently risky opportunities, they will not always be successful in identifying an opportunity that the firm can take advantage of. On the other hand, significant understanding is obtained in a very short time regardless of the outcome because only three months is spent on the process.

Segment your market and choose the target market you will serve— from that all else flows. Chapter 7 provides a quantitative methodology, STUP, for determining how to segment the market, helping choose the best target market for the firm, gaining understanding about the needs and problems that target most wants solved, and then positioning the firm's new product in the eyes of the consumer. Although these steps are fundamental to new product success, they frequently are glossed over or completed intuititively. This chapter provides all the techniques needed for the team to complete this process on their own. This is another technique that smaller firms also will find most useful.

A rose by any other name is still a rose. Chapter 8 clarifies "What's in a name?" and provides a product naming methodology for effectively naming products and avoiding naming errors such as the Chevy "Nova" and Pontiac "Fiero" (no go and ugly old woman, respectively, in Spanish). The chapter presents a six-step process to develop an effective new product name, one that does not infringe on other product names and that may be protected legally from others' using it. A number of useful brand Web resources are also provided in this chapter.

Chapter 9 leads you through a process for building assumptions-based new product forecasts. Assumptions-based forecasts start with identifying the target market for a new product and defining the overall potential size, if everyone bought the new product. They then identify different factors that can be expected to reduce the size of the market (such as incomplete distribution coverage), to repeatedly partition down the overall market into smaller, more realizable sales, until finally the expected market size is obtained. These models can be developed rather rapidly, and updated over time as improved information is obtained. Again, this is a tool that can utilized regardless of the size of your organization.

3

Applying Trade-off Analysis to Get the Most from Customer Needs

Nelson Whipple
Director, Resource Systems Group, Inc.

Thomas Adler
President, Resource Systems Group, Inc.

Stephan McCurdy
Director, Resource Systems Group, Inc.

To build profitable products around the customer, the product team must not only identify the range of needs that the customer experiences, but also understand how the customer prioritizes those needs. The product team must make trade-offs when deciding which needs to address given available resources, and a clear, quantitative structuring of the trade-offs that customers are willing to make can greatly help the product team to focus and make those decisions.

Unfortunately, quantitative customer research to measure trade-offs has typically been addressed by techniques such as conjoint analysis and discrete choice modeling that are considered to be expensive, to require special training, and to require special software to execute them. These factors contribute to the perception that trade-off analysis should only be applied late in the development process (rather than at the front end), and that such analyses are outside the capabilities of the product team.

In recent years, however, a new research tool has emerged that makes trade-off analysis of customer priorities more accessible to the product team at more phases of the product development process. This technique, alternatively known as maximum difference scaling (MaxDiff) and best-worst scaling (BWS), does not require formal training in statistics, does not depend on special software to execute it, and does not require a PC- or Web-based survey. The budget and time resources required are more similar to a typical customer survey than to a typical conjoint analysis. Although MaxDiff was primarily developed to improve how survey questions are designed and analyzed (especially for international markets), it has also evolved into a tool that simplifies the way customer trade-offs can be measured, opening up more opportunities to gather these valuable insights (for a discussion of market research issues addressed, see Cohen 2003).

In trade-off analysis, customers are given a list of items relating to concepts that the product team wants to understand (needs, features, etc.), and they are asked to choose between alternatives. From these customer responses, a number is derived for each item that represents how much customers value the item relative to each other item. Analyzing these numbers, the product team can rank the items, but, more importantly, can understand the relative differences in value from item to item. In addition to knowing the top three customer needs, for example, the team will also know whether the second and third ones (and fourth and fifth ones) matter at all compared to the higher-ranked items.

This chapter will provide all the information you need to design, execute, analyze, and interpret customer need priorities using MaxDiff without any software, other than a spreadsheet program such as Excel. Although MaxDiff can apply to a wide variety of issues for which you may wish to quantify customer priorities, this chapter will focus on how to apply it to customer needs analysis. This chapter will walk through the steps in a quantitative needs analysis that uses MaxDiff, including generating the data needed to draw conclusions and explaining how to do the following:

◆ Analyze and interpret customer need priorities in the context of overall product decisions.

◆ Design and write a survey that uses MaxDiff.

◆ Calculate importance values.

◆ Sort needs into categories based on importance.

◆ Develop charts that put the customer priorities in the context of the competitive environment.

◆ Develop a scoring system to prioritize opportunities.

Although special software can be used during some steps of this process, the example illustrates manual methods that can be applied using tools you already have (plus the information in this chapter).

To discuss each step of the process outlined in Figure 3-1, an example from an industrial equipment study that focuses on 15 need statements will be used. The manufacturer in the example makes and sells equipment for businesses, and it is planning its next product update. The manufacturer completed *Voice of the Customer* work to identify over 100 unique customer needs, and then narrowed the list down to 15 key needs that it wished to prioritize further. The reduced

FIGURE 3-1. Overview of key steps in analysis of customer need priorties.

list of needs focuses on those involving maintenance and repair, the equipment operator, environmental issues, fuel economy, and machine performance.

RESEARCH DESIGN AND EXECUTION

The most successful products meet basic customer requirements and differentiate by meeting important needs better than competition. A quantitative survey using trade-off methods will help to identify those basic requirements and opportunities to differentiate by prioritizing customer needs and understanding how well these needs are met today. The typical components of a trade-off study are as follows:

◆ A list of items (needs, benefits, product features, etc.) to be tested
◆ An experimental design to structure the trade-off questions
◆ A survey that applies the items to the design
◆ A process for administering the survey questions to respondents
◆ A statistical procedure for estimating the priorities from the answers given in the survey.

The survey should also include background information to classify customers, as well as ratings of current competitor performance on each of the items.

In the industrial equipment example, the list of items are 15 needs statements (see Table 3-1), the experimental design is applied via look-up tables, the survey is administered on paper, and the priorities are estimated using a manual calculation. The industrial equipment study shows how to execute the work for 15 needs. A subset of the actual experimental design look-up tables is included in this chapter, but a complete set of look-up tables for analyzing anywhere from 5 to 15 needs can be found at: *www.rsginc.com/pdma/toolbook3/design_tables*.

TABLE 3-1.
List of Customer Needs from Industrial Equipment Example

1. Actively manages traction for maximum performance.
2. Automatically helps a less skilled operator perform up to the level of a good operator.
3. Automatically tracks and schedules its regular maintenance.
4. Can accurately self-diagnose problems and report the information in a clear and easy-to-understand format.
5. Can be easily reconfigured and set-up.
6. Can display easy-to-use electronic versions of its service manual, operator manual, and parts catalogue on its monitor.
7. Can easily maneuver and work effectively in tight spaces.
8. Can effectively operate in extreme conditions, and warns the operator if conditions are reaching stability limits.
9. Can run on reduced emission fuel types without sacrificing performance or durability.
10. Controls can be easily adjusted to comfortably fit each individual operator.
11. Is designed to be extremely easy and quick to clean.
12. Is exceptionally quiet.
13. The machine greatly reduces the amount of dust generated during operation and transport.
14. The machine has the ability to automatically perform repeated tasks.
15. Uses only half as much fuel as today's equipment to perform the same amount of work.

Finalize List of Customer Needs

The first step is to finalize the list of customer needs you will analyze. Assume that you already have a list of customer needs, perhaps from Voice of the Customer work like that described by Gerald M. Katz in *PDMA ToolBook 2* (Katz 2004), although you may not have narrowed them down to 15 yet. If you have not yet done that, there are several ways to focus on a narrower set of needs, some of which build on existing knowledge. Before narrowing the needs, you may want to have the following available for reference:

◆ Existing data or knowledge based on customer needs from previous efforts

◆ Established value propositions or strategic objectives for your company or product area (these are stated strategies for how you will differentiate your offerings relative to competition to win customers)

◆ Customer needs affinity diagram

◆ List of features you have brainstormed that correspond with specific needs

These tools will help you understand which needs are most likely to have an impact on the product's success and therefore should be included in the trade-off analysis. Table 3-2 lists steps to take to finalize the list of customer needs.

An affinity diagram like the type shown in Figure 3-2 can be used to group needs together (see Katz 2004). Similar needs (as determined by the product team or, better yet, a group of customers assembled for this purpose) are placed together into sets. Each set can then be named by a member of its set that best summarizes the group or a by new statement developed as a summary. For example, the five needs in the far left side of Figure 3-2 can be summarized as "Machine Performance." If the needs list is too long, it can be reduced by substituting the summary statements for the master list of all detailed statements. Figure 3-2 demonstrates that the 5 needs just below the "Industrial Equipment" box might be used if you wanted to reduce the list of needs from 15 to 5. (In the diagram, labels have been used instead of statements due to space constraints.)

Develop the Sample Plan

Before putting the needs statements into a survey, you must decide which customers to interview. This decision can sometimes influence the content of the survey. You must also identify if there are any key subgroups of interest that should be included. Set targets for numbers of interviews overall, possibly by subgroup. This is called a *sample plan*.[1] Although perhaps an obvious

---

[1] For an expanded discussion on sample planning, including templates for sample and segment planning, please visit *www.rsginc.com/pdma/toolbook3/sampleplan*.

TABLE 3-2.
Steps to Take to Finalize the List of Needs

Evaluation Steps	What to Look For
Revisit the affinity chart.	• Are all the groupings distinct? Perhaps these could be further consolidated. • Consider whether you can still meet your goals by focusing on the summary needs rather than all of the needs listed in the diagram.
Review any existing data to which you might have access on customer needs that concern this product.	• Are there some basic needs that are already well understood that don't need to be measured again? • Which needs will provide new customer insights if they are better understood?
Review any value propositions that have been established by your company or for your product area.	• Are some needs less critical to supporting established value propositions for your company or product area?
Consider the scope of your effort.	• Does your current initiative address a product line or an individual product? If you are making decisions for a product line, consider which needs provide the most leverage in addressing multiple products. • Can each need be addressed by a technology that will exist within the timeframe you are considering for this initiative? Consider removing any that will not be met until beyond your timeline. • Do some of the needs fall outside of what you can address with your effort?
Focus on the voice of the customer.	• Do any statements reflect technical features that are not customer needs? • Make sure needs statements are phrased in the customer's language.

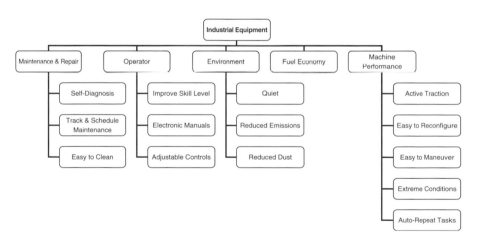

FIGURE 3-2. Example affinity diagram.

point, it is important to recognize that the results you gather will reflect the make-up of the people you interview. If you choose to interview only your own customers, you will not gain insight into how to attract new customers who may have different needs. Conversely, if you interview only prospects, you will not gain any insight into how to retain current customers. The sample plan will help you to manage and control these influences.

The composition of the sample you survey will influence the results, because different customers will prioritize needs differently. High-volume purchasers may differ from small-volume purchasers. Consumers in California may differ from consumers in the Northeast. Large companies may have different priorities than small companies. Customer priorities may differ according to each company's key applications of your product. At the individual level, users may have different priorities from those responsible for maintaining the product or from the purchase decision maker. If certain types of customers are systematically excluded, their unique need profiles will not influence the results. If certain types of customers are more heavily represented in your sample than in the market, their need profiles will more strongly influence the results that you get relative to what might actually happen once the product is in the market.

To develop an appropriate sample plan, you must have a clear idea of who you want to survey and how you can represent them appropriately, given the proportions of different key subgroups.

You must also plan to get enough completed surveys to be confident in the results. The following must be decided:

◆ Scope of the market to address—the whole market or strategic segments? (i.e., respondent qualifications)
◆ Whether to analyze subgroups of the selected market (i.e., quota groups)
◆ How many interviews to conduct

To support these decisions, consider the parameters of the product development effort as they relate to market objectives and identify and quantify the key market segments that will be important to the product's success.

For different types of surveys, there are different ways to determine your minimum sample size. For MaxDiff, unfortunately, a set of rules for determining the necessary sample size has not yet been established (Flynn et al. 2006). However, some guidelines from another set of trade-off methods, conjoint analysis (Orme 2006), can be borrowed. In his book *Getting Started with Conjoint Analysis: Strategies for Product Design and Pricing Research*, Brian Orme discusses rules of thumb for establishing sample size in different kinds of studies, distinguishing between research that is intended to be a robust market analysis and research that is intended as investigational to develop hypotheses. For a robust market analysis (one with statistically significant results), he recommends a total of 300 surveys. If you have quotas for subgroups and wish to compare them, he recommends 200 surveys per group, and the total number of surveys would increase accordingly. If your goal is to generate hypotheses and to understand generally how the market prioritizes needs, he suggests that 30 to 60 may be adequate.

Table 3-3 summarizes the possible ways to determine sample size based on the quota groups you may have and your goals for the overall and subgroup analyses. If you or your team do not have quota groups and want a robust analysis, target a minimum of 300 surveys. If you have quota groups and want a robust analysis of those groups, target a minimum of 200 per group, and the

TABLE 3-3.
Determining Sample Needs

Overall Goal	Goals By Quota Group	Minimum Recommended Surveys
Robust analysis	(No quota groups)	300 total
Robust analysis	Robust analysis	200 X each subgroup to be analyzed
Robust analysis	Directional analysis	50 X each subgroup to analyzed; ensure a total of 300 overall
Directional analysis	(No quota groups)	50 total
Directional analysis	Directional analysis	50 X each subgroup to be analyzed

overall target will result from that calculation. If you only want a directional analysis at the subgroup level, target 50 per group, but make sure that the overall total reaches 300. You can accomplish this by raising the targets in each group or including qualified people who are not in one of the quota groups. If you have quota groups and only need a directional analysis overall, you do not need to make sure that you have 300 overall. A directional overall analysis without quota groups should target a minimum of 50 surveys.

In the equipment example, the manufacturer was mainly interested in a robust understanding of large and very large companies, plus three industry subgroups. Thus, the overall sample size was establish as 600, driven by the three industry subgroups that required a minimum of 200 completed surveys each (see Table 3-4). Other factors could have driven this target to a lower number. For example, if it was felt that it would be difficult to survey some subgroups due to cost, timing, and the total number of companies available to interview, that could influence a lower target to be set or some subgroups to be omitted. For example, if there were not many very large companies available to survey, the target for these companies might be set conservatively, below 200. These might be analyzed as part of the overall market analysis, but not as a separate subgroup.

However, one might choose to look at results from a smaller group in spite of the sample size, especially if the analysis was conducted in the early stages of the product upgrade process. In this case, the team might not want to expend the resources to conduct a completely robust analysis. They might require only directional input on what was generally important to the market, and so might not expend as many resources to get information from a larger sample.

When you recruit respondents to take the survey, you will first make sure that they are qualified by asking them to answer questions from a short survey called a *screener*. Recruiting can be accomplished by telephone, by intercepting people (e.g., at a mall or a trade show), by e-mail, or even by regular mail. If you acquire a list of potential respondents, make sure that you know the guidelines for using that list, if any, with respect to how, when, and why you

TABLE 3-4.
Sample Plan for Industrial Equipment Example

Industrial Equipment Sample Plan: 600 total completed interviews

Qualification Description	Qualification Definition
Company size	• Own at least 40 pieces of "common"equipment
Title/Decision maker	• Owners/executives for companies in Large segment • Managers and above at Very Large companies • Have "significant" input into the selection of one equipment manufacturer over another

Quota Description	Quota Definition
Company size	• 200 Large companies (minimum) • 200 Very Large companies (minimum)
Industry	• 200 Industry A • 200 Industry B • 200 Other Industries

are allowed to contact the people on it. If you are going to intercept people, make sure that you have permission from the mall, trade show, or other venue. (If you hire a company that manages recruiting, it will know how to handle these situations.)

Figure 3-3 shows an example screener for the industrial equipment study, including a greeting, with questions addressing the main qualifications and quotas. This example is administered via telephone; the qualification questions precede the quota questions in order to minimize the average length of time needed to screen someone.

When a screener respondent qualifies to take the survey and accepts, make sure to assign a unique respondent ID that can be used to match the screener responses to the survey responses.

MAXDIFF TRADE-OFF DESIGN AND ANALYSIS

Design and Draft MaxDiff Survey

After developing the sample plan and the screener, the next step is to draft the survey. The survey will collect three essential kinds of information:

1. Importance of each customer need (e.g., trade-off using MaxDiff)
2. Perceptions of competition on each need
3. Demographic and other background information that helps to put a context around the answers (e.g., customer or non-customer, purchase volume, experience with key competitors, etc.)

Hello, may I please speak with (NAME FROM SAMPLE) or the Owner/President or the Director of Equipment Acquisitions.

My name is _____ and I'm calling on behalf of (name of company conducting research). We're doing a research project on industrial equipment and would like to include your opinions. Please let me assure you we are not selling or advertising anything, nor will this be used for direct sales purposes. Your name will not be associated with your answers in any way. We are solely interested in your opinions for the purposes of product development.

Considering the following four types of equipment (READ THE FOUR TYPES), which of the following best describes **your** role when your company is deciding from which manufacturer to acquire a piece of equipment? Would you say you.... (**READ LIST**)

1. Are the sole decision-maker
2. Have a significant amount of input into the decision
3. Have some input into the decision (ASK FOR REFERRAL)
4. Have little or no input (ASK FOR REFERRAL)
5. (DO NOT READ) Don't know/Refused (THANK AND TERMINATE)

Still, considering the following four types of equipment (READ THE FOUR TYPES), IN TOTAL, approximately how many pieces of equipment does your company currently own? Please include all pieces that your company has that were acquired new through a purchase or lease.

• None (THANK AND TERMINATE)
• 39 or fewer pieces (THANK AND TERMINATE)
• 40 to 99 pieces
• 100 pieces or more

Next, I'm going to read a list of areas or industries. Please stop me when I read the area your company is **primarily** involved in. If your company works heavily in more than one of these areas, please choose the one that best describes your primary line of work.

1. Industry A
2. Industry B
3. Industry C
4. (**DON'T READ**) Other (specify)_____

FIGURE 3-3. Example screener based on industrial equipment sample plan (administered via telephone).

First, the MaxDiff trade-off will be discussed from design through analysis. Then, the discussion will return to the survey design to discuss competitive perceptions.

The MaxDiff trade-off section consists of a series of questions that look like the example in Figure 3-4.

Creating the questions requires the needs list, an *experimental design* that determines how to structure the questions using the needs, and the question itself. How you phrase the question will determine what people think about when they answer and how you interpret what they say.

The question should supply a more specific context than simply asking the customer to state what is most and least important, because the lack of context leaves open the issue, "Important for what?" In Figure 3-4, the customer is asked to indicate what would be of "most or least value to their company." For this product team, the importance is benchmarked as the extent to which its products help add value to the customer's business; such considerations influence the question wording. Answers might be different if the question asked, "What would be most important in helping you to retain skilled operators," or "What would be most important to you when purchasing your next piece of equipment."

If the equipment manufacturer's strategy was focused on differentiating versus competition based on operator satisfaction, the former question might be appropriate. The latter question is inappropriate because it focuses on near-term purchases, and this may cause the customer to think too much about what is feasible in the near term. When designing for the long term, one doesn't

Which ONE of the following would be of the MOST value and which ONE would be of the LEAST value for your company's equipment in the future?		
Most Valuable		Least Valuable
☐	Actively manages traction for maximum performance.	☐
☐	Can be easily be reconfigured and set-up.	☐
☐	Can effectively operate on steep inclines, and warns the operator if conditions are reaching stability limits.	☐
☐	Is designed to be extremely easy and quick to clean.	☐

FIGURE 3-4. Sample MaxDiff question.

want the customer to say that something is unimportant simply because he or she does not think it is feasible or forthcoming.

Notice that this list of needs may look similar to a list of features. However, in this case, the needs and benefits are embedded in the feature descriptions so that the value to the customer is clear. The list was developed from customer interviews using their words rather than internally, using company lingo, so there is high confidence that the list of statements is meaningful to the market.

To design the sequence of questions, one needs to match the needs to an experimental design. The experimental design tells us (1) how many questions (experiments) to ask, (2) how many needs to show in each question, and (3) which needs are shown in what order. It also reveals how many different versions of the survey are needed. The goal is to accurately determine the relative importance of each need without overburdening the customer with too many questions or questions that are too complicated. To do this, the questions must be designed in a way that satisfies the statistical requirements while optimizing the task from the customer's point of view. To meet these requirements, there must be different versions of the survey, each with a different set of questions.

Tables 3-5a and 3-5b show 2 of the 10 experimental designs required to study 15 needs. Because there are 10 designs, there will be 10 different surveys, each with a different set of MaxDiff questions (as mentioned previously, these can be found on the Web at the address given), but identical in all other ways. Table 3-5a shows one of the 10 unique designs: It has 11 MaxDiff questions (rows numbered 1 to 11), and each question will show four needs statements (the four columns labeled "Needs Statement #"). The cells where the question rows and need statement numbers intersect each contain a number corresponding to a need statement that will be shown in the question.

Applying Table 3-5a to the industrial equipment example, the first question would look like Figure 3-5a. Reading across the needs statements for Question 1, the first need to show is #3 ("Automatically tracks and schedules its

TABLE 3-5a.
Survey Version #1 of 10 for 15 Needs Statements

Survey Version #1

Question #	Needs Statement #	Needs Statement #	Needs Statement #	Needs Statement #
1	3	2	6	12
2	6	15	14	13
3	15	10	1	5
4	2	5	4	11
5	10	6	11	8
6	8	13	9	7
7	14	1	8	2
8	1	9	10	12
9	5	3	13	9
10	12	7	15	4
11	11	7	3	14

TABLE 3-5b.
Survey Version #2 of 10 for 15 Needs Statements

Survey Version #2

Question #	Needs Statement #	Needs Statement #	Needs Statement #	Needs Statement #
1	4	8	10	3
2	14	12	9	5
3	7	2	5	6
4	1	3	7	15
5	9	4	6	1
6	13	14	4	10
7	4	9	2	11
8	8	11	12	1
9	2	10	3	14
10	5	15	8	9
11	11	13	2	15

regular maintenance" from Table 3-1), the second need is #2 ("Automatically helps a less skilled operator perform up to the level of a good operator" from Table 3-1), and so on. The order of the needs list and how the statements are numbered (Table 3-1) does not matter as long as the same numbers correspond to the same needs statements for each question generated using the experimental design tables. The statements are placed into the questions exactly in the order shown in the design table, with the statement in the leftmost column listed first and the number in the rightmost column listed last. When the entire design (i.e., each of the 10 tables) is implemented, each statement will appear the same number of times in each position, in the same number of questions, and with each other statement an equal number of times. These properties are important to maintain in order to prevent bias in the results, and applying the experimental design accurately helps to achieve this.

Which ONE of the following would be of the MOST value and which ONE would be of the LEAST value for your company's equipment in the future?			
Most Valuable		Least Valuable	
☐	Automatically tracks and schedules its regular maintenance.	☐	(3)
☐	Automatically helps a less skilled operator perform up to the level of a good operator.	☐	(2)
☐	Can display easy-to-use electronic versions of its service manual, operator manual, and parts catalogue on its monitor	☐	(6)
☐	Is exceptionally quiet.	☐	(12)

FIGURE 3-5a. First MaxDiff question from Table 3-5a using the needs list in Table 3-1.

Which ONE of the following would be of the MOST value and which ONE would be of the LEAST value for your company's equipment in the future?			
Most Valuable		Least Valuable	
☐	Can display easy-to-use electronic versions of its service manual, operator manual, and parts catalogue on its monitor.	☐	(6)
☐	Uses only half as much fuel as today's equipment to perform the same amount of work.	☐	(15)
☐	The machine has the ability to automatically perform repeated tasks.	☐	(14)
☐	The machine greatly reduces the amount of dust generated during operation and transport.	☐	(13)

FIGURE 3-5b. Second MaxDiff question from Table 3-5a using needs list in Table 3-1.

Figures 3-5a and 3-5b show the first two questions of the first survey that the experimental design for the industrial equipment analysis generates. Each question is written by following the process of reading a statement number from Table 3-5a, then placing that statement in the proper position in the question corresponding to the row number. (The numbers in the margin to the right of the question show the number of each need from the master list. This is included to make it easier to code the data later). Another 9 questions are generated in addition to those in Figures 3-5a and 3-5b for a total of 11 questions. This entire process is repeated nine more times to generate a total of 10 unique sets of 11 questions.

Each set will be placed in a separate survey, and each person taking the survey will see only one set of 11 questions. Each of the 10 versions should be administered an equal number of times across the sample. In the industrial equipment example, a total of 600 completed surveys are targeted, so each version of the survey will be administered to 60 people (600 completes divided by 10 versions). In practice, due to the logistics involved in recruiting people to take the survey, it is not always possible to administer each version an exactly equal number of times. For example, 60 complete interviews for each version (10 percent) might be targeted, but the end result might end up with 8 percent to 12 percent for some versions, for example. Although not perfect, this outcome is still acceptable. The ideal is to have each need statement appear the exact same number of times overall, in each position, and with each other statement; however, the results are valid as long as the percentages are close to the expected values.

Implement and Administer Survey

When producing the printed surveys, print a code on the survey that identifies the version so you can easily tally how many of each version were completed. Also, assign each respondent a unique code so that it is easy to track individuals throughout the process. You can use the ID from the screening process or another unique ID as long as you have other information that will link the survey back to the screener.

The industrial equipment survey is a paper survey, but other methods can be used. The survey itself can be administered on paper, as an electronic form that is e-mailed, on a CD, or online. Telephone is not a good option for a MaxDiff survey because respondents are asked to consider several items at once when making choices, and this is difficult to conceptualize over the phone. It may happen that someone suggests that a salesperson should bring the survey to their customers and administer it (this is a particularly popular suggestion from salespeople). In general, this is a poor idea, as the presence of the salesperson inevitably leads to biased answers. The methods mentioned here do not require someone to administer the questions; each type can be self-administered by the respondent.

Finally, it is a good idea to test the survey with a control group before administering it to the whole target group. The main goals of the test are to make sure that customers interpret the needs the way that you intended and that the questions are implemented so that they can be easily understood and answered. Show them the list of needs again, and ask them to explain each one and point out any that require extra concentration to understand. You can also test whether the group you have selected via your sample plan is defined correctly. When you debrief the customers, ask them to describe their role in the purchase decision for your product. Their answers should tell you whether the screening process has accurately identified the group you intended to reach.

Ideally, the control group test would be conducted with a small group (5 to 10) of customers who are debriefed afterward. Changes (e.g., to the wording, the survey layout, or screening/sampling) are discussed among the product team before deciding if they are merited. If customers cannot be used for the pretest, customer surrogates can be used as a last resort (such as salespeople, customer service reps, or market research people who have experience with customers).

Process Data and Calculate Importance Values

As you receive completed surveys, tally how many of each version of the survey are completed overall and by segment. When all the completed surveys are tallied, each version should have been completed a roughly equivalent number of times.

The completed answers to the MaxDiff questions will look like Figure 3-6 and must be entered into an electronic data file for analysis. First, set up a spreadsheet like the one in Table 3-6. Make a column for the respondent ID (Resp ID), a column for the question number (Q#), and one column for each of the 15 needs statements. Each respondent's MaxDiff data will be represented in 11 rows. Place the respondent ID in the first column in each of the first 11 rows, and, in the second column, number the rows from 1 to 11; these will correspond to the 11 MaxDiff questions. Each row will contain the answers to one of the MaxDiff questions, and 11 rows will represent one complete survey. When a need is chosen as "Most Important," it will receive a "1" in its column; when "Least Important," a "−1". Needs that were displayed but not selected should receive a "0"; this will be important if you want to calculate an average score (for example, to compute a confidence interval). Excel will count the 0 as part of the average and will exclude empty cells.

Table 3-6 shows the complete set of MaxDiff answers for respondents 1,000 and 1,001. To determine the relative importance for each need, simply sum the values in each Need Statement column. The resulting numbers (or points) represent the relative importance of each need, and range from negative to positive. The aggregate importance scores for respondents 1000 and 1001 are shown in Table 3-7. Each need is shown an average of three times per respondent, so the 6 next to need #15 (fuel economy) suggests that both respondents rated it as "Most Important" every time it was shown. At the opposite end, it looks as though need #1 (actively manages traction) was almost always picked as "Least Important."

Note that because the number of the need statement was included in the margin of the question, one doesn't need to know what version of the survey

Which ONE of the following would be of the MOST value and which ONE would be of the LEAST value for your company's equipment in the future?			
Most Valuable		Least Valuable	
✔	Automatically tracks and schedules its regular maintenance.	☐	(3)
☐	Automatically helps a less skilled operator perform up to the level of a good operator.	☐	(2)
☐	Can display easy-to-use electronic versions of its service manual, operator manual, and parts catalogue on its monitor.	✔	(6)
☐	Is exceptionally quiet.	☐	(12)

FIGURE 3-6. First MaxDiff question with responses.

TABLE 3-6.
Complete MaxDiff Data for Two Respondents

Resp ID	Q#	1	2	3	4	5	6	7	8	9	10	11	12	13	14	15
1000	1		0	1			−1							0			
1000	2						−1								0	0	1
1000	3	0				−1					0						1
1000	4		1		0	0						−1					
1000	5						−1		1		0	0					
1000	6							0	1	−1				0			
1000	7	−1	1						0							0	
1000	8	−1									0	0	1				
1000	9		1			−1			0					0			
1000	10				0		−1						0				1
1000	11		1					0				−1				0	
1001	1		−1	0			1						0				
1001	2						0								0	−1	1
1001	3	−1				0					0						1
1001	4		0		1	−1						0					
1001	5						1		0		0	−1					
1001	6							−1	0	0				1			
1001	7	−1	0						1							0	
1001	8	−1									0	0	1				
1001	9			−1		0				0				1			
1001	10				0		−1						0				1
1001	11			1			−1					0				0	

each respondent saw in order to do the coding. The data can be entered into the spreadsheet directly from the completed survey without referring back to the experimental design tables.

Table 3-8 shows these results for all 600 industrial equipment respondents in the "Raw Score" column. Need statement #15 is far and away the leader, with 1,612 points; the second most important need, #7 (maneuverability), is well behind with 1,036. The "Pct." column is the average of the "Raw Scores"; it is labeled "Pct." because the average of the scores also represents the percentage of the total possible score that each need could have achieved. Need statement #15 achieved 89.6 percent of the total possible high score (100 percent would mean everyone chose it as "Most Important" every time it was shown). At the bottom end, need statement #6 ("Can display easy-to-use electronic versions of its service manual, operator manual, and parts catalogue on its monitor.") achieved 66.4 percent of the total possible *worst* score.

The next column, "95% CI," is the confidence interval around the average at the 95 percent confidence level. To understand if the difference in importance values between two needs is statistically significant, calculate a confidence interval for each value. To calculate a confidence interval for the final score, follow these steps (functions and notation refer to Excel functions):

Compute an average (mean) for each need column [in Excel, = Average(number1, number2,...)]. In the formula, "number1, number2..." represents the range of rows in your data set. In the industrial equipment example,

TABLE 3-7.
Importance Scores for Two Respondents

15. Uses only half as much fuel as today's equipment to perform the same amount of work.	6
3. Automatically tracks and schedules its regular maintenance.	3
8. Can effectively operate in extreme conditions, and warns the operator if conditions are reaching stability limits.	3
12. Is exceptionally quiet.	2
13. The machine greatly reduces the amount of dust generated during operation and transport.	2
2. Automatically helps a less skilled operator perform up to the level of a good operator.	1
4. Can accurately self-diagnose problems and report the information in a clear and easy-to-understand format.	1
10. Controls can be easily adjusted to comfortably fit each individual operator.	0
6. Can display easy-to-use electronic versions of its service manual, operator manual, and parts catalogue on its monitor.	−1
9. Can run on reduced emission fuel types without sacrificing performance or durability.	−1
14. The machine has the ability to automatically perform repeated tasks.	−1
5. Can be easily be reconfigured and set-up.	−3
11. Is designed to be extremely easy and quick to clean.	−3
7. Can easily maneuver and work effectively in tight spaces.	−4
1. Actively manages traction for maximum performance.	−5

there are 600 respondents with 11 rows each, or 6,600 rows, so the formula will be expressed as "=Average(C2:C6601)." Excel will count the "0's" in the average and ignore the empty cells. In the example, there are a total of 600 respondents who saw each need three times, so each raw score is divided by 1,800.

Calculate a 95 percent confidence interval: In Excel,

♦ =1.96*STDEV(number1, number2. . .)/SQRT(COUNT(number1, number2. . .)). If the first need statement is represented by column C and the data start in the second row, the confidence interval is given by =1.96*STDEV(C2:C6601)/SQRT(COUNT(C2:C6601).

♦ Subtract the number computed in Step 2 from the mean computed in Step 1. This gives you the lower bound of the confidence interval.

♦ Add the number computed in Step 2 to the mean computed in Step 1. This gives the upper bound of the confidence interval.

The confidence interval for need statement #15 is 2.0 percent; technically, this means that there is a 95 percent probability that the true average of need statement #15 is within 2.0 points of 89.6 percent. The confidence range for need statement #15 is 87.6 percent to 91.6 percent. Any other need with a confidence interval that overlaps that interval is not considered to be different from need statement #15; if the intervals do not overlap, they are considered to be different.

For easier communication, it is appropriate to rescale these numbers to range from 0 to 100, as in the column labeled "Final Score." To rescale, first

TABLE 3-8.
Final MaxDiff Results for Industrial Equipment Case

Need Statement	Raw Score	Pct.	95% CI	Final Score	95% CI	Grade
15. "Uses half as much fuel…"	1612	89.6%	± 2.0%	100	± 1.3	A
7. "Can easily maneuver…"	1036	57.6%	± 3.6%	79	± 2.3	B
2. "…helps a less skilled operator…"	980	54.4%	± 3.6%	77	± 2.3	B
8. "…operate in extreme conditions…"	884	49.1%	± 3.7%	74	± 2.3	B
1. "Actively manages traction…"	748	41.6%	± 3.7%	69	± 2.4	B
9. "…reduced emission fuel w/o sacrificing…"	664	36.9%	± 3.6%	66	± 2.3	B
14. "…able to automatically perform repeated tasks…"	608	33.8%	± 3.7%	64	± 2.4	B
12. "…exceptionally quiet…"	212	11.8%	± 3.5%	50	± 2.3	C
4. "…self-diagnose problems and report…"	160	8.9%	± 3.1%	48	± 2.0	C
3. "…tracks and schedules its maintenance…"	64	3.6%	± 1.8%	45	± 1.1	C
10. "Controls can be easily adjusted…"	−232	−12.9%	± 3.5%	34	± 2.2	D
5. "…easily reconfigured and set-up…"	−348	−19.3%	± 3.3%	30	± 2.1	D
13. "…reduces dust during operation and transport…"	−684	−38.0%	± 3.5%	18	± 2.3	D
11. "…extremely easy and quick to clean…"	−1028	−57.1%	± 3.0%	6	± 1.9	D
6. "…displays e-versions of service manuals…"	−1196	−66.4%	± 2.5%	0	+ 1.6	D

subtract the lowest value from the highest value (e.g., $1,612 - (-1,196) = 2,808$). Now, subtract the lowest value from the value for each need; for example, subtract $-1,196$ from 748 for need statement #1 ("Actively manages traction for maximum performance") to get $1,944$. Now, divide this number by the largest difference ($2,808$) to get a score of 69. Repeat this calculation for each need to complete the Final Score column.

To convert these boundaries to the 0 to 100 scale, multiply the confidence interval for the percentage (from the first "95 percent CI" column) by 100 and divide by the range of percentages (from the "Pct." Column). For example, the confidence interval for need statement #1 is 100 times 3.7 divided by 156 (89.6 minus −66.4), or 2.4. Table 3-8 shows these results in the second "95 percent CI" column.

If the sample size supports it, you can analyze segments. First, select the customers that are in the segment you wish to analyze (e.g., males, companies with more than 50 employees, non-users, etc.) and run the calculations for those people. A confidence interval can be computed for each need within each segment to see if the results are statistically significant within the reduced sample.

Analyze Customer Priorities

The most universal tool for analyzing how customers prioritize needs is the Pareto Chart, a bar graph in which the bars represent importance values. These are sorted from most important to least important so that the key customer

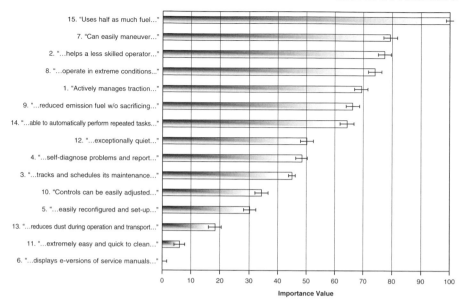

FIGURE 3-7. Final MaxDiff results for industrial equipment case.

needs can be identified, distinguishing the "vital few" from the "useful many" (see Figure 3-7).

In this example, 95 percent confidence intervals are computed and displayed so that statistically significant differences are apparent. If the confidence intervals for two bars overlap (e.g., 7 "Can easily maneuver and work effectively in tight spaces" and 2 "Automatically helps a less-skilled operator perform up to the level of a good operator"), the values are not considered to be statistically different. If these do not overlap (e.g., 7 and 12 "Is exceptionally quiet"), the two are considered to be different.

To simplify the discussion of the results, the needs can be assigned letter grades as follow:

A—a member of the vital few

B—a key driver

C—a potential segment driver

D—useful, but not a driver

Grading is somewhat arbitrary, but must be based on the importance values and must be done consistently. Methods to assign grades include:

◆ A range of values for each grade (e.g., A = 80–100, B = 60–79, C = 40–59, D = 0–39)

◆ A grade for each quartile (A = top 25 percent of needs, B = top 26–50 percent of needs, etc.)

◆ According to confidence level, if confidence intervals are computed (e.g., all needs that overlap the confidence interval of the highest ranked need get an A, all those that overlap the confidence interval of the highest ranked need after the As get a B, etc.)

The last column in Table 3-8 shows letter grades for the industrial equipment results based on the first method. For use in presentations and documents, grades can also be color-coded for easier comprehension.

If you have calculated importance values within customer segments, you can also grade them in the same manner. This will tell you whether any of the needs that were graded lower for the overall market show up as more important for specific segments. Although it is unlikely that any needs that are Ds for the overall market would show up as As or Bs within a segment, you may find a few Cs or Bs that are graded higher within a segment. This analysis prevents you from ignoring any needs that are "must haves" for important customer groups. It is important to try to look at as many different segments as the data will support so that you can be satisfied that you have not missed any critical needs.

In addition to analyzing the segments for differences, also analyze them for commonalities. Needs that are universally considered to be As and Bs could indicate the key components of a product platform or requirements, while the differences might indicate potential product variations.

The composite results for the overall market suggest a product built around fuel economy. Reviewing the segment profiles may reveal different underlying patterns of needs that suggest other alternative concepts. These segment-driven concepts can be carried forward for further evaluation.

COMPETITIVE PERFORMANCE RATINGS

Survey Design

Now that the MaxDiff process has been discussed, it's time to return to the survey design and discuss how to incorporate the competitive performance ratings. If you know where competitors perform today on customer needs, you can target *open space*—product spaces with important needs that are not currently or likely to be served by the competition. To design these questions, you must decide which competitors to ask the respondents to rate and how to phrase the question.

Figure 3-8 shows competitive rating questions from the industrial equipment example. In this case, two competitors were rated (identified here as Company A and Company B). Asking respondents to rate two competitors rather than one has a couple of benefits. First, it increases the total number of ratings you will have available to analyze. Second, it encourages the respondents to put more thought into the ratings task because they will make

Given what you know about each of the companies below, how likely is each one to meet the following need in the next few years? Please circle a number below where "10" means that the company is extrremely likely to meet this need and "1" means that they are not at all likely to meet this need.

machine actively manages traction for maximum performance											
	NOT AT ALL LIKELY								EXTREMELY LIKELY		
Company A	1	2	3	4	5	6	7	8	9	10	Don't Know
Company B	1	2	3	4	5	6	7	8	9	10	Don't Know

machine can be easily reconfigured and set-up											
	NOT AT ALL LIKELY								EXTREMELY LIKELY		
Company A	1	2	3	4	5	6	7	8	9	10	Don't Know
Company B	1	2	3	4	5	6	7	8	9	10	Don't Know

machine is exceptionally quiet.											
	Meets this need VERY POORLY								Meets this need VERY WELL		
Company A	1	2	3	4	5	6	7	8	9	10	Don't Know
Company B	1	2	3	4	5	6	7	8	9	10	Don't Know

FIGURE 3-8. Example competitive rating questions.

comparisons across the two competitors. The presence of a second competitor provides a richer context which tends to jog memories and stimulate minds.

First, you must choose which competitors will be rated. With a paper survey, the competitors must be prespecified and printed on the page; in other words, they cannot be customized based on the respondent's experience. The competitors could be represented in the survey by their names or by a description (e.g., "your current brand," "the leading competitor"). If you use actual names, one of them should be your company's name so that you can know where you stand. The other name could be the market share leader (or the second competitor, if you are the leader) or a competitor that you feel will be your strongest competition for selling this product. If you use your company's name in the survey, the second company should not be identified descriptively (such as "strongest competitor") because showing your name next to a generic name will give away that you are sponsoring the survey. (If you are collecting perceptions, it is a good idea not to reveal the sponsor of the survey, as that will bias the results.)

The disadvantage to naming specific competitors is that the respondent may not be familiar with them, since the competitors' inclusion is based on your preference, not the respondent's experience. One way to address this issue is to make familiarity with one of the competitors a qualification in the screening process. Add a question to the screener that asks them to rate their experience with the named competitors and disqualify anyone who does not pass a minimum threshold for either competitor. If a respondent is not familiar with *your* company, you will need to decide whether you still want his or her input. If you are looking to enter a new market, for example, you may want to know what needs those customers have even if they cannot rate your company's performance.

The survey should include a question that asks for the respondent's experience with each key competitor so that you can evaluate their ratings accordingly. Some examples of experiences to ask about include the following:

◆ Current customer, past customer, non-customer

◆ Purchased the brand, evaluated it for purchase, have used it in the past, are familiar with it, are not familiar with it

◆ Primary brand used; use the brand, but is not primary; don't use the brand

◆ Most frequently purchased brand in past three months, etc.

A common approach is to ask customers to rate only the product they currently own. The main drawback to this approach is that customers tend to be fairly satisfied with the products they have, so the ratings are often not very differentiated across customer groups. Also, this approach provides no information on what noncustomers think about your product, so you can't get a sense of the size of the gap you need to bridge in order to win them over.

It is a good practice to have customers rate at least one other competitor that they know in addition to the one they use. This will make it possible to assess the gap that you need to overcome to win new customers and identify the weaknesses that competitors might have outside of their customer base.

The ratings questions themselves can take many forms, but the example in Figure 3 8 is a common type that works well. The question asks how well each competitor is positioned to meet a particular need. In this example, a 1 to 10 rating scale is used. Many researchers prefer an odd-numbered scale (e.g., 5-, 7-, or 9-point) so that customers can select the center of the scale if they don't know or have no opinion. Others argue that an even-numbered scale is better precisely because people can't sit on the fence and choose a middle point. There is a lot of debate on this issue; overall, relative to the importance measurement issues, the exact design of the rating scale for competitive performance is not a major concern.

The final survey section includes the background data that are used to classify people, analyze known segments, and profile particular response patterns, for example, customers who think *durability* is more important than *reliability* or that *tastes great* is more important than *less filling*. To design these questions, revisit your sample design discussion and decide which information is important to capture. These can include demographics, corporate profiling information, title or position, volume of consumption, relationship with competitors, and so forth.

Process and Analyze Competitive Performance Ratings

The first step in processing the competitive performance ratings is to make sure that you have matched the ratings to the right competitors in your data file. If your survey listed the actual names of competitors, then the ratings are already matched up properly. If you used descriptive wording such as "Your primary supplier" in the rating question text, then you need to recode the data so that the ratings can be analyzed for each competitive product. To do this, you need to refer to the question in the survey that identifies which supplier matches the description in the question. For example, if the rating questions ask to rate "Your primary supplier," there should be a question in the survey that asks who is their primary supplier. In the data file, create one field for each rating for each supplier mentioned and place the corresponding ratings in those fields. If Respondent 1000 considers Competitor A as their primary supplier and Respondent 1001 considers Competitor B as theirs, then 1000's ratings for "Your primary supplier" are copied into the fields for Competitor A and 1001's into the fields for Competitor B.

There are some alternative approaches to analyzing competitive ratings, such as computing an average, determining the percentage of ratings that are at the top of the scale, or subtracting the ratings at the bottom of the scale from those at the top. Averaging the ratings can sometimes flatten the data so that differences are not as apparent, but this process takes into account the full scale that was used. Focusing on the top part of the scale (e.g., on the 8s to 10s of a 10-point scale) does not consider differences in the lower part of the scale to be meaningful, but sharpens the differences in the data by emphasizing the strongest opinions. The results are reported as the percentage of the sample that rated the competitor at the top of the scale. A variation of this is to subtract the percentage of ratings that were at the bottom of the scale from the percentage of ratings that were at the top, producing a net rating.

To get a clearer sense of your strengths and weaknesses, take the ratings for all other competitors (using any of the previously described methods) and subtract it from your ratings. Positive results will indicate your strengths and negative ones indicate weaknesses.

Analyze Customer Priorities and Competitive Performance

At this point, the list of needs has been prioritized from the customer's perspective, some logical concepts to test later have been identified based on segment differences, and strengths and weaknesses in competitive performance have also been identified from the customer's perspective. However, conclusions about what actions to take cannot be drawn without further analysis. The customer may be king, but even a king can't have everything, and the information developed so far helps you to understand the trade-offs that a customer is willing to make. To know where the opportunities are, a more complete profile must be built around each important need, especially with respect to gaps versus competition.

Quadrant analysis is a common tool for analyzing the importance of needs relative to where competitors perform on them (Katz 2004) and is straightforward to implement. In the adaptation of Katz's example in Figure 3-9, the

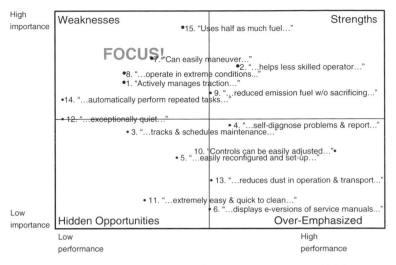

FIGURE 3-9. Quadrant analysis adapted from Katz, 2004.

results for the industrial equipment example have been charted. The importance scores are based on the overall market average, and the performance scores are based on ratings of Competitor A relative to competition.

For each need statement, a performance score has been calculated by averaging all the ratings for Competitor A and subtracting the average rating for Competitor B. (If you collect perceptions on more than one other competitor, subtract the average for all other competitors combined.) The vertical line in the middle is drawn through 0, representing performance that is the same as competition.

The placement of the cross hairs is somewhat arbitrary, so the placement of needs into the quadrants is also arbitrary. Consider the following alternatives:

◆ Draw them so that half the needs are on each side of the line (left/right, above/below).

◆ Draw the importance line to separate the As and Bs from the Cs and Ds.

◆ If you have performance by competitor, calculate the gap between your performance and your competitors', and draw the line through 0 so that strengths are to the right and weaknesses are to the left.

Using these methods to draw the cross hairs will make the charts easier to interpret because the placement of the lines will be meaningful to you.

Using the performance data for the rest of the competitors, with a little more effort you can expand on the quadrant chart with a method called *open space* analysis. Open space analysis helps you to build a differentiated value proposition for your product by illustrating which needs are *owned* or *closed* by competitors and which are open, ready to be filled with a differentiated offering (these are also referred to as *unmet needs*). In open space analysis, the needs are sorted into four categories (see Table 3-9).

TABLE 3-9.
Definitions of the Four Open Space Categories

Openness Category	Competitive Situation
Co-owned	• Top performers have no headroom to improve • No clear leader among top performers
Owned	• Top performer has no headroom to improve • Top performer has a significant lead over all other competitors
Led	• Top performer can still improve its perception • Top performer has a significant lead over all other competitors
Open	• Top performers can still improve their perception • No clear leader among top performers

Needs are categorized via significance testing of the performance data to establish which differences across competitors are significant and if there is headroom for a leader to improve. A competitor owns, co-owns, or leads a need if the lower bound of its performance rating confidence interval (calculated as it was for the importance rating data) does not overlap the upper boundary for competitors with lower average ratings. For needs that are *owned* or *co-owned*, there is no opportunity to differentiate (unless you are already the owner). For needs that are *led* by another competitor, there is still headroom to establish differentiation, but differentiation requires not only substantially improving your performance, but also surpassing an established leader. Needs that are *open* provide the clearest opportunities to differentiate.

To determine whether there is headroom on a need requires some judgment. The idea is that a competitor is so highly rated that it would be impossible to achieve a higher rating that would be statistically different from the current leader. For example, suppose a competitor had an average rating of 9.5 on a 10-point scale with an upper bound of 9.75. It would be nearly impossible to achieve a score with a lower bound that exceeded the competitor's upper bound. In this case, there is no headroom to improve on the current leader's position, unless the leader's performance declines or your product radically changes how customers perceive how their needs are filled. For example, the introduction of a fax machine radically changed how well customers thought overnight delivery met their needs, and the introduction of e-mail changed the perception of the fax's ability to meet similar needs. However, such situations are uncommon.

Opportunities become clearer when they are matched with the importance values in an open space chart (see Figure 3-10). In this chart, the needs are plotted according to importance within the openness category to which they belong. If a competitor owns, co-owns, or leads a particular need, its name is appended to the need label. Analyzing this chart lets you understand your

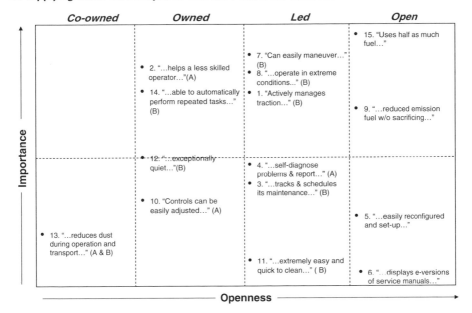

FIGURE 3-10. Example of open space chart for two competitors (A & B).

current sources of advantage, potential opportunities to differentiate, and which needs you cannot own (unless you have a disruptive action to take). The goal of this analysis is to identify potential winning value propositions for your product development effort based on customer needs. You want to select the needs that your product will lead or own, and you need to identify enough of them to overcome—not just to match—the value proposition established by competitors as defined by the needs they lead and own. You may identify several promising value propositions, in which case you can develop a set of criteria for scoring them relative to each other.

The implications of the open space analysis are somewhat different for product updates versus disruptive new offerings. The latter may change the rules of the game so that a need that is currently owned by a competitor may become vulnerable to the new offering. When using open space analysis, the product team must consider whether proposed introductions are disruptive enough to dislodge an entrenched competitor with respect to the needs it owns.

In the industrial equipment example in Figure 3-10, two competitors are displayed, Competitor A and Competitor B. Competitors A and B co-own need statement #13 ("...reduces dust during operation and transport..."); there is no room to improve and no way to gain an advantage (unless the performance of one competitor declines). Each competitor owns two needs outright. Competitor A owns need statements #2 ("...helps a less skilled operator...") and #10 ("Controls can be easily adjusted...") while B owns #14 ("...able to automatically perform repeated tasks...") and #12 ("...exceptionally quiet..."). In these areas, the lower-rated competitor cannot realistically expect to surpass the leading competitor because the overall level of performance is high.

From this chart, the most likely needs to target would be need statement #15 ("Uses half as much fuel...") and #9 ("...reduced emission fuel w/o sacrificing..."), which are both open and relatively important. Of the two, #15 is much more important, making it attractive to target. Need statement #7 ("Can easily maneuver...") is also an important one, and there is room to improve on current offerings. However, Competitor B already leads the market in fulfilling this need, and it may be more difficult to overtake an existing leader than to establish leadership on another need that does not have a leader. Using the information in the open space chart, Competitor A can start to analyze issues that will help it to build a stronger value proposition for its product.

Note that in the quadrant chart (Figure 3-9), need statements #13 and #9 are positioned near the center of the chart from left to right, indicating that Competitor A and Competitor B perform similarly to each other on these needs. However, the open space chart adds another key piece of information: both competitors are performing as well as possible on need #13, but need #9 is open, indicating that it is possible to establish leadership on this need.

The open space chart can be used to group needs together. For example, the "owned" and "co-owned' spaces show that Competitor A and Competitor B each control the same number of needs, and that each of these sets are of similar importance overall. Competitor A must choose some other needs in order to differentiate its value from Competitor B's. Improving on need statement #4 ("...self-diagnose problems & report...") might be a good move because Competitor A already has a lead, the need is relatively important, and there is room to improve. Competitor B, however, leads on five needs, indicating that Competitor A must differentiate on more than one additional need. The best candidates will come from the needs that are classified as open. Following this sort of reasoning, Competitor A can group together different sets of needs that might create a differentiated advantage and subject those to further scrutiny.

Draw Conclusions and Prioritize Actions

So far, the customer needs analysis has produced the following key tools and deliverables:

- ◆ List of needs with corresponding features that could address each one
- ◆ Affinity chart
- ◆ Grades for each need
- ◆ Quadrant analysis chart
- ◆ Open space chart
- ◆ Potential product concepts or needs grouping (from segment analysis and open space analysis)

TABLE 3-10.
Example of a Needs Profile
Environmental Issues

	Grade	Open Space Category	Our Capability	Best Competitor Capability
9. "...reduced emission fuel w/o sacrificin..."	B	Open	Can do it now	Have technology
12. "...exceptionally quiet..."	C	Owned by B	Technology exists	Can do it now
13. "...reduces dust in operation & transport..."	D	Co-owned	Can do it now	Can do it now

	Proposed Features	Key Segments
9. "...reduced emission fuel w/o sacrificing..."	[list features]	• Very Large
12. "...exceptionally quiet..."	[list features]	• None
13. "...reduces dust in operation & transport..."	[list features]	• None

In order to draw final conclusions, it is valuable to take this information and use it to develop profiles for each potential value proposition or concept. Table 3-10 offers a schematic example of a needs profile that summarizes the key learning of the effort so far.

The profile summarizes one group of needs from the affinity chart. The lower table has the list of potential features that could address each need and the key segments that have high value for that need (if any). The upper table contains the importance grade for each need and the open space categorization (which notes who leads or owns the need, if applicable). The last two columns result from analyzing the features that address each need and drawing conclusions about how readily they can be met by you and by the most capable competitor. The following classifications are used:

◆ Has the technology and can implement it now
◆ Has the technology but is not ready to implement it
◆ Does not have the technology, but the technology exists
◆ The technology does not exist

Documenting the capabilities helps to assess the immediacy of the opportunity and the risk of not acting first on important needs. When all this information is considered together—customer priority, competitive position, and readiness to address—the product team can begin to make its own trade-offs regarding which needs to target.

The information in these profiles is also useful to help score the value propositions that result from the open space analysis. In this step, the team defines a set of internal criteria and a set of market criteria. Internal criteria might include factors such as readiness to implement, cost to implement, and fit with corporate value proposition. Market criteria might include customer importance, ability to address strategic segments, and degree of competitive advantage or threat. Each of these criteria is rated on a five-point scale, where each point is given a meaningful, measurable label. The criteria are assigned weights relative to each other, and a weighted average score is computed for internal criteria and market criteria.

When scoring the needs, it is up to each product team to define the criteria they want to use to evaluate their likely success. They must also decide how to weight these criteria relative to each other, and this can be an iterative process as weights and criteria are adjusted after new plots have been produced. Each iteration of testing the criteria and their relative weights helps the product team to become more aware of its own decision making process. Examples of criteria for internal scoring include:

- ◆ Cost to implement
- ◆ Clarity of actions that would result from targeting the need
- ◆ Fit with marketing and sales process
- ◆ Time frame for results
- ◆ Fit with other development efforts

Examples of criteria for market scoring include:

- ◆ Relevance to established target segments
- ◆ Market coverage
- ◆ Potential to differentiate from key competitors
- ◆ Credibility with the market (will the market believe you can deliver it?)
- ◆ Sustainability of differentiation

Using these types of criteria, the product team can then analyze the customer needs relative to their own capabilities and strategies in order to target individual needs or bundles of needs that will have the highest likelihood of success.

Continuing the industrial equipment example, these criteria are plotted against each other, as in Figure 3-11. The line through the middle is placed at the discretion of the team. In this case, the team has grouped the needs together according to the affinity chart. The product team evaluated the market opportunity for each need in terms of its importance to customers and the potential to establish leadership on it. Needs that are open receive a higher score, and those that are owned by someone else receive a lower score. Because of this, two needs that have similar importance could be considered by the

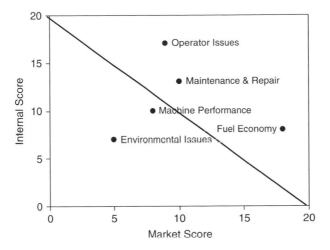

FIGURE 3-11. Example of scoring needs.

team to represent different levels of opportunity because there might be more opportunity to lead on one set (e.g., a set of needs that tend to be open) than on another, (e.g., a set of needs that tend to be led or owned by a competitor). Fuel economy—which is both open and important—is clearly the best opportunity as far as the market is concerned, but internal considerations may make it less attractive.

With respect to the internal measures, fuel economy does not score very highly relative to the other need categories. Competitor A currently does not have strong capabilities in this area, and it would take a great deal of resources and organizational commitment to excel here. Operator issues are where their strongest internal capabilities lie today, but the market opportunity is not so great based on the relative lack of importance of some of these issues. The scoring chart does not provide a definitive solution, but it provides a systematic way to discuss the alternatives in the context of internal and market realities.

Timing and Budget

This type of project can be executed in about a month to six weeks, with data collection as the main driver of the timeline. It could take longer if the design is complex (e.g., a lot of different quotas), the respondents are hard to find and recruit, or a slower methodology is used for fieldwork (e.g., mail or customer site visits). Time may also increase if you need to assemble a lot of information in order to make decisions on the sample planning. The project could take less time if the survey is programmed and administered on a computer or the Web. Web-based interviewing in particular will decrease the time needed for data collection and data handling, as all the data would be available immediately. A typical breakdown by main activity:

Activity	Week #
Project, survey and sample design	1
Recruiting and survey administration	2–4
Data processing & importance estimation	2–5
Charting & segment analysis	5
Opportunity scoring & conclusions	6

This project can be executed by one person, with support for some of the activities. The project leader will design the survey, request and review the market information to make the sample plan, receive updates on the data collection progress, oversee the charting process, design the opportunity scoring, and lead the internal discussion of the results. Project support would include: gathering data for the sample planning; creating the different versions of the survey; acquiring a list[2] of potential respondents; recruiting people to take the survey; administering the survey; paying respondent incentives; data processing; estimating importance; and creating charts.

The out-of-pocket cost largely depends on your sample plan and how you plan to administer the survey. Typical out-of-pocket data-collection expenses will include the cost of a list, respondent incentives, and recruiting costs. Purchasing a list could cost from a few hundred dollars up to $5,000, depending on the type of respondent targeted. If you use an e-sample provider or a data collection agency that provides a list, these costs will be built into their cost per interview (CPI). Data collection for research on products sold to a mass market of consumers will cost less than for products sold to people who are difficult to identify or hard to reach (like executives). CPIs could range from less than $5 for a consumer study to $100 or more when people are harder to find and recruit.

With respect to administering the survey, your costs will differ depending on the method you choose. For a paper survey, you have the cost of printing the surveys, getting them to the respondents, and getting them back. If you can administer the survey via computer or online, you can hire someone to program and host it for you (about $3,000 or less for this type of survey).

SUMMARY

The accurate prioritization of customer needs is critical information for successful product development, and recent advances in market research techniques make trade-off analysis of customer need importance accessible to product teams. Trade-off analysis is superior to importance rating approaches because it reduces bias, increases customer interest in the task, and forces customers to say what they are willing to give up in order to get what they need. MaxDiff introduces a way to measure customer trade-offs that is easier to use, more

---

[2] For tips on customer lists, please visit *www.rsginc.com/pdma/toolbook3/tips*.

affordable, and more flexible than traditional conjoint or discrete choice methods. For many applications, product teams can execute a MaxDiff analysis without any specialized software or outside help.

These tools and techniques will not tell you what products to build. However, they will help you to make decisions by providing a structured approach for making fact-based decisions. In addition to helping you make decisions, they will help you to articulate your own decision-making process.

REFERENCES

Supporting information for this chapter can be found at *www.rsginc.com/pdma/ toolbook3*. This includes experimental designs to analyze for five 5 to 15 needs, detailed discussion of sample planning with planning templates, and additional tips.

Cohen, Steven H. 2003. *Maximum Difference Scaling: Improved Measures of Importance and Preference for Segmentation*. Sequin, Washington: Sawtooth Software, Inc.

Cohen, S. H., and L. Neira, 2003. "Measuring Preference for Product Benefits Across Countries: Overcoming Scale Usage Bias With Maximum Difference Scaling." ESOMAR 2003 Latin America Conference Proceedings. Amsterdam: The Netherlands.

Colbourn, Charles J., and Jeffrey H. Dinitz. 1996. *The CRC Handbook of Combinatorial Designs*. CRC Press.

Flynn, Terry N., Jordan Louiviere, Tim J. Peters, and Joanna Coast. 2006. "Best-Worst Scaling: What It Can Do for Health Care Research and How to Do It." CenSoC Working Paper No. 06-001.

Griffin, Abbie, and John Hauser. 1993. "The Voice of the Customer." *Marketing Science*, 12 (1), Winter: 1–27.

John, J. A., and E. R. Williams. 1995. *Cyclic and Computer Generated Designs,* 2nd ed., London: Chapman and Hall.

Katz, Gerald. 2004. "The Voice of the Customer." *The PDMA ToolBook for New Product Development*. Hoboken, NJ: John Wiley & Sons.

Katz, Gerald. 2001. "The 'One Right Way' to Gather the Voice of the Customer." *PDMA Visions* 25 (2), October.

Louiviere, Jordan J., and Towhidul Islam. 2004. "A Comparison of Importance Weights/Measures Derived from Choice-Based Conjoint, Constant Sum Scales, and Best-Worst Scaling." CenSoC Working Paper No. 04-003.

Orme, Bryan. 2006. *Getting Started with Conjoint Analysis: Strategies for Product Design and Pricing Research*. Madison, WI: Research Publishers LLC.

Sawtooth Software, Inc. Software for Designing Experiments, *www.sawtooth software.com*

Sawtooth Software, Inc. 2005. "The MaxDiff/Web System Technical Paper." *www. sawtoothsoftware.com*.

4

The Slingshot: A Group Process for Generating Breakthrough Ideas

Anne Orban
Director, Discovery & Innovation, Innovation Focus Inc.

Christopher W. Miller
Founder & Chief Executive Officer, Innovation Focus Inc.

INTRODUCTION

This chapter provides a detailed explanation of the Slingshot process for generating breakthrough ideas. The Slingshot is a group process that uses four different types of participants and two different processes in continuous time to achieve breakthrough ideas. The four types of participants are prosumers (defined later), consumers, project team members, and a moderator/facilitator. The two processes are focus group and creative-problem-solving sessions. This chapter discusses what makes a Slingshot different and useful, when to use a Slingshot, how to implement a Slingshot, and keys to success and possible pitfalls.

This detailed explanation of the Slingshot group process is addressed primarily to team leaders who would make the decision to use a Slingshot and be responsible for its implementation. The chapter will focus primarily on a Slingshot's use in a discovery-phase project at the front end of a product development process. Sufficient detail will be provided so that a Slingshot can be selected and used effectively in other contexts when a problem needs a high-quality solution set in a short amount of time.

The Slingshot process emerged from a British Airways (BA) product development team working to upgrade its business class service. A focus group with frequent business class users was followed by a debriefing session. Several of the participants happened to be both frequent trans-Atlantic business-class customers and skilled professionals in product development. One of these participants was also included in the extensive debriefing creative-problem-solving session immediately after the focus group. This person was "slingshot" between participating in a consumer experience and then employing her professional product development background in the debrief session. In the debriefing session, she challenged the BA team's conventional thinking and came up with a breakthrough idea leading to very satisfactory results for the final outcome in seat design. The innovation in process became formalized as the Slingshot group process and the person acting in the consumer and professional product development team role was christened a *prosumer*.

What Makes a Slingshot Different and Useful?

There are three characteristics that make a Slingshot different from and more useful than freestanding implementation of either of its component processes in the task of developing a high-quality solution set:

1. *Introduction of the prosumer in both consumer and creative roles.* Prosumer participation contributes depth to understanding the consumer experience. Prosumers also provide high-level creative input from a product development professional, as well as impartial and informed challenge to team biases and assumptions from an outside peer.

2. *Purposeful development of creative tension.* Creative tension helps uncover breakthrough ideas when prosumers and project team members are slingshot from the consumer experience and into the creative idea-generator role.

3. *Close proximity in time of a qualitative research experience and a creative-problem-solving session to optimize creative tension.* Anecdotal evidence supports the value of the close proximity of listening to the voice of the customer and then processing it immediately in a creative-problem-solving session, whether on the same day or consecutive days.

The relatively short time needed to plan and complete a Slingshot also makes the tool useful. A basic Slingshot can be accomplished in three to four weeks when the topic is not too specialized. The degree to which the topic area is highly specialized will impact the amount of time necessary to obtain participation from the right prosumers and consumers—key components of a successful Slingshot.

Finally, a Slingshot is useful because third-party costs for doing the process are moderate, given the value of its output. The estimated cost for a basic daylong Slingshot with a moderator/facilitator, one focus group with 10 recruits, and two prosumers, is estimated at $9,000. The components in this estimate include:

◆ Focus group and/or other facility rental (video & audio recording)
◆ Ten consumers (recruiting fee and incentives)
◆ Two prosumers (honoraria)
◆ Catering

PROSUMERS

A prosumer for a Slingshot is an individual who can play both the consumer and the product development professional role. Candidates for the prosumer role can be individuals from other divisions of a large enterprise, noncompeting product development practitioners from other enterprises, academics, and consultants. They can be from a range of disciplines, including market research, marketing, design, logistics, manufacturing, engineering, science, economics, general management, project management, finance, IT, and supply chain.

In a Slingshot, a prosumer has three roles (Figure 4-1):

1. Connected consumer
2. Skilled idea generator
3. Impartial challenger

Prosumer as Consumer

In the consumer role, prosumers are invited to immerse themselves in the consumer experience. In this way prosumers develop the intrinsic motivation to care about finding solutions to consumer problems. To be effective in the consumer role, prosumers must have enough characteristics in common with the prequalified focus group participants to blend with them. Similarly, prosumers are expected to provide reaction and insights to topics introduced by the moderator in the same way as the prequalified target-market consumers.

Prosumer as Skilled Idea Generator

In a creative-problem-solving session, prosumer immersion in the consumer experience is coupled with content expertise and creativity skills that purposefully develop creative tension. Creative tension plays a significant role in stretching the prosumer to offer breakthrough ideas.

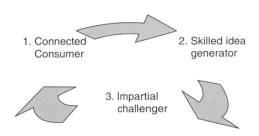

1. Connected Consumer → 2. Skilled idea generator

3. Impartial challenger

FIGURE 4-1. Three roles for the prosumer.

Prosumer as Impartial Challenger

A prosumer can act as an impartial challenger to stretch the project team because he or she is not part of the team's business context and dynamic and yet the prosumer has a deep understanding of the team's challenge. In terms of team dynamic, a prosumer does not have the same filters or biases as project team members. Neither does a prosumer share assumptions inherent in that company's culture. In terms of the team's business context, a prosumer is uninhibited in offering breakthrough ideas because a prosumer is not constrained by internal responsibilities related to developing and implementing any of the ideas.

Characteristics of a Prosumer

A prosumer in a Slingshot is an individual with the following characteristics:

- ◆ No conflict of interest
- ◆ Ability to sign a confidentiality agreement
- ◆ Consumer connection with the topic area
- ◆ Knowledge relative to the topic
- ◆ Experience with new product development
- ◆ Creative-problem-solving skills
- ◆ Good interpersonal communication skills

Appendix 4-1 provides tips to prospective prosumers on how to wear two hats in a Slingshot process.

CREATIVE TENSION

Amabile (1998) identifies three core components of individual creativity in the business context as *expertise* (technical, procedural, intellection knowledge); *creative thinking skills* (the flexibility and imagination with which individuals approach problems); and *motivation* (especially intrinsic motivation—the inner passion to solve a problem). Creative tension is a term to describe the state of mind that is the result of combining individual content expertise with creative thinking skills and intrinsic motivation to want to solve a problem in the context of a Slingshot process. A Slingshot's topic and objectives provides the focus of a specific business context for the application of creative tension. In the Slingshot process, project team members and prosumers are active participants in the purposeful development of the creative tension state of mind.

Creative tension = a *function* of (Individual content expertise + Creative thinking skills + Intrinsic motivation) in a specific business context.

The Slingshot purposefully develops creative tension in two ways. The first is by immersing individuals with topic expertise and creative thinking skills in a specific consumer/customer context to create intrinsic motivation to solve their problems. The second way is to immediately harness these individuals' topic expertise, creative thinking skills, and inspired intrinsic motivation in a problem-solving process designed to yield breakthrough ideas for the specified business topic.

CLOSE PROXIMITY OF FOCUS GROUP AND CREATIVE-PROBLEM-SOLVING SESSIONS

The rationale for collecting voice of the customer/consumer data is well supported in studies and articles such as in the *2003 PDMA Comparative Performance Assessment Study* (*http://www.pdma.org/cpas*). In that study, *understanding* the customer was ranked in the top two boxes as key factors driving new product development success, while *misunderstanding* the customer was ranked in the top two boxes as factors causing failure in new product development. Similarly, Cooper (1999) argued that product developers will continue to introduce failures if customer input is missing. Alam (2005) observed that better strategies were needed to effectively interact with customers to obtain necessary input. Mixing prosumers with consumers in focus group sessions is one way to improve this interaction.

Following a focus group with a creative-problem-solving session is a natural flow for developing breakthrough ideas. The rationale for the back-to-back use of these two processes derives from the value of creative tension as a springboard for breakthrough idea generation. With the voice of the consumer resonating in the heads of project team members and prosumers (who continue to carry the torch for the consumer), a skilled facilitator has lots of creative tension with which to stimulate breakthrough thinking in the follow-on creative-problem-solving session.

WHEN TO USE A SLINGSHOT

Project leaders should consider using a Slingshot when a project topic needs to yield a set of quality ideas in a short amount of time within a limited budget using first-hand consumer insights. A Slingshot is an effective tool in the Discovery phase of product development to identify product opportunity gaps by offering project team members easy exposure to consumers' needs and then immediately focusing them on turning the insights gathered into product ideas. It can be used to jump start the work of product modifications and extensions by gathering information through a developmental focus group that investigates adequacies and deficiencies of a current product and immediately applies that information to solicit ideas for a second generation product. Applied in the Development phase, a Slingshot can be an effective way for a project team to harness consumer input on concepts or prototypes that can immediately be put to use to refine those concepts and prototypes.

Harvesting the value of a Slingshot depends on paying attention to all components of the process. There are 10 steps in implementing a Slingshot:

1. Set the topic and objectives
2. Select the moderator/facilitator
3. Assemble the project team
4. Set up the logistics for the session
5. Screen and recruit prosumers and consumers
6. Develop the focus group discussion guide
7. Develop the creative-problem-solving process
8. Conduct the focus group(s)
9. Conduct the creative-problem-solving session
10. Document and disseminate results

Project leaders can use the basic methodology of a Slingshot in a variety of circumstances, as the basic process can be expanded and repeated depending on the topic and objectives. The details of each step will be described and illustrated with a case study.

INTRODUCING THE CASE STUDY EXAMPLE

In the discovery phase of product development, a Slingshot may be selected as a component of the research plan in a multistage front end of innovation process such as *Hunting for Hunting Grounds*.™ (see the *PDMA ToolBook 1*, Chapter 2). The case study that will be used to illustrate all the steps in the Slingshot process is an example of that circumstance.

With all the interest in preventing childhood obesity and improving the health and wellness of children, the pediatric nutrition division of a pharmaceutical company wanted to determine what nutritional products it could develop to improve the health outlook for this important cohort. The research phase had to be accomplished in two months in order to complete the project to meet the gate review timetable set by senior management.

Step 1: Set the Topic and Objectives

Setting the topic and clarifying the objectives for a Slingshot is the first step in successful implementation. The topic and objectives for a Slingshot will depend on the business context for which it has been selected. In general, strategic intent should drive project definition and project definition should drive process selection.

The project leader is the person responsible for ensuring that the Slingshot is focused on the right topic and that the objectives are clear. The topic of the Slingshot will determine the specifications for screening prosumers and consumers into the process. Setting the objectives for the Slingshot is equally important. They will determine the content of the discussion guide and the process design for the creative-problem-solving session. A well-written topic and set of objectives should clarify the following:

- ◆ Target market
- ◆ Problem to be understood
- ◆ Desired outcomes

Setting the Topic and Objectives in the Case Study Example

In the case study, the strategic intent was to ensure that the company had explored all possible ways of making a positive contribution to the significant national public health challenge of childhood obesity. The project leader selected a multistep front end of innovation process with exploratory research as a component. The Slingshot process was one of a number of processes selected as part of the exploratory research agenda which included focus groups with children and ethnographies in homes with households representative of the target population. The target population age range of 1 to 14 years was selected because parents and schools are perceived to have significant control over the nutritional choices for that age group. The topic for the Slingshot process has uncovered opportunity gaps in pediatric nutrition for improving the health and wellness of children ages 1 to 14 years. The desired outcomes for the Slingshot were to identify nutritional product gaps for the target population and to develop at least 20 concepts for new products and services with potential to fill the gaps.

Step 2: Selecting the Moderator/Facilitator

One skilled person can perform the roles of both focus group moderator and creative-problem-solving facilitator. Alternatively, two people may be recruited—one with the moderator skill set and the other with the creative-problem-solving skill set. The person(s) to fulfill the roles of moderator and facilitator may be recruited from within the company or from outside. If the person recruited for the role is internal to the company, then it is very important for that individual to be perceived as neutral, objective and impartial in relations to the topic, to project team members and especially to the team leader.

The candidate for moderator must be able to accomplish all steps in preparing for and conducting the focus group, including developing a screener to obtain people qualified as being within the target market, writing the discussion guide to obtain the relevant consumer input during the focus group, and effectively managing participant discussion.

The skill set required in the facilitator role for the creative-problem-solving session includes experience with a range of tools and techniques for managing group dynamics, eliciting insights and ideas, and developing and prioritizing ideas for next steps.

Selecting the Moderator/Facilitator in the Case Study Example

In the case study, the project leader recruited a seasoned person external to the firm who had both moderator and facilitator skill sets. The continuity of one person doing both roles streamlined project management and reduced the briefing time necessary for these roles. From the moderator/facilitator's perspective, involvement in both components of the Slingshot enhanced familiarity with the content and rapport with the team. This increased the person's effectiveness in both roles.

Step 3: Assemble the Project Team

Assembling the project team is another responsibility of the project team leader. There is considerable agreement about what constitutes an effective team for product development in terms of number, type of participants, functions, gender, information processing strength, and topic experience. Similarly, there is agreement that effective team members must have enough time allocated to do the work of the team and that co-location improves communication and overall effectiveness.

It is generally accepted that a project team works well with 6 to no more than 12 members (Rees 1997). Team members should represent a range of functions appropriate to the project topic, a range of experience inside and outside the company, and a mix of gender and age. It is also possible to put together teams that are diverse in terms of problem-solving preferences, creativity, and personality that can be identified using various psychometric tools.

Assembling the Project Team in the Case Study Example

In this case study, there were 12 members of the project team. The team was led by a senior manager of marketing and business development. There were two other members from marketing and business development, a representative from packaging research and development, two consumer product managers, and one team member from marketing research. From the research side of the house there were five scientists. Seven of the team members were men and five were women. Several of the team members were new to the company. The team leader appointed a non–team member with significant project management skills as coordinator for logistics.

The team leader and a senior member from the marketing and the science side were full-time on the project. They constituted the core team and were co-located for the duration of the project. The other members of the project team had 50 percent of their time allocated to the project but did not co-locate.

Step 4: Set up the Logistics for the Session

The project team leader is responsible for determining the geographic location for a Slingshot. Multiple locations for a Slingshot can be selected if the project is national or international in scope. In making any location decision, the project leader will be guided by a number of factors, including the perceived value of a location in fulfilling the Slingshot's objectives, the impact of the location on the overall project budget and team members' availability, the likely incidence and availability of the kinds of prosumers and consumers that are needed to ensure success, and the desirability of having a choice of appropriate market research facilities. When the location decision is made, then the detailed logistics for the Slingshot can be handled by the project leader collaborating with the moderator/facilitator and assisted typically by a project manager or logistics coordinator.

The focus group discussion must take place in a typical market research facility that provides a front room for the moderated discussion and a back room with one-way glass and a sound system for project team members to see and hear the front-room discussion. Each session should be both audio- and videotaped, so that the team and other stakeholders can review the content in the future. As for scheduling the focus group and the creative-problem-solving session, the project leader will confer with the moderator/facilitator concerning time slots convenient for getting participation from the desired consumers and prosumers. If consumers can be recruited for a morning discussion, then the creative-problem-solving session can follow in the facility in the afternoon. If the desired consumers and prosumers can best be recruited for a late afternoon or early evening focus group, then the creative-problem-solving session must take place the next day. It could take place in the same facility or in another convenient location.

The Slingshot process can include more than one focus group session. Table 4-1 offers agenda formats for Slingshots with one and two focus groups. The project leader will determine the total number of focus groups based on overall project considerations, Slingshot topic and objectives, budget, and time available from project team members.

Setting up the Session Logistics in the Case Study Example

Given the national scope of the project, the project leader wanted an East Coast, West Coast and Midwest location. The objectives for the Slingshot necessitated access to medical and health professionals more likely to be found in large cities with public and private universities, hospitals, schools, and with

TABLE 4-1.
Agenda options for slingshots.
Option 1a: One-Day Slingshot Agenda with One Focus
Group Session

Time	Task
8:30–10:30	Focus Group
10:30–11:00	Break
11:00–12:30	Creative Problem Solving Session
12:30–1:15	Lunch
1:15–2:30	Creative Problem Solving Session
2:30–2:45	Break
2:45–4:00	Creative Problem Solving Session

Option 1b: One-Day Slingshot Agenda with Two Focus
Group Session

Time	Task
8:30–10:30	Focus Group 1
10:30–11:00	Break
11:00–1:00	Focus Group 2
1:00–1:30	Lunch
1:30–2:45	Creative Problem Solving Session
2:45–3:00	Break
3:00–5:00	Creative Problem Solving Session

Option 1c: One-Day Slingshot Agenda with Two Focus
Groups and Two Creative Problem Solving Sessions

Time	Task
8:30–10:30	Focus Group 1
10:30–11:00	Break
11:00–1:00	Creative Problem Solving Session
1:00–1:30	Lunch
1:30–2:45	Focus Group 2
2:45–3:00	Break
3:00–5:00	Creative Problem Solving Session

Option 2a: Slingshot Agenda for Consecutive Days

Time	Task
Day 1	
7 pm–9 pm	Focus Group
Day 2	
8:30–10:30	Creative Problem Solving Session
10:30–10:45	Break
10:45–12:15	Creative Problem Solving Session
12:15–1 pm	Lunch

TABLE 4-1.

(*continued*)

1:00–2:45	Creative Problem Solving Session
2:45–3:00	Break
3:00–4:00	Creative Problem Solving Session

Option 2b: Slingshot Agenda for Consecutive Days with Two Focus Group Sessions

Time	Task
7 pm–9 pm	Focus Group

Day 1

4:30–6:30	Focus Group
6:30–7:00	Break
7:30–9:30	Focus Group

Day 2

8:30–10:30	Creative Problem Solving Session
10:30–10:45	Break
10:45–12:15	Creative Problem Solving Session
12:15–1:00	Lunch
1:00–2:45	Creative Problem Solving Session
2:45–3:00	Break
3:00–4:00	Creative Problem Solving Session

a range of enterprises in the nutritional areas of interest such as organic foods. The locations chosen were Boston, Chicago, and San Francisco.

Step 5: Screen and Recruit Prosumers and Consumers

Consumers and customers will be needed to participate in the focus group research. Customers are those who buy the company's products. Consumers are those who have the problems the company is trying to solve with their products, but who may buy from someone else, buy a different type of product, make rather than buy a solution, or go with the problem unsolved because no one has "the right" solution. If you only talk with customers, you will only know about the needs of those who are already happy with your products.

To start with, the project leader must understand whether the Slingshot topic needs an exploratory focus group or a developmental focus group. The difference is that in an exploratory focus group, the task is to better understand consumer/customer experiences related to the topic and to probe to identify gaps in current available products and services. In a developmental focus group, the task is to understand consumer/customer reactions to concepts and/or prototypes.

There are some rules of thumb to guide project leaders in establishing screening specifications for all focus group participants:

◆ If it is an exploratory Slingshot focus group, then the emphasis is on consumers who also have the expertise necessary for exploring the topic.
◆ If it is a developmental Slingshot focus group, then recruit a mix of consumers and customers who are familiar with the competitive range of currently available products and services.

Screen and Recruit Prosumers

Prosumers will be needed as participants in both the focus group research and in the creative-problem-solving debrief. Prosumers need to be recruited based on the following criteria:

◆ Relevant content knowledge
◆ Degree of competition or potential collaboration
◆ Relevant consumer experience
◆ Amount of product development experience
◆ Creative-problem-solving skills
◆ Interpersonal skills

If a company requires prosumers to sign a confidentially agreement, then it is important to determine, at the outset, whether the prosumer candidate is able to do that.

Prosumer candidates can be found from within other divisions in a company, in the membership of the Product Development & Management Association, in professional organizations, in speaker rosters from professional conferences, from researching trade and academic publications, and from networking at professional meetings.

The first task in screening and recruiting prosumers is to identify necessary topic expertise and to develop a screening instrument with questions that reflect the breadth of capabilities desired, such as those covering the six criteria just listed. The project team leader's responsibility to establish the acceptable score range for all components of the prosumer screener. The range of acceptability will vary from Slingshot to Slingshot, depending on the perceived competitive sensitivity of the project and the importance of prosumer content knowledge and skill set.

The next task is for the project team to identify a lot of possible candidates. Team members can then contact prosumer candidates and have the qualifying conversation based on the template. Results of the screening conversation will help the project leader identify a short list to contact for securing participation of the required number of prosumers in the Slingshot. As a rule of thumb, two prosumers are optimal for participation in one focus group in which there are

TABLE 4-2.
Prosumer screening template.

Date: Project Name: Objectives for Slingshot: Prosumer Candidate: Brief Biography: (one paragraph)	Screener's Name: Project Topic:
Relevant content knowledge 1 2 3 4 5 Low High	Comments:
Ability to sign confidentiality agreement 1 2 3 4 5 No Yes	Comments:
Degree of competition with company OR Degree of potential collaboration 1 2 3 4 5 Low High	Comments:
Direct consumer experience 1 2 3 4 5 Low High	Comments:
New product development experience 1 2 3 4 5 Low High	
Creative problem solving experience 1 2 3 4 5 Low High	
Interpersonal communication skills 1 2 3 4 5 Low High	
Prosumer candidate should score between X and Y points to be considered for this Slingshot (team leader determines range) **TOTAL POINTS:**	
Ask to participate in Slingshot Team Leader's Signature	YES NO

six to ten consumers. The prosumer screening template in Table 4-2 can be used to guide the interview with prosumer candidates.

Screening and Recruiting Prosumers in the Case Study Example

In the Boston Slingshot location, four prosumers were recruited who represented a mix of business professionals and academics with relevant health and nutrition credentials. The ratio of 4:4 rather than the guideline 2:6 was chosen

TABLE 4-3.
Prosumer description.

Prosumers Used in the Slingshot (Boston)

Tom: Strategy consultant to companies in global nutrition and natural foods products marketplace. Academic degrees in food engineering, chemical engineering, and an MBA.

Bob: Co-founder of a natural foods products consulting company providing assistance in supply chain and distribution, strategic planning and organization development. Previous experience as VP sales and corporate development of fast growing natural food company.

Paul: Senior business development manager for the materials technology group of a contract R & D and small-volume manufacturing company with extensive knowledge of innovation process and tools commonly used to accelerate innovations to market. Member of the PDMA. Advanced degrees in chemical engineering and materials science. Parent of two children under age of 8.

Christina: Assistant Professor of Nutrition at a major research university. Her research efforts focus on the interaction among exercise, diet, body composition, and bone health using longitudinal studies and lifestyle interventions starting early in life. Principal investigator on two large children's studies focusing on role of calcium and exercise to increase bone density and muscularity in children first through third grades.

---

because of the complexity of the topic and the need for a range of technical expertise. Table 4-3 describes the prosumers who participated in the Slingshot in Boston.

Screen and Recruit Consumers and/or Customers

The purpose of the Slingshot and the type of people who will provide meaningful input will determine the screening specifications for a Slingshot focus group. In general, the screening instrument will want to consider specifications that also include a range of topic-appropriate demographics such as age, gender, household characteristics, race/ethnicity, and location. The exact type and number of people to be recruited for participation in a Slingshot focus group will vary based on whether the purpose of the discussion is exploratory or developmental.

In a focus group with an exploratory purpose, the emphasis is on learning as much as possible from appropriate content experts and allowing them plenty of time to interact with each other and the topic in the time allotted. Therefore, a small group of content expert consumers/customers is preferable for exploratory focus groups. As noted in the Slingshot format options in Table 4-1, more than one focus group can be included in a Slingshot process, which permits in-depth discussion with more experts in the exploratory context.

In a focus group with a developmental purpose, the emphasis is on getting as many relevant perspectives as possible based on participants' experience with the topic. The British Airways example that originated the Slingshot is an example of a focus group with a developmental purpose. It started with several potential concepts for redesigned seating for the range of purposes that seats are used during lengthy flights. To be included in the focus group discussion about new business class seat concepts, participants had to be customers and consumers who frequently made the trans-Atlantic crossing in business class for business purposes. More than one focus group can be included in a Slingshot process that permits breadth of input from eligible participants in the developmental context.

Screening and Recruiting Consumers/Customers in Case Study Example

In the case study, the purpose of the Slingshot focus group research was exploratory. Each Slingshot focus group needed to recruit four participants who in their professional roles were consumers (not necessarily the company's customers) of nutritional products for children. All participants had to be trained professionals in some aspect of childhood health and nutrition, either as a pediatrician, pediatric nurse, or dietician. Finally, each professional had to have some experience addressing the needs of children with diabetes, weight management or obesity problems. Participants who screened into the Slingshot focus group in Boston are described in Table 4-4. The questions, recruiting specifications grid, and nondisclosure agreement used to recruit consumers for the case study focus group discussions are in Appendix 4-2.

Step 6: Develop the Focus Group Discussion Guide

The discussion guide in a Slingshot provides the framework for learning about the topic area from the perspective of consumers and customers. Whether the purpose of the discussion is exploratory or developmental, the outline and flow of discussion guides tend to be more similar than different. The terms *exploratory* and *developmental* refer to the primary purpose of the focus group. An exploratory focus group's purpose is to explore an area in as

TABLE 4-4.
Case study consumer profiles.

Exploratory Focus for Pediatric Obesity and Childhood Nutrition Consumers used in Slingshot (Boston)

Pam: Registered Dietician, diabetes educator, 20+ years experience. Parent with a child aged 14.
Margaret: Registered dietician on campus working full time looking after 200+ students. Helps with meal planning, problems in obesity in sedentary lifestyle of visually impaired children.
Lakshmi: Pediatrician at an obesity clinic. Parent of two children ages 3 and 7.
Roberta: Registered dietician in a weight management clinic. Parent with children ages 9 and 12.

much breadth as possible. It has implications for the content of the discussion guide, and the kinds of stimuli introduced. The case study is an example of an exploratory focus group. A developmental focus group is one with the purpose of adding depth and understanding to a beginning concept or prototype. The discussion guide will typically involve presenting the concept or prototype for unprompted reaction followed by probes around perceived value, possible issues, and improvements.

The typical concerns for developing a discussion guide are shown in Table 4-5. Three areas in the discussion guide—homework, stimuli, and deconstructing the topic—require some explanation:

1. Prework that is done at home and brought to the focus group facility for the session is important for helping participants focus on the topic area, as well as for helping the moderator generate rich discussion. For example, participants can keep diaries for the week before the focus group(s) that record activities and thoughts relevant to the topic area that otherwise might be forgotten or go unmentioned. Participants using materials at home may assemble collages—combinations of visual images, words, and tables— relevant to the topic. These can reveal values and frames of reference that might otherwise be hard to express in words only in front of strangers. (Figure 4-2 is an example of a collage prepared for the case study topic.)

2. Stimuli refer to things that represent the Slingshot topic. These include things such as product or service physical prototypes, samples of snack mixes, visual representations of product/service concepts, descriptions of concepts, mock-ups of Web sites, and software programs.

TABLE 4-5.
Guidelines for developing a discussion guide.

Mechanics	Preparing participants for the contexts/situations of inquiry	Designing probing questions
◆ Keep the group size between 6 and 10 ◆ Allow a full two hours for discussion ◆ Cover fewer areas in greater depth	◆ Require homework that helps unlock the social and emotional dimensions of experience and need (especially if purpose is exploratory research) ◆ Create appropriate stimuli for use in the discussion (especially if purpose is developmental research)	◆ Deconstruct the topic to promote in-depth understanding of issues/opportunities ◆ Probe to reveal tacit knowledge about functional, social and emotional dimensions of topic area ◆ Use how, where, and why open-ended questions ◆ Incorporate homework and/or stimuli

FIGURE 4-2. Collage representing a prosumer's perception of nutritional challenges for children one to fourteen years of age.

Introducing stimuli to participants can be an integral part of the discussion guide in the focus group session.

3. Deconstructing the topic area refers to efforts within the focus group discussion to uncover, understand, and explore all aspects of the topic in one or a range of contexts. For example, understanding the opportunities to improve service for business class flyers across the Atlantic requires deconstructing all aspects of the experience from the decision to make the trip to collecting any checked baggage and leaving the airport, and all the steps in between.

Developing the Focus Group Discussion Guide in the Case Study Example

In the case study, the moderator and the project team leader collaborated on developing a framework for managing discussion of the scope and complexities

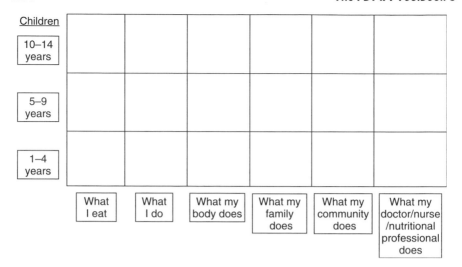

FIGURE 4-3. Chart prepared for case study topic analysis discussion.

of the topic. The discussion framework was visually represented by a multi-celled chart with three consumer age segments on the y axis and six nutrition contexts on the x axis. (See Figure 4-3.) The Slingshot case study focus group discussion guide is detailed in Table 4-6.

Step 7: Develop Creative-Problem-Solving Session Process

The creative-problem-solving session immediately follows the focus group and is designed to collect insights and generate new concepts to accomplish the Slingshot's objectives. The project team leader and facilitator should agree on the amount of time necessary to achieve the desired results. A creative-problem-solving session should use a minimum of two hours but can last as long as a day. It is the responsibility of the facilitator to design a process using appropriate tools and techniques to get results in the time allotted. (For a sample process see *PDMA ToolBook 2*, Chapter 17.)

A pattern of divergent and convergent exercises is typical of a successful creative-problem-solving session. The selection of tools and techniques depends on the objectives of the Slingshot. For example, a facilitator could choose a brand pyramid (de Chernatony 2001) exercise to help a marketing team develop communication strategies for a product launch; a morphological analysis (Kahn 2001) to drive idea generation at the intersection of selected market needs and technological capabilities; or an analogical reasoning (Sifonis et al. 2003) exercise to stimulate thinking about new paradigms.

A process plan contains information about timing, tasks, and materials necessary to accomplish the tasks. It must also factor in the need to create a record of the discussion and outcomes. Basic materials for every group *creative-problem-solving session include a flip-chart stand, pads, different*

TABLE 4-6.
Discussion guide in the case study example.
Participant homework: All focus group participants were asked to create a collage using visual images, words, tables etc. to represent the nutritional challenges for children in the USA today.

Time	Topic
0:00	Introduction—Task, participants, people behind glass, video recording.
0:05	◆ Have topic analysis chart prepared and hang on wall (See Figure 4-3)
	◆ Hang prepared spider diagram/mind map outline on wall (See Figure 4-4)
	◆ Put the collages on the wall for all to see. (See Figure 4-2)
	◆ Ask participants to present and explain their collages
	◆ Facilitate discussion at the intersection of the consumer age segmentation and nutritional contexts and ask participants to use Post-it® notes to identify nutrition strategies for doing the right thing, barriers, and enablers
	◆ Probe to understand opportunity gaps in nutrition strategy for children Solicit input from the perspectives of Health Care Professional and Parent. Note: Children's perspectives will be explored in kids groups later
	◆ Probe to understand what's happening, physiologically, socially, and within the family dynamic in terms of nutrition habits and behaviors
0:55	◆ Move to focus on obesity
	◆ Ask participants what they think are the core issues.
	◆ Put each core issue in the center of a sheet of chart paper and ask participants to contribute to developing mind maps/spider diagrams branching out from each core issue
1:30	Break
1:40	◆ Look for synergies or where the two discussions overlap—ask the group to identify 8–12 points of synergy
	◆ Look for gaps or where the two discussions diverge—ask the group to identify 8–12 points of difference
2:55	◆ Thanks and Adjourn the Focus Group Session

colored markers, different colored Post-it® notes, paper and pens for participants, blue masking tape to hang paper and enough wall space to do that, different colored dots for voting, and paper formatted for writing concepts. Table 4-7 links some creative-problem-solving tasks with tools and techniques that facilitators can employ.

Developing the Creative-Problem-Solving Process in the Case Study Example

In the case study, the moderator/facilitator designed a four-hour process to create a rich set of insights and learnings and to turn those into breakthrough

TABLE 4-7.
Matching process tasks with creative problem solving tools and techniques.

Creative Problem Solving Session Tasks	Tools & Techniques
Recording observations	◆ Working alone exercise using Post-it® notes then sharing them with the team ◆ Facilitator writing on chart paper
Connecting and clustering observations	◆ Mind Map/Spider diagram (Wycoff 1991, 1995) (Figure 4-4)
Generating insights derived from observations	◆ Brain writing (Van Gundy, 1988) ◆ Headlining nugget of an insight with "I wish"
Listing top likes from consumers' observations	Heartstorm™ exercise: brainstorming for positive emotions (Miller, 1997)
Listing top dislikes from consumers' observations	Thunderstorm™ exercise: brainstorming for negative emotions (Miller, 1997)
Sorting and categorizing ideas	Clustering/Bucketing exercise
Selecting and prioritizing	Dot voting exercise
Developing ideas into beginning concepts	Concept sheet format (see Appendix 4-3 for an example)
Advice to the project team from prosumers	Writing a letter to the team which could include strategic insights, comments on powerful differentiators, and what is still to be developed and more.

ideas and concepts. Selected creative-problem-solving tools and techniques included divergent and convergent techniques of wish brainstorming and dot voting, as well as processes for concept writing and synthesizing learnings. Table 4-8 outlines the plan. Appendix 4-3 provides a template for developing beginning product concepts.

Step 8: Conducting the Focus Group

Prior to starting the Slingshot focus group, the moderator will coach project team members and prosumers on the need to acknowledge and manage their biases and assumptions. To help manage biases and assumptions and to promote active listening, the participants in the back room will be asked to value what they don't know, suspend judgment, and adopt an attitude of unconditional positive regard for the people in the front room and what they say. The project team members in the back room will also be instructed on how to send notes into the moderator during the focus group session for new lines of inquiry or for additional probing questions.

The moderator will instruct participants in the back room on how to record the discussion in the front room, asking them to focus on observations and insights (See Figure 4-5). Observations are the raw material for generating insights. Observations include verbatim reporting of verbal comments and tone

TABLE 4-8.
Slingshot creative-problem-solving process plan in case study example.

Time	Task
00:00	◆ *Remind prosumers that they are now assuming the role of project team member*
	◆ *Review task and desired outcomes*
00:02	◆ *Ask prosumers and project team members for key insights and learnings, number and write them on sheets of chart paper for all to see*
	◆ *Based on insights and learnings, facilitate identifying components of a nutrition strategy for ages 1–14 and scribe on chart paper*
	◆ *Use wish brainstorming to generate ideas for products and services for each of the components of the nutrition strategy*
01:00	◆ *Using dot voting (number of dots for each participant should approximate 10% of all numbered ideas on the wall), ask team to place dots beside the ideas that interest and intrigue them the most*
01:10	◆ *Ask each person to develop a selected idea as a concept using the concept sheet format (see Appendix 4-3)*
	◆ *Report out concepts to the group and write any builds as new concepts*
	◆ *Place concepts on chart paper and place on the wall for all to see*
01:40	◆ *Review nutrition strategy components, discuss any gaps*
	◆ *Vote to prioritize nutrition strategy components*
	◆ *Use action brainstorming to drill down into selected areas for more ideas*
	◆ *Use dot voting (same 10% rule for number of dots for voting) to identify the ideas that interest and intrigue them the most*
	◆ *Write up selected ideas as concepts*
02:15	*Break*
02:30	◆ *Report out concepts to the group and write any builds as new concepts*
	◆ *Add concepts to chart paper and place on the wall for all to see*
03:00	◆ *Use dot voting (same 10% rule for number of dots for voting) to cast an advisory vote on all concepts displayed on the wall*
03:10	◆ *Ask prosumers to write and read out a letter of advice to the project team based on the Slingshot*
	◆ *Ask each project team member to write a statement of priorities for the project team and read to group*
03:20	*Review the results of the Slingshot and next steps for the concepts*
03:30	*Thank and Adjourn*

of voice, and description of how participants physically handled any stimuli and related body language. Insights are the ideas inspired by the observations within the business context of the topic.

The moderator also briefs prosumers on how to handle their consumer role. Before inviting the consumers to enter the discussion room, the moderator

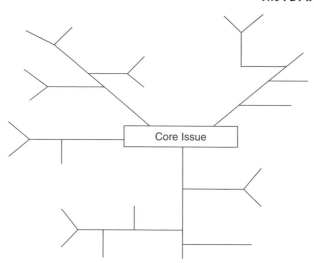

FIGURE 4-4. Chart prepared to illustrate mind map/spider diagram skeleton in case study.

will ask the prosumers to join the consumers in the waiting room so that they can be ushered into the meeting space together.

In the focus group session, a skilled moderator (1) ensures that all the consumers contribute to the discussion, (2) stimulates productive interaction among all participants, (3) manages time so that all areas of the discussion guide are covered, and (4) integrates additional back-room questions as seamlessly as possible.

Conducting the Focus Group in the Case Study Example

The stage was set for an in-depth discussion from a diverse set of highly informed perspectives. The moderator had hung on walls in the focus group room flip-chart sheets of paper prepared for the topic analysis (see Figure 4-3) and for the mind map/spider diagram (Figure 4-5) exercise. The table had Post-it® notes, pens, and markers ready for the participants to use in the stimulus exercises. The focus group consisted of four prosumers and four consumers. The discussion lasted for three hours—somewhat longer than is typical for a consumer focus group, which is usually 90 minutes to two hours. The longer amount of time was necessary because of the complexity of the topic and the desire to take advantage of the expertise and experience of the eight participants. The moderator had a flip-chart stand and pad in the room to use as necessary for stimulating and documenting discussion. Step 6 and Table 4-6 contains details about the discussion guide.

FIGURE 4-5. Front room Slingshort focus group discussion as viewed from back room via closed circuit television.

Step 9: Conducting the Creative-Problem-Solving Session

In the creative-problem-solving session a skilled facilitator helps project team members and prosumers achieve breakthrough ideas. Effective facilitation involves:

◆ Managing the dynamics of the group to ensure that everyone participates.

◆ Coaching prosumers for the transition from consumer to team member role.

◆ Implementing the process plan with all its divergent and convergent group processes in the time available.

◆ Making sure that there is a complete record of the session discussion and its outcomes.

The creative-problem-solving session begins after consumers have left the facility. Project team members bring their notes and join prosumers around the table in the front room (see Figure 4-6). The facilitator reviews the process and restates the overall task and desired outcomes. Then the facilitator introduces the first of the debrief exercises to the group and continues through the process plan to the final exercise. During the session, the facilitator checks in with the team leader to get any feedback that might require adjustments to the process. At the end of the session, the facilitator reviews the outcomes. The team leader will thank prosumers and comment on next steps. The team leader and facilitator will collect all documentation generated during the session for digitizing and delivery in hard and soft copy for next steps.

FIGURE 4-6. Slingshot creative problem solving session.

Conducting the Creative-Problem-Solving Session in the Case Study Example

In the case study, after the focus group session, the facilitator coached pro-sumers for their role in the creative-problem-solving session. They were told that they are not responsible for implementation and to use that to manage any reflex to self-censor. They were also asked to follow the facilitator's lead to stretch beyond the threshold of acceptability. (See Appendix 4-1 for more advice to prosumers on how to wear two hats).

The session started with a review of the task and desired outcomes. Prosumers and consumers were then asked to offer key insights and learnings. The facilitator went around the room as many times as possible in 30 minutes. All insights and learnings were numbered and scribed onto flip-chart pages, and these were hung on the walls for all to see. Participants noted that it was useful to be able to see the insights and learnings visually, as it helped stimulate content for the next exercise. Based on the insights and learnings, the team was asked to write components of a nutrition strategy on large Post-it® notes and attach them to chart paper on the walls for all to see. The facilitator then asked the team to use a brainstorming format starting with "I wish," "We could," "What if," and "How to," followed by the nugget of the idea (7 to 10 words) to generate ideas for products and services for each of the nutrition strategy components.

The facilitator introduced voting with dots to identify those ideas that had the most interest for participants. Different colored dots were used for prosumers, marketing, and scientific participants because the project leader wanted to know what each group thought was most interesting. The facilitator then introduced the concept sheet format. (See Appendix 4-3.) Each participant was asked to select two ideas to write up as concepts. Each concept was presented and discussed for builds and placed on chart paper for all to see.

The facilitator then asked the team to take a step back to see what gaps, if any, there were in the components. There were gaps in education strategies that the team leader decided were not useful to pursue at this point. With that discussion completed, the facilitator asked the team to use three dots to vote and prioritize the components of most importance to the company. Only those components that got at least three votes were to be used for the next exercise. The team leader was asked to select ideas in each of the prioritized components to develop using action brainstorming (Miller 1997). These ideas were voted on, and each participant selected one to develop as a concept. After the break, participants presented their concepts to the whole group. This step also included facilitated discussion for builds and connections. These concepts were also displayed on the wall. Each participant was then asked to use five dots to identify their top five concept choices.

As a final exercise, each prosumer was asked to write a letter of advice to the project team and each team member was asked to identify his/her top priority for action going forward. After the Slingshot session, all the notes were digitized and top vote-getting breakthrough concepts were entered into an electronic database for use in the next stage of the front end of innovation process. (See, for example, ToolBook 1, Chapter 2.)

Step 10: Document and Disseminate Results

Documenting results is an important component for any process. A Slingshot process can be documented in various ways, such as:

- Digitizing the notes made by participants in the back room
- Digitizing all ideas and concepts (with votes) produced during the creative-problem-solving session
- Archiving the video and audio recordings made during the focus group
- Creating a searchable video/audio database of the focus group discussion
- Asking the moderator to write a report of the focus group discussion
- Putting all concepts into a database

Documenting the output of a Slingshot is guided by the project's need to create a database for further use and the need to disseminate the process outcomes.

The purpose for which a Slingshot was chosen will determine the distribution list for disseminating results. Such dissemination is the responsibility of the project leader. All the various stakeholders in the Slingshot project should receive the results. Team members and others who have actively participated in the Slingshot process are stakeholders in the tactical results and need to receive a report of the outcomes for use in next steps. Other types of stakeholders are those in the company who need to know the results to work out any strategic implications. Reports—whether for tactical next steps or for strategic deliberations—benefit from being written in ways that demonstrate understanding of the requirements of the stakeholders.

Documenting and Disseminating Results in the Case Study Example

In the case study, concepts generated from each Slingshot session were digitized and stored in a secure electronic database. All project team members had access to the database. These persons were asked to review and rank all the concepts for next steps in the Hunting for Hunting Grounds™ process (PDMA ToolBook 1, Chapter 2). The information gathered in the Slingshot sessions helped shape development of two new business platforms, each with a portfolio of concepts for the deliberations of strategic decision-makers.

SLINGSHOT CHECKLIST

The checklist in Table 4-9 is designed for the project team leader who is ultimately responsible for the implementation of a Slingshot.

Slingshot Pitfalls to Avoid

It is always important when choosing to use a Slingshot to communicate a clear understanding of why the Slingshot process has been selected and the expected outcomes. So, obvious pitfalls to avoid are failure to adequately understand and communicate process, and failure to clarify deliverables from the process.

An all-too-human pitfall is the proclivity of project teams to overestimate the value of a high-profile prosumer candidate and to underestimate the value of a less well-known, lower-profile and equally qualified prosumer candidate. A similar pitfall is when project teams want prosumer candidates who are an exact match for their project. Bring fresh and challenging perspectives to the Slingshot process is important, so consider prosumers who have core competencies tangentially related to the topic area. For example, in a Slingshot for a manufacturer who wanted to explore diaper-changing contexts, it was hard to convince the team to recruit the male designer of Black & Decker's successful Snake Light, although he had two children under the age of three. This team just did not see men as diaper changers. As it turned out, not only had the designer had a lot of first-hand experience as a diaper changer, but his creation of the snake light had been prompted by the insight that what was needed was a third hand to deliver focused work light when both hands were already occupied. It always seems like a third hand is needed in diaper changing as well.

TABLE 4-9.
Slingshot Check List

Slingshot Process Task	Who is Responsible	By When
• Define project, budget and recruit team		
• Identify reasons to use a Slingshot		
• Make decision to use a Slingshot		

• Clarify topic, objectives, and format(s) for documenting output
• Determine details including
  • Number of focus group sessions
  • Total number of prosumers
  • Total number of consumers
  • Number of locations
  • Which locations		

• Decide if moderator/facilitator role to be done in-house or outsourced		
• Select moderator and facilitator		
• Determine date and location		
• Determine total number of prosumers • Screen & recruit prosumers • Collect non-disclosure agreements		
• Develop screener with recruiting specifications • Select market research company to do the recruit • Screen and recruit consumers • Ensure adequate facility spaces for all Slingshot needs • Manage all aspects of using the market research company and it's facility		
• Develop focus group discussion guide • Approve discussion guide • Disseminate guide to team members		
• Design creative-problem-solving process • Approve process • Communicate it to project team members and prosumers		
• Manage all aspects of logistics and catering for focus group and creative-problem-solving session • Send all necessary details to team members and prosumers		
• Conduct focus group (s) • Document discussion		
• Conduct creative-problem-solving session		
• Document and disseminate results		
• Implement next steps		

Slingshot Keys to Success

There are three keys for accomplishing a successful Slingshot:

1. Detailed preparation and planning
2. Selecting a skilled moderator and facilitator
3. Implementing the process with rigor, discipline, and flexibility

All successful projects require detailed preparation and planning. The most important details to get right include:

◆ Ensuring that a Slingshot is the right process tool for the task
◆ Recruiting qualified prosumers and consumers
◆ Designing a discussion guide to extract real value from participants' experience
◆ Designing a creative-problem-solving process to optimize the purposeful development of creative tension

Aligning important tasks with the skills and personal characteristics of individuals is a key to success for any undertaking. Some of the important attributes for a moderator/facilitator include:

◆ Process design skills
◆ Ability to acquire language appropriate to topic area
◆ Skill to probe and challenge without alienating participants

The right mix of rigor, discipline, and flexibility is a key to the success of any undertaking. The art of a successful Slingshot process comes from knowing when to insist on process rigor and discipline and when to be flexible, and that knowledge comes with practice.

APPENDIX 4-1 HOW TO WEAR TWO HATS AS A PROSUMER IN A SLINGSHOT PROCESS

It is important to the effectiveness of prosumers that they understand what to expect and how to manage themselves in both components of a Slingshot.

Wearing the Consumer Hat

In the consumer focus group component of the Slingshot process, prosumers need to be able to engage effectively as a consumer in the focus group.

USEFUL TIPS:

- ◆ Follow the moderator's lead. A skilled moderator will work to elicit and develop the consumer participants' experience and fold the prosumer seamlessly into the discussion.
- ◆ Make a mental effort to manage your own filters and relax into the consumer mindset.
- ◆ Time share and encourage dialogue with other consumers.

Wearing the Problem-Solving Hat

In the creative-problem-solving session, prosumers need to manage the reflex to self-censor by offering ideas that they know how to make work.

USEFUL TIPS:

- ◆ Prepare to experience creative tension and to use it as a springboard.
- ◆ Follow the facilitator's lead to stretch beyond threshold of acceptability.
- ◆ Remember that you are not responsible for implementation.

APPENDIX 4-2 CASE STUDY FOCUS GROUP SCREENER

Participant's Name:	*Home Phone:*
Address:	*Work Phone:*
Recruited By:	*Date:*
Confirmed By:	*Date:*

Hello, my name is _____ from (INSERT NAME OF FACILITY). We are conducting a survey today regarding nutrition and obesity and would like to ask you a few questions. Please be assured that this call does not involve sales of any kind.

1. *First, previous surveys have shown that people who work in particular fields may have different attitudes toward certain products than others do. For this reason, we need to know the occupation of each income earner in your household. (RECORD RESPONSES BELOW. DO NOT READ LIST.)*

Self: _____
Spouse/Partner: _____

RECRUITER: DO NOT RECRUIT ANYONE FROM THE FOLLOWING:

- ◆ *An advertising agency*
- ◆ *A marketing or marketing research firm or department*
- ◆ *A public relations or promotions firm*

2. *Of the following industries, which one most closely relates to your field of work?*

Agriculture, Fisheries, Forestry	TERMINATE
Banking, Financial, Insurance	TERMINATE
Medical	CONTINUE
Manufacturing	TERMINATE
Construction	TERMINATE
Utilities	TERMINATE
Wholesale trade	TERMINATE
Retail trade	TERMINATE
Accommodations and food service	TERMINATE
Arts, entertainment, recreation	TERMINATE
Public administration	TERMINATE
Technology	TERMINATE
Service	
Please list: _____	TERMINATE, UNLESS IN THE MEDICAL FIELD
Other	
Please list: _____	TERMINATE, UNLESS IN THE MEDICAL FIELD

3. *Given your work in a medical field, what kind of position do you hold?*

Administrative staff	TERMINATE
Surgical staff	CONTINUE
Nonsurgical staff	CONTINUE
Laboratory	TERMINATE
Retail	TERMINATE
Other please specify _____	TERMINATE IF NOT TARGET (SEE GRID BELOW)

RECRUITER SPECS:

April 14 from 5:30 to 7:30	*At least one from each specialty: Pediatrician, pediatric nurse, registered dietician (Recruit 5 total)*

4. *Which position do you hold?*

Anesthesiologist (of any kind)	TERMINATE

Nurse (of any kind)	CONTINUE
Orderly (of any kind)	TERMINATE
Doctor (of any kind)	CONTINUE
Technician (of any kind)	TERMINATE
Dietician/Nutritionist	CONTINUE

5. *What type of Nurse/Surgeon/Dietician are you?* _____
     MUST SAY PEDIATRICIAN, PEDIATRIC NURSE OR DIETI-
     CIAN WORKING WITH CHILDREN

6. *What age ranges do you work with?*

0–1	TERMINATE
1–10	CONTINUE
11–14	CONTINUE
15 +	TERMINATE

7. *Do you work with children in the 1 to 14 age range on a daily basis?*

Yes	CONTINUE
No	TERMINATE

8. *Do you work with children or counsel children on any of the following areas?*

Health and Nutrition	CONTINUE
Weight Management/Obesity	CONTINUE
Diseases	TERMINATE IF THIS IS ONLY AREA
Family Issues	TERMINATE
*Other*_____*(describe)*	TERMINATE IF NOT NUTRITION- OR WEIGHT-MANAGEMENT RELATED

Now for something a little different: What is your favorite movie? (RECORD BELOW)

_____

What might you tell someone to convince him or her to see it? (Probe)

_____

     MAKE SURE RESPONDENT IS ARTICULATE. IF HE/SHE IS
UNABLE TO DESCRIBE MOVIE IN A WAY THAT MAKES IT SOUND
INTERESTING, OR CANNOT SAY MORE THAN A FEW WORDS
ABOUT IT, THEN TERMINATE.
     We would like to invite you to participate in a new product development discussion around child nutrition and weight management/obesity. This will be a facilitated brainstorming process with other professionals around new nutritious food products. We would like to offer you $_____ for your time.
     Would you be interested in participating in this effort?

Yes......	CONTINUE
No.......	TERMINATE

◆ *Advise participants about the nondisclosure agreement they will be asked to sign.*

◆ *Inform them that they will be required for three hours and that the session will be recorded audiovisually.*

◆ *Tell them a follow-up call will be made on X date*

Recruiting Specifications

April 14 from 5:30 to 7:30	At least one from each specialty: Pediatrician, pediatric nurse, registered dietician (Recruit 5 total for 4 to show)		
	Pediatrician	*Pediatric Nurse*	*Dietician*
April 14 from 5:30 to 7:30			

Sample Nondisclosure Agreement

Dear Participant:

We will be paying you $_____ to discuss and/or create ideas for our client. In consideration of our client you agree to respect the secrecy of any strategies, concepts, or ideas discussed or generated as a result of the session and not to disclose such ideas or concepts to any third party. In addition, you agree that any ideas that are generated as a result of your participation in these meetings will become the property of our client, and, if necessary, upon request you agree to execute such documents as may be necessary for our client to obtain, maintain, or exercise its ownership rights.

Participant's Signature	
Participant's Name (print or type)	

APPENDIX 4-3: CONCEPT FORM TEMPLATE

Concept Name: _____
Who worked on this: _____

DESCRIBE THIS CONCEPT IN ONE SENTENCE:
WHAT IS THE IDEA?

HOW WOULD THE IDEA WORK?

Pluses:
+
+
+

Concerns:
–
–
–

REFERENCES

Alam, Intekhab (Ian). 2005. "Chapter 16: Interacting with Customers in the New Product Development Process." In *The PDMA Handbook of New Product Development*, 2nd Edition. Ed. Kenneth B. Kahn, Associate Editors, George Castellion and Abbie Griffin. Hoboken: NJ: John Wiley & Sons.

Amabile, T. M. 1996. "How to Kill Creativity." *Harvard Business Review*, 76 (5).

Chen, H. K., D. Bommarito, and C. M. Sifonis. 2003. "Business Process Innovation Through Analogical Reasoning." Paper presented at the annual Innovation Convergence Meeting, Minneapolis, MN, September.

Cooper, R. G. 1999. "The Invisible Success Factors in Product Innovation." *Journal of Product Innovation Management*, 16 (2): 115–133.

de Chernatony, Leslie. 2001. *From Brand Vision to Brand Evaluation*. Oxford: Butterworth Heinemann, in Association with the Chartered Institute of Marketing.

Sifonis, C.M., Chen, F.H.K., & Bommarito, D.E. 2003. "Analogical Reasoning and Its Application to the Business Process." *GM R&D Center Contract Report CR-03/20/VDR*. Troy, MI

Kahn, Kenneth B. 2001. *Product Planning Essentials*. London: Sage Hill Publications.

Miller, Christopher W. 2002. "Chapter 2: Hunting for Hunting Grounds." In *The PDMA Toolbook for New Product Development*, edited by Paul Belliveau, Abbie Griffin, and Stephen Somermeyer. New York: John Wiley and Sons.

Miller, Christopher W. 2005. "Chapter 17: Getting Lightning to Strike: Ideation and Concept Generation." In *The PDMA Handbook of New Product Development*, 2nd Edition. Edited by Kenneth B. Kahn, Associate Editors, George Castellion and Abbie Griffin. Hoboken: NJ: John Wiley & Sons.

Miller, Christopher W. 1997. *The Focused Innovation Technique: A Creative Problem Solving Process*. Ed. Linda S. Crill. Lancaster, PA: Innovation Focus.

Perry, Barbara, Cara L. Woodland, and Christopher W. Miller. 2004. "Chapter 8: Creating the Customer Connection: Anthropological/Ethnographic Needs Discovery."

In *The PDMA Toolbook 2 for New Product Development*. Ed. Paul Belliveau, Abbie Griffin, and Stephen Somermeyer. New York: John Wiley and Sons.

Rees, Fran. 1997. *Teamwork from Start to Finish*. San Francisco: Pfeiffer—An Imprint of Jossey-Bass Inc.

Sifonis, C. M., A. Chernoff, and K. Kolpasky. 2006. "The Innovation Pipeline: Analogy as a Tool for Communicating about Innovation." *International Journal of Innovation and Technology Management*.

VanGundy, Arthur B. 1988. *Techniques of Structured Problem Solving*. New York: Van Nostrand Reinhold Company.

Wycoff, Joyce. 1995. *Transformation Thinking*. New York: Berkley Publishing Group.

Wycoff, Joyce. 1991. *Mindmapping: Your Personal Guide to Exploring Creativity and Problem-Solving*. New York: Berkley Publishing Group.

Integrating User Observations with Business Objectives to Drive Product Design

Larry Marine
Principle User Experience Architect Intuitive Design

Chad A. McAllister Ph.D
Solutions Consultant, LexisNexis

New product development (NPD) fails at an alarming rate, with most companies finding a single market winner for every 10 attempts (Cooper, Edgett, and Kleinschmidt 2004). How can new product ventures improve their likelihood for success? One way is to incorporate user observations in a cohesive process as a means of discovering the specific innovations that will make a product a market winner.

Other books, including previous *PDMA ToolBooks*, have described how to conduct Voice of the Customer (*ToolBook 2*, Chapter 7) and ethnographic research (*ToolBook 2*, Chapter 8). This chapter extends those concepts to describe a process for integrating actionable business objectives and user observations to identify more successful product opportunities. Product managers in all industries will find that this process improves their product and service designs. This process has been applied to commercial, consumer, and enterprise products and services, and has been used in the development of games, medical devices, manufacturing processes, e-commerce Web sites, software applications, and many others. In all, over 250 products have been successfully created with this process, with each focused on achieving the stated objectives and meeting users' needs.

OVERVIEW OF THE CHAPTER

This chapter was written for product managers looking for ways to improve their success rate with NPD—more market winners with fewer attempts. It is also valuable to product managers charged with improving an existing product and is applicable in both B2B and B2C environments. In addition, business and marketing managers who need to expose their organizations to improved processes for designing highly successful products will find this chapter a useful tool to add to their toolbox.

The chapter briefly describes several key reasons for why you should integrate user observation research with your business objectives, then covers the six steps of the process for creating a product design based on an initial business objective and user observation research. The steps include: (1) define the objectives, (2) plan for conducting observations, (3) conduct the first round of observations, (4) create the priority matrix, (5) conduct the second round of observations, and (6) design to the tasks. Each step is illustrated using an actual case study that applied the process to solve a *knowledge management* problem for a large government agency. The chapter concludes with several common pitfalls to avoid when using this process.

WHY START WITH A BUSINESS OBJECTIVE?

Novice and Olympic marksmen both have one thing in common: they ready themselves, aim at the target, and then fire their weapon. This simple *ready, aim, fire* process is used over and over to help them hit their target. The same process is discussed in business: the importance of defining a target, taking aim, and executing to hit the target. Although this is an obvious process, product development often appears to follow a different process: ready, *fire*, aim. Instead of first aiming at a specific business objective, such as capturing a new market or expanding sales in a current market, and doing those things that are necessary to hit the target, NPD efforts too frequently begin with the inception of a novel idea or technology and then attempt to build a business around the resulting product.

Failing to aim first at the business objectives and doing what is necessary to achieve the objectives contributes to the gap between companies with high NPD success rates and companies that have lower success rates. One example of this gap is that the most successful companies only need 4 ideas to be developed to generate a market winner instead of the nearly 10 ideas that less successful companies need (Adams and Boike 2005). Companies that are not aiming at specific targets must have more ideas in development in order to generate a winner, with the consequence of wasted effort and higher NPD costs.

WHY OBSERVATIONS MUST BE PART OF AN NPD PROCESS

Observing prospective users of a new product in their environment provides an understanding of their needs and produces innovative inputs for a successful product design. Observation studies, or ethnographic research, are necessary because they produce the insights into the users' real needs. Interviews, focus groups, and surveys tend to expose different information than can be found through observations. These tools derive reactive comments to an existing product or service design, while observations tend to identify proactive insights that lead to emerging product opportunities. The following are just a few good reasons why user observations are better for informing new product design efforts:

Users Tend to Describe Problems within the Limits of Their Own Perceptions

Users tend to define their problems within the boundaries of their understanding of the technology. When users are asked about their needs, they often identify incremental ideas based on their experience. Observations help identify bigger, revolutionary changes that alter the competitive landscape. Because most users tend to describe their problems in terms of existing products and experiences, typically they share suggestions for improving an existing design, not innovations that dramatically improve a product.

Users May Say One Thing and Do Another

It is common for users to ask for one thing but really need another (Gallivan and Keil 2003). They may discuss areas where they had the most problems, which may not accurately represent their real needs. Observing the users typically identifies that users have habituated to a particular work-around and fail to describe that as a problem. Thus, designing to their stated needs may merely automate their current frustrations, while user observations will point out that the frustrating task could be removed altogether. For instance, many software users describe the need for a report generator, when in fact they take those reports and then transpose the numbers to a charting and graphing tool. What they really need is to have the product generate graphs and charts, skipping the reports.

Users Have Difficulty Expressing What They Want

A common characteristic of human behavior is the difficulty of articulating what we want before we see it or are exposed to scenarios in the context of the problem. Requirements are difficult to determine because users are unsure what is possible, have trouble describing the problem, or do not sufficiently understand the problem (Kazmierczak, Dart, Sterling, and Winikoff 2000). Users often describe their problems in terms of a solution, leaving the interviewer to reverse engineer the solution to define the problem. Observing users removes these obstacles inherent in human behavior and places the responsibility of requirement creation with the design team.

EXAMPLE: THE PERFECT TRACTOR

Product designers conducted research in Germany to improve a farm tractor (*ToolBook2*, Chapter 8). A focus group of farmers was asked about their tractors. One farmer responded that his tractor was perfect and he emphatically requested that the next model remain unchanged. During an interview in the same farmer's home, he reiterated his position that the tractor was perfect as-is. The designers asked to see the tractor. He then proudly showed them his *perfect* tractor, which he had personally customized with over 20 modifications. Only after observing the farmer's tractor did the design team have a better appreciation for what the farmer considered to be the perfect tractor. Relying only on what the farmer said would have produced very misleading research results.

STEP 1: DEFINE THE OBJECTIVES

Business Objectives

A good business objective provides the reason and guidance for creating a product, enhancing a product, or extending a product family. Many business objectives are too general to be of help to NPD teams, who end up struggling unnecessarily to understand what objective they are supposed to achieve. Also, far too many business objectives are defined *after* creating the product, clearly reflective of a ready, *fire*, aim approach. A simple way to tell if a well-defined business objective exists for a product is to ask each team member to write the project's single most important business objective on a piece of paper and then discuss what they wrote. Don't be surprised if everyone has a different concept of the key objective. Differences can then be discussed and a common business objective agreed to (Davis 2005).

Keep in mind that business objectives typically represent one of two basic goals:

1. Increase revenue
2. Decrease costs

Business objectives must also be observable, measurable, and have a finite timeline. Consider the following robust and clear business objectives:

◆ Increase sales to existing customers by 25 percent over the next three years.
◆ Penetrate an adjacent vertical market, gaining 20 percent market share in the first year.

◆ Convert 10 percent of the competitors' customers to be our customers within one year.

◆ Decrease support costs of an existing product by 25 percent with the next version.

Contrast those objectives to these common, yet weaker and less helpful, objectives:

◆ Leverage existing technology.

◆ Open a new market.

◆ Create a world-class product.

Notice that the stronger objectives have clearly defined, observable, and measurable targets, and they provide more definition of the product's market and intended users than the weaker objectives.

TIP: FOCUS ON A SINGLE BUSINESS OBJECTIVE   An executive management team discussed specific objectives, and together agreed on a course of action. After the meeting, the executives all returned to their functional areas—sales, marketing, business development, product development—where execution necessary to meet the objective became skewed based on the needs, perspective, and politics of each executive. Even when objectives are clearly understood and actionable, each party involved may choose another course of action. Minimize this problem by creating a laser-like focus on a single, clear business objective, like a skilled marksman who aims at one target.

KNOWLEDGE MANAGEMENT CASE STUDY: BUSINESS OBJECTIVE

A government agency in Washington, D.C., asked for assistance in designing a knowledge management (KM) system to help them reduce the amount of effort it took to answer questions from congressional staffers. The goal was to avoid reinventing the wheel and allow agency employees to leverage work they have done in the past for other staffers. The first step was to clearly define the problem. KM means many things to different people, but it generally involves processes to collect, store, access, retrieve, apply, and create knowledge. After sufficiently exploring the dimensions of the problem and coming to a common understanding, the business objective was discussed as reducing the amount of time and effort required to identify the answers to the questions from congressional staffers. A constraint was that the solution must not require an inordinate amount of development effort since the organization did not have internal resources for writing software, nor funds to develop a large custom-developed system.

The stated objective then became: Decrease the amount of effort to answer questions by 50 percent. Additionally, the project must cost less than $100,000 and be implemented within 18 months.

Marketing Objective

With a clear business objective for a product, the next step is to identify the key marketing objective. For the purposes of the process discussed here, the goal of the marketing objective is to identify the source of customers for the proposed product. Marketing objectives tend to fall into three basic customer categories:

1. Deepen existing relationships with customers.
2. Source customers from competitors.
3. Open or expand a market.

In many cases, the business objective contains sufficient detail to determine the correct customer category. For example, if the business objective is "Convert 10 percent of the competitors' customers to be our customers within one year," then the customer category is "source customers from competitors." If the business objective does not provide the guidance to make this determination about the market, then the business objective should be reexamined or a marketing objective created that clearly paints a picture for the future of a product and its impact on the business. For instance, if the business objective is simply to increase sales by 25 percent in the next year, then the marketing team might recognize that the greatest opportunity for that increase is in expanding an immature market base.

KNOWLEDGE MANAGEMENT CASE STUDY: MARKETING OBJECTIVE

Since this was an enterprise project intended to support users within the government agency's organization that were already using various tools, it is fairly obvious that the marketing objective was to deepen existing relationships. The stated objective then was to achieve 80 percent adoption by the intended users within one year of implementation. Although that may be a formidable goal for a consumer or commercial market, it is a reasonable goal for the captive audience of this enterprise domain.

STEP 2: PLAN FOR CONDUCTING OBSERVATIONS

The focus of user research in this process is on visiting users in their environment and observing their work. Observation studies discover what the proposed product must provide for the users in order to satisfy them—the value proposition. The emphasis on observations stems from years of product development experience that found the self-reporting bias inherent in asking users what they want results in a distorted view of their needs, missed opportunities, and ultimately, products that provide far less value than was possible. Instead of relying on users to say what they want, observe what they need. If observations are not part of the NPD process, it is difficult to identify the user needs that significantly increase the value of the product.

> **EXAMPLE: INSPIRATION FOR THE PALM PILOT**
>
> By observing users, we see different problems, problems that create new product opportunities, problems that users are unlikely to identify in a survey or interview. Before launching the Palm Pilot, the product designer observed people using the current generation of electronic personal organizers. He noted that these products were optimized for getting information into the organizer, but users more frequently needed to retrieve information from the organizer. This observation resulted in the idea to create an organizer optimized for information retrieval. To do this, they observed users' detailed use of the different personal organizer functions and found that the calendar, address book, to-do list and memo functions were the most frequently used. The Palm team thus created buttons for each of these functions that instantly access the information contained within the function. Information input would rely more on synchronization with a personal computer. Prior to the introduction of the Palm Pilot, the personal organizer market was saturated and owned primarily by Sharp and Casio. Within a few years of its introduction, the Palm Pilot had transformed the personal organizer into the PDA. The Palm Pilot PDA stimulated a market that was thought to be saturated and flat, even outselling TV sets for several years.

Two Rounds of Observations

For the best results, plan to conduct two sequential rounds of user observations. The first round provides a high-level perspective about what is important to the users. Once the initial discoveries have been identified and analyzed, they are prioritized to focus the second round of observations. Experience shows that the initial round of research identifies opportunities that are different from the current product concept, warranting a pause for the business and marketing stakeholders to reconsider their options. Once the business, marketing, design, and technical teams have reviewed and assessed the opportunities, the next round of observations is constructed to learn specifically about the highest priority user tasks that will shape the product design.

Who to Observe

Before conducting observations, determine which users are appropriate for your marketing objectives. Users generally fall into one of three categories: existing users, users of competitors' products, and new users. Utilizing Table 5-1, determine the appropriate user classification for the product development project, based on the business and marketing objectives. Although this might seem obvious, many teams frequently observe the wrong users. For example, observing existing users to discover ways to improve a product to appeal to new users is a common error.

IDENTIFY USER ROLES Begin by identifying the specific user roles affected by the product, such as end user, decision maker, support person, evaluator, administrator, etc. More than likely, the initial user roles defined prior to

TABLE 5-1.
User observations matrix

		Business Objectives	
		Increase Revenues	Reduce Costs
Marketing Objectives	Deepen Relationships	Existing Users	Existing Users
	Source Customers	Competitors' Users	Not Applicable
	Open New Markets	New Users	Not Applicable

conducting the observations will change rather significantly after the first round of observations, but it provides a starting point.

AVERAGE, NOT EXPERT USERS   Expert users do things very differently than the average users, so avoid observing expert users, unless they are the target user group. The goal is to identify and observe the typical or representative user. Remember, even expert users were novices at one time.

HOW MANY USERS   A common question is how many users must be observed. Since this research is not intended to derive statistically significant results, it is typically successful to observe between three to six users of each user role or type in each of the two rounds of research (Nielsen and Landauer 1993). This is expected to uncover between 80 percent and 90 percent of the potential user needs. Of course, this is predicated on the notion that representative users are observed. Only observe enough users to provide a clear understanding of the users' needs, goals, and tasks.

RECRUITING USERS   Finding representative users can prove to be difficult, more so in B2B products than in consumer-oriented B2C products. Based on the classification of users to observe—existing users, new users, or competitors' users—different techniques will be effective.

1. *Existing and competitors' users:* For existing users and competitors' users, your company's marketing and sales teams are an excellent source for recruiting users. An advantage of relying on the sales team is that they are likely to have developed relationships with representative users, and this makes it much easier to establish the observation appointment. Some participants are initially concerned about being observed, for both privacy and company proprietary reasons, and established relationships can help assuage these concerns.

2. *New B2C users:* When new users must be observed, market research firms can help find users, assist with scheduling, and handle other logistical issues. They are skilled at screening prospects to ensure a representative user sample, setting reasonable schedules, addressing last-minute scheduling issues, and so on. They typically have a contact database of representative participants on file and can successfully recruit outside of their database, and given that this is their focus, they

are not sidetracked by having other responsibilities to attend to, as most sales teams are.

3. *New B2B users:* Although finding new users for B2B products can be a bit more challenging, market research firms, industry professional groups, professional conferences, and trade shows are useful sources for identifying candidate users. Further, your sales team may be a good resource for suggesting representative users.

B2B and B2C users can also be recruited through various organizations, such as AARP, schools, business groups, and manufacturing organizations. There are also other means of locating users, including media ads, flyers, and Internet lists, such as *CraigsList.com*. These alternate methods are often more successful for finding participants that represent the general B2C population and less so at finding very specific participants in niche B2C product domains or B2B products.

KNOWLEDGE MANAGEMENT CASE STUDY: WHO WAS OBSERVED

The researchers who were creating reports to help answer questions from congressional staffers were first observed to understand what they needed a KM system to do, which was the original goal for the product. Through the observations, other types of users were discovered, such as facilitators, who gave the researchers their tasks and provided the necessary resources, and the reviewers who critiqued the content to ensure compatibility with standards and security issues. Facilitators were often mid-level managers within the agency or at research firms contracted by the agency. The reviewers were either managers or other researchers, usually working within the agency. No more than six users of each user type, with varying degrees of task experience, were observed. For each user type, the observers were confident that the critical information was collected because little to no new information was discovered during the fifth or sixth user observation of each type.

How to Observe

Conducting observation studies requires listening, discerning, watching, and examining skills. A few guidelines follow:

- ◆ *Make the user comfortable with the process.* Before conducting observations, share with the participating users how the findings will be protected and anonymity will be assured. For example, aggregating all observations and limiting access to collected data is a means of protecting the participant's identity. Also, avoid overwhelming the user by limiting the number of observers to just one or two—three at most. These observers should remain out of direct sight, usually slightly behind and to one side, so as to make accurate observations without impeding the user in any way.

◆ *Be unobtrusive.* While the observers may need to prompt the user or ask clarifying questions, observers must resist the impulse to talk or otherwise interrupt the user as they perform their tasks. Position each observer with clear views of the work area, without getting in the way of the user. As a matter of practice, plan on noting questions as they arise and wait until after the observation is over to ask specific questions.

◆ *Catch the users performing their normal tasks.* The best results occur when the observations are not forced. However, some tasks are so infrequent that users may have to be prompted to perform the desired tasks. Asking users to recall the last time they had to perform the task and to follow the same scenario is one good way to prompt them. Another way to prompt the user is to start asking a few questions regarding what events, actions, or triggers often cause them to perform their tasks, and then ask them to follow a scenario as if one of those triggers occurred.

◆ *Expect interruptions.* Interruptions are a natural part of many tasks, coming in the form of ringing telephones, arriving e-mail messages, knocks at a door, crying babies, and more, depending on the environment. For many products, it is important to understand how tasks are impacted by interruptions and provide a means for users to easily continue their task after returning from the interruption. Try to understand how the user resumes the task after returning from an interruption.

◆ *Avoid asking or expecting users to talk while they work.* In an effort to feel more at ease with the observation process, many users feel compelled to think aloud, explaining what they are doing. While this may seem useful, it can cause users to be unusually conscious of their task. If users are thinking too much about the task instead of going about it as they normally would, the observations may be significantly affected. Before starting observations, remind users to perform their work as they normally do without unduly narrating their actions.

◆ *Ask clarifying questions later.* The process begins with a simple introduction to ease the users' fears and anxieties about the process, but should move quickly to observing the users and their tasks. When the observations have been collected, plan to review the findings with the user for about 15 to 45 minutes to ask any questions identified during the observations.

◆ *Work backward to improve task efficiency.* By working backward from the users' desired end result, it is easier to understand what tasks the users perform, as well as identify inefficiencies and potential opportunities for improvements. By knowing the user's end result, trace the task backward to identify the original trigger for starting the task. That trigger may be related to another task that needs to be followed backward to identify its task, and so on. Eventually, a series of tasks are identified that end with the desired result and begin from an original trigger. Most often, improvements involve some process reengineering. The objective is to create a set of tasks that link together to achieve the desired outcome. Ideally, each task should result in an outcome that appropriately triggers the next task. In conducting this analysis, try to identify opportunities for eliminating one or more tasks.

◆ *Conduct at least three observations.* No fewer than three observation sessions should be conducted with each user type to ensure that a useful composite perspective of the users' tasks is created. Anomalies can be missed or misidentified if only one session is conducted. If inconsistencies are observed or significant questions remain after conducting three sessions, conduct additional sessions. Practical experience indicates that more than six sessions per user type are rarely needed.

◆ *Record the observations.* A video record of the observation is useful for sharing the information with team members who did not participate in the observation. It is also a helpful aid in reviewing any unclear observation points. It is imperative to explicitly request and receive written permission to record any observation. This is especially important when observing children, medical patients, or other special classes of participants. Also, people in their work environment are often reluctant to grant permission to make any recording in the workplace. When audio or video recordings are not possible, the observers must be prepared to rely on their notes and any artifacts that they can collect, such as blank forms or charts. Many users are willing to share "sanitized" artifacts, with any proprietary information blotted out, so expect to collect these as they help tell the story along with the observer's notes.

◆ *If used, make recordings valuable.* A key problem to address in video recording an observation is to make sure that the camera captures the entire work environment. Users often utilize many various artifacts that are not usually part of the product, but are part of their task. For instance, most desks where people use a computer have several piles of information. Each pile has a different meaning to the user, and each likely impacts the user's tasks differently. A video camera should be able to capture the user's interactions with all of the artifacts used during a task. This may mean that instead of setting the camera on a tripod, one of the observers may have to control the camera. If the user is mobile while performing the task, then video recording is further complicated and requires the videographer to always be mindful of capturing the relevant information.

KNOWLEDGE MANAGEMENT CASE STUDY: HOW USERS WERE OBSERVED

Users were asked to continue their normal work to find answers to questions from congressional staffers. All of the users had at least one question that they were recently given that still required significant information gathering efforts. Two observers watched each user as they engaged various tools to find, collect, and organize information.

Each observation session lasted about an hour and started with a few key questions to set the stage: `` How did you receive the research request,'' and ``What will your work look like when it is done?'' The users were observed as they searched for information to respond to the research request. At the end of each observation session, the observers asked users clarifying questions for up to 45 minutes. As an example, some users were asked, ``We noticed that you don't save your e-mail attachments in your various project folders; can you help us understand this?'' The common answer was that it took too much time and impeded the users ``flow'' when working on a project. The analysis of this answer, combined with other information from the observations, suggested that a traditional KM tool that required an additional effort by the users to classify and store information would impose an undue burden on users who were focused on retrieving information. A design that could automatically classify information without significant user involvement would offer greater value without interrupting the users' workflow.

The customer would not allow the observations to be audio or video recorded, nor could pictures of artifacts be taken. Given these constraints, the observers took careful notes of their observations and answers to clarifying questions and also asked for and received *sanitized* copies of key artifacts, such as a research request form.

Table 5-2 provides an example of the type of notes collected when observing one of the researchers.

TABLE 5-2.
Sample User Observations Notes

Observation Notes
Observed: John S. – Research analyst
He spends more time than he would like writing each report. Given the short timeframe with many ad hoc requests, the depth of the data suffers since more time must be allocated to writing the content.
The current process and system is not very adaptable to last-minute changes.
Because of the time it tasks to research, analyze, draft, and edit each cyclic report, reports are often up to two years out of date by the time they are completed.
Because so many models include inflated numbers (gaming) in anticipation of budget cuts, the resulting calculations are always somewhat suspect. This is especially true when an analysis is combined with others and extrapolated over a longer timeframe.
Each program has their own "black box" modeling – it's hard to see what tweaks they've introduced, because it's buried in the appendix (which no one reads – no time).
He might engage in some peer-review processes to ensure valid calculations.
All analysis for 11 programs needs to be immediately available. Want to see a clear path from detailed to strategic levels of documentation.

TABLE 5-2.
(continued)

The current process is so flexible as to create problems and inhibit reuse of artifacts.
There is little learning and sharing across departments.
Credibility might be related to confidence in the reports.
Most reports are too long to read or review on-line.
Once the review is done, he needs to email the report to appropriate recipients or
an announcement of its completion and location on the shared drive.
Saving the report or interim artifacts means knowing where to put it.
Uses email as a primary document storage mechanism.
Reviews email for high priority items and deadlines.
Gets mobile email on Blackberry, limited integration with other desktop apps (e.g.
calendar, Daylite).

Where to Observe

Where to observe users is not always as obvious as one would think. If the research is being conducted for a new product, users need to be more broadly observed than if the research is to aid the design of the next version of an existing product. In the case of new products, observe users performing similar tasks with an analogous product. For instance, observe people using airline ticket machines to develop a new sporting event ticket kiosk. In both cases, it is imperative to observe users as close to their task environment as possible.

Even inviting users into a lab that is set up "just like" their office, home, car, and so on is insufficient because it is not their actual environment. By removing the user from their environment, external factors that play an important part in the users' task are also removed. For example, naturally occurring interruptions and distractions will not be observed in a lab. Observing users perform a task may mean following them from one location to another. If users begin a task at a desk, continue the task in a vehicle, and complete the task in a warehouse, important information will be gained by observing them at each task location.

KNOWLEDGE MANAGEMENT CASE STUDY: WHERE USERS WERE OBSERVED

All of the users performed their research tasks from their own desks. Consequently, the observations were scheduled during normal working hours and conducted in the users' offices. The observers positioned themselves to view a user's work area but did their best to remain out of the way, even if the users moved from their desk to a file drawer or bookcase to find something related to the research task. Often, observers positioned themselves directly behind the users so that they could view the users' computer screens and workspace.

What to Observe

Observing is more than just watching people—it is also watching and analyzing tasks. Tasks are the actions, steps, and process flows that users perform in order to achieve a desired result. For instance, the task of determining what salespeople need to do over the next month to ensure that they make their bonus may include reviewing which key customers they must visit over the next four weeks. This task may have subtasks that include reviewing how many times they have seen each key customer and determining which ones need to be visited again, then planning those visits.

The goal is to understand how users think about their objectives and the tasks they perform to achieve those objectives. The information users need to succeed at their tasks and the difference between what information they are likely to have and what information they are not likely to have, yet need, should be identified. Good observations include tasks that are "outside" of the current or expected product usage model or environment. When designing the next version of an existing product, it is important to observe what drives the user to use the product and what happens when they are done. In other words, the entire task environment must be observed.

LOOK FOR TASKS, NOT FEATURES   Observations should be about the users' tasks and goals. Do not leap from observations to product features too quickly or the real needs of the user will not be sufficiently investigated and understood.

TIP: LEAVE THE FEATURE BANDWAGON   Much of product development, and the mindset of many product managers, is more concerned with features and less with value to the customer. Recall that while most organizations find one successful product in every ten attempts, the best organizations find success in one in every four attempts. The best way to supercharge the probability of product success is making this critical shift in thinking—from features to tasks. Concentrate on what users need to accomplish and not the features they may use in accomplishing it. Features will flow from a deep understanding of tasks and the problems users need to solve.

NOTE HOW AND WHEN USERS INTERRUPT THEIR OWN TASKS   One of the key things to look for when observing the order of the tasks and subtasks is where it seems that users take a mental pause. This gives an indication of where a subtask ends and another begins. Each subtask can be thought of as a cognitive chunk. For instance, if anyone were to ask for a phone number, they do not want to be interrupted in the middle of writing it down. That is because a typical phone number requires pretty much the maximum amount of working memory that we have available for a task—the well-known seven items plus or minus two items guideline. If the task is interrupted, they will likely lose some part of the phone number and need to hear it again. This working memory limitation is crucial in understanding how users perform tasks. This is especially important when attempting to reengineer a task with the introduction of a new product. The new product may avoid the necessity to remember information, but you must take care not to impose an additional cognitive task requirement that exceeds a person's working memory limit. Given this notion that users pause their tasks at predictable points and cannot be interrupted at others, it is necessary to note what pauses occur and to identify what cues suggest is a reasonable place to pause the task.

IDENTIFY USERS' OBJECTIVES OR OUTCOMES   Identify the desired outcome the user is trying to achieve. Knowing the endpoint helps to better understand the steps users take to get there. Also, note what form the outcome takes. For example, if a user generates a report, that does not mean the report is the desired outcome. Inevitably, a report is used to complete some higher order task, such as to make a decision or invoke an action. The observations should strive to determine the end state of the task, not an end state of a feature, product, or service.

LOOK FOR COMPLETE TASKS, BEGINNING TO END   Too often products include capabilities that satisfy part of a task but still require the user to perform additional tasks outside of the feature or product in order to achieve their desired goal. The inability to solve the complete task erodes a product's value, resulting in missed sales and opening the door for a competitor to recognize how to improve on the product. For instance, a network system-monitoring product may require users to run reports on various system functions to find anomalies in the network's performance. Instead of requiring users to run the reports to identify problems, the product could do most of the work and alert users when a problem is encountered. The monitoring product could even automatically perform some of the repair actions for the user.

IDENTIFY WHAT TRIGGERS AN ACTION OR TASK   Pay particular attention to what events occur to cause the user to begin a task, or resume after an interruption. These triggers may be an event, such as a warning; a person, such as a manager requesting a status; or something altogether different. It also helps to understand the importance or priority of these triggers. For instance, if the user is interrupted in one task by a trigger for another task, then the interrupting trigger has a higher priority. Attempt to identify what makes the new task more important than the current task. Also, try to determine how well the user understands the desired outcome of each trigger.

RECORD THE ORDER IN WHICH USERS PERFORM TASKS   One reason observations are so powerful is that they identify the tasks and subtasks that users fail to describe when asked. Do not be surprised to find that users are not always efficient in what they do. Often the process they follow may be inefficient, or even backward. These are indications that some part of the task encourages an inefficient task flow. This can be an opportunity to suggest a best practices approach to a task.

NOTE WHEN USERS THINK THEY ARE DONE   In addition to the cognitive pauses that occur in tasks, identify when users think they are done with a task or key subtask. It is important to note what cues exist to suggest to the user that they have completed the task. The product design should take advantage of these cues and should not inadvertently eliminate them.

COLLECT ARTIFACTS SUCH AS FORMS, PRINTOUTS, NOTES   Users may create and maintain artifacts to provide cognitive cues that aid in the performance of their tasks. Some common artifacts are forms, checklists, sticky notes, a dog-eared user guide, controls marked by permanent marker, and taped over controls. Examples of all these artifacts need to be collected to aid in understanding the users' perception of their tasks. These artifacts indicate where users have learned that they cannot reliably rely on their own memory to maintain an adequate level of success in the task. These artifacts help the user unburden their working memory. Photographs can be useful, but keep in mind that not all environments allow photographs to be taken. In such cases, capture clear notes of the artifacts.

KNOWLEDGE MANAGEMENT CASE STUDY: WHAT WAS OBSERVED

The tools, artifacts, and processes users employed in completing their tasks were observed. Users collected and organized various forms of information to compile a report that answered the research questions they were asked. They most commonly searched for information on the Web, in their e-mail folders, local computer folders, remote computer folders, physical file cabinets and bookcases.

Who Should Observe

Who conducts the observations is just as important as who to observe. Ideally, the observations should be conducted by a team of two to three people, which is led by a behavioral observational specialist, such as a usability professional or ethnographer, who is familiar with ethnographic and behavioral observation techniques. A professional observer will have the skills and training necessary to attribute specific user thought processes to the observed tasks and behaviors. This provides a deeper understanding of the users' tasks and needs.

USE CROSS-FUNCTIONAL TEAMS   Observations benefit from using observers with different backgrounds, which provides multiple perspectives of the users' tasks and creates better opportunities for innovation. By using a skilled professional observer, other team members will quickly grow in their ethnographic abilities through experience. Additional team members can include people with more of a business background, an understanding of technology, or experience with the problem domain. This may include product managers, business analysts, marketers, technical writers, and technical developers. However, technical representatives typically are less skilled at understanding user observations and may not necessarily be good choices for observers, though that varies by person and product domain. Nonetheless, it is sometimes helpful to have a technical representative participate in a few observations to help address issues that arise in the development cycle regarding user tasks and to establish credibility of the research with the development team.

LIMIT THE USE OF LESS SKILLED OBSERVERS   Less-skilled observers tend to observe users' actions without understanding why the users do them. Such observations provide little more than time and motion data and rarely provide the insight needed to help the project succeed. Who should observe is based on a combination of skill and personality, but each team should have a minimum of one skilled and experienced observer. Of course, it may be necessary for the skilled professional observer to provide some initial training and guidance to other less experienced team members, but much of the skill is gained through actual observational experience.

SHARE AND SYNTHESIZE NOTES   When several different teams participate in the observation effort, plan to include sessions where the teams meet to share their insights and raise additional questions. This ensures that each of the observation teams are equally prepared to observe and notice potential insights.

KNOWLEDGE MANAGEMENT CASE STUDY: WHO OBSERVED

The observation team consisted of three people. Two experienced behavioral observational ethnographers with a background in cognitive psychology and user interface design were often accompanied by a manager from the agency. The manager provided a different perspective and insights that were unfamiliar to the ethnographers. Prior to conducting the user observations, the observation team reviewed the user roles that had already been identified by the government agency and discussed the types of tasks and artifacts they may observe. This helped them to begin thinking more deeply about the problem domain. They also reviewed the ground rules for conducting observations, such as not asking questions or interrupting the users while they are performing their tasks.

STEP 3: CONDUCT FIRST ROUND OF OBSERVATIONS

In the first round of observations, look for new discoveries, not for validations of proposed product designs. It helps to approach observations with the expectation of discovering something entirely novel. Observations should produce *ah-ha* experiences, not *that-is-what-we-thought* experiences. Some ah-ha discoveries may be more novel than others, but they are all useful, and almost always elevate your product above the competition.

Although not intuitive, the key to the success of the first round of observations is to avoid focusing too closely on what is seen. Look for general tasks and activities, especially those that users are doing that could be done better by the product under design. If the users are using another product, take special note of the tasks the users perform that are not necessarily part of their normal task, but are more to serve the needs of the product. This is an opportunity to improve the users' tasks and, again, to avoid automating current frustrations.

KNOWLEDGE MANAGEMENT CASE STUDY: CONDUCTING OBSERVATIONS

Existing users of current systems within the organization were targeted for observation. Several key user roles were identified and observed. Each role was very different, but could easily be done by the same person at different times. The three key user roles were:

1. *Topic researcher:* Responsible for conducting the research, sifting through the data, and producing the draft report.
2. *Report reviewer:* Responsible for verifying the accuracy of the information, ensuring consistency with established guidelines, and finalizing the report.
3. *Team facilitator:* Responsible for giving the topic researchers their tasks and providing the necessary resources for the project.

A particularly interesting result was noting how many different information repositories were used and where they were located, including local computer folders, remote computer folders, e-mail folders, department intranets, and hard copy paper reports. Users also were likely to use incomplete artifacts, such as spreadsheets and tables, from other unfinished research projects. Moreover, given the existence of information barriers, such as network firewalls between some departments and other security measures, access to the various repositories was just as varied, yet everyone needed access to all of the information.

Additionally, the variety of the internal storage structures and paradigms each person employed to organize their information was noted. Some users used a broad and shallow approach—many folders with little content in each—while others used a narrow and deep approach—few folders with large amounts of content in each. This suggested that a highly constrained organizational structure, as found in some KM systems, would impose a barrier to user adoption of the product.

Clearly, a traditional KM tool was not the right solution for this problem. Users needed a better search mechanism instead of a KM tool. This fact alone justified the investment in the user research. Without the research, this project would have resulted in an expensive tool that did not meet users' needs.

Depicting the Observations

The results of observations are captured with a generalized task flow diagram that indicates the major tasks and subtasks performed by users. This flow diagram does not need much detail, just enough to convey the observed task flows, key artifacts, triggers, and outcomes. It may also help the process by creating user profile caricatures that help the design team remember the user roles they observed. For example, the caricature of Browsing Betty could be created for a user who wanders serendipitously through several retail

Web sites before finding something that seems interesting, as opposed to Specific Sam, who knows exactly what he needs, where it is, and how much it costs.

KNOWLEDGE MANAGEMENT CASE STUDY: GENERALIZED TASK FLOW

Figure 5-1 is the generalized task flow for the KM project, which illustrates the high-level information learned from the initial observations of the typical researcher. A caricature of Hurried Henry was used to describe the nearly frenzied nature of researchers to find the information they need as quickly as possible. The task flow shows that researchers perform three basic sets of tasks that each involve several subtasks. One factor to note, also, is that these users worked on several projects simultaneously. They were almost always under a time constraint.

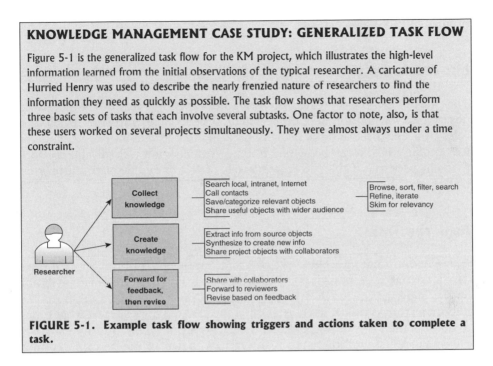

FIGURE 5-1. Example task flow showing triggers and actions taken to complete a task.

STEP 4: CREATE THE PRIORITY MATRIX

The results from the first round of observations provide information to help refine the business directions of the product. The results are insights into the needs and priorities of the users in light of the business objectives and technical constraints. In this process, a priority matrix is created that quantifies the products' potential desirability, profitability, and feasibility. The matrix also helps engineering teams understand what to build, and the business and marketing groups to understand the engineering constraints, such as time and resources.

The prioritized task matrix indicates the tasks to focus on during the second round observations. Do not be surprised if a new business objective or opportunity is discovered as a result of the observations and prioritizations. New ideas should be expected to evolve from these initial observations. A common hesitancy with this approach is a reluctance to act on the new ideas because preconceptions about the product design have already taken shape. Oftentimes, the information learned from the initial observations tends to contradict or conflict with otherwise widely held industry notions of the domain, which makes people question the observations. When conflicts are found between what is common (or reflected in a competitor's product) and what observations reveal, an opportunity exists to provide great value—a distinct and defensible competitive advantage. Instead of choosing to follow convention and making "me-too" products, act on the conflict to push a product ahead of competitors.

List the Tasks

The priority matrix is created by listing the tasks, not features, which were observed in the first round of observations. It helps to organize this list by user role, grouping a set of tasks (and subtasks in complex environments) under a user profile. Also, it can be helpful to reiterate a task if it is applicable to more than one user role. Table 5-3 is an example of what a simple matrix listing the tasks looks like.

Rank the Tasks

A three-point ranking scale works best to determine the priority for each task. Any task that is on the list is, by default, important, although many will not be the highest priority. A task that is ranked as a 1 is not necessarily less important than a task ranked as a 3, but the ranking does suggest what order to address the tasks during product design. Remember, the priority matrix is only a guideline, and there are many factors that may not be captured in a ranking system that may affect the final decisions. Adding more granularity to the ranking scale does not really help as much as it makes it more difficult to assign a value to each task.

A task-ranking meeting should be scheduled to include those that performed the observations, the business and marketing project stakeholders, and the design and development stakeholders. By meeting together, the tasks can be more easily explained, misunderstandings can more quickly be addressed, and everyone is provided an opportunity to be part of the process.

USERS' RANKINGS: DESIRABILITY   In the User Experience column of the matrix, the actual observers should indicate the priority of the tasks with respect to how desirable the task is to the users' overall objectives. The rankings should be provided by the actual observers prior to conducting the prioritization meeting with the rest of the team. The rankings used are as follows:

> 3: Primary task and most desirable
> 2: Secondary task and modestly desirable
> 1: Tertiary task and less desirable

TABLE 5-3.
Priority Matrix with Tasks

Function	User Experience	Business Potential	Technical Feasibility	Overall
Find previous authors/users				0
Become aware of files, reports, artifacts				0
Determine relevancy of information				0
Collect artifacts in project "folder"				0
Contact relevant experts/authors				0
Search shared drives				0
Search local hard-drives				0
Search multiple intranets (cross firewalls)				0
Search email folders				0
Search the internet				0
Search pay sites and groups				0
Search hard copy libraries				0
Share useful artifacts with others				0
Keep some content private				0
Share some content locally (group/dept)				0
Compile selected data				0
Create working docs				0
Distribution 1's	0	0	0	
2's	0	0	0	
3's	0	0	0	

User experience: Task immediacy in meeting user's objectives 3: primary task and most desirable; 2: secondary task and modestly desirable; 1: tertiary task and less desirable
Business potential: Impact of user's task on business potential/profitability 3: high impact or most profitable; 2: medium impact or moderately profitable; 1: low impact or less profitable
Technical feasibility: Difficulty of delivering task completion 3: easy or low risk; 2: moderate or medium risk; 1: hard or high risk

BUSINESS AND MARKETING: PROFITABILITY   In the Business column, the business and marketing stakeholders determine the impact of the users' task on the business objectives, such as impact on potential profitability. Oftentimes, this process helps teams reassess their objectives, even helping to refine objectives or ways to achieve an objective. The rankings used are as follows:

> 3: High impact, or most profitable
>
> 2: Medium impact, or moderately profitable
>
> 1: Low impact, or less profitable

DESIGNERS AND DEVELOPERS: FEASIBILITY   In the Technical Feasibility column, design and development stakeholders rank the difficulty of solving for the tasks, accounting for risk and effort. It helps for the design and development teams to discuss the tasks prior to the prioritization meeting to give them time to consider the impact and ask clarifying questions as necessary. The rankings used are as follows:

> 3: Easy or low risk
>
> 2: Moderate or medium risk
>
> 1: Hard or high risk

Remember that this matrix is only a tool, but it provides extremely useful guidance. While it lacks detail, it collects and organizes the necessary information for making decisions in a very efficient manner.

Equal Distribution of the Rankings

This priority matrix is most useful when an equal number of tasks are ranked as 1s, 2s, and 3s in each of the three columns. Remember that the rankings signify the immediacy of a task, not the importance of it. All of the tasks are important; otherwise, they would not be on the list. Prioritizing the tasks allows the design and implementation of the product to be phased over time, as necessary, given the resources available. The tasks ranked the highest will be addressed first.

One method that has worked fairly well is to rank all the tasks, determine the number of 1s, 2s, and 3s, and then review the task rankings again, adjusting the scores to get an equal distribution of the scores. It is also easier to work on the matrix one column at a time. Although this requires at least two passes through each column, the ensuing discussion can be very helpful. Another method is to "give" each group assigning rankings a similar number of 1s, 2s, and 3s to use. For example, if 12 tasks need to be ranked, those doing the rankings know they should rank four tasks as 1s, four tasks as 2s, and four tasks as 3s even before the ranking process begins. Regardless of which method you use, this is not an easy process the first time you try it, but remaining faithful to the process provides the best results.

Score the Tasks

Once the ranking has been completed, simply add the scores for each task. The combined scores of the three factors—user desirability, business profitability, and technical feasibility—will range from 3 to 9 with the higher scores representing the most likely tasks to design first. The best possible result of the prioritization would be a user ranking of 3 (most desirable), 3 for business (most profitable), and a 3 for feasibility (easiest or lowest risk to produce), resulting in a final score of a 9. Typically, the matrix results in a fairly wide distribution between 4 and 8.

The next step is to highlight the higher-priority items in the matrix that are doable within the given time and resource constraints. This helps focus the subsequent research and design activities on the highest-priority items. This also provides a roadmap for future evolutions of the product. Knowing what will likely be required in a subsequent version makes it a little easier for the design team to leave room for the next set of features in the design. Table 5-4 illustrates what a completed priority matrix with the highest priority tasks highlighted looks like.

KNOWLEDGE MANAGEMENT CASE STUDY: CREATING THE PRIORITY MATRIX

The tasks for the priority matrix come from the generalized task flow, shown in Figure 5-1. Although this may not be specific enough to create a design from, when combined with the observation notes, the task flow provides enough information to complete the priority matrix.

Table 5-4 is the completed priority matrix with just the tasks listed for one user type, the researcher, in this case. This example simply lists all observed tasks for this one user type, but it is also common to list the tasks grouped by each user type identified in the observations (even if some of the tasks are repeated for each type) and to score each user separately. Thus, the matrix could be organized to group the tasks for each observed user type, researcher, facilitator, and reviewer. Each project is different, but making one long list is usually easier the first time.

This matrix was created by first identifying the priorities from the user's perspective. This was iterated until an equal distribution of 1s, 2s, and 3s was produced. Once the user rankings were completed, rankings for the business column were created in a similar manner. The process was then repeated for the technical feasibility column.

The last step to complete the matrix was to add the rankings for each task to produce a composite ranking. Note the wide distribution of scores between 5 and 8 in Table 5-4. After analyzing the composite rankings, the development team was confident it could complete all tasks with a composite rank of 6 or higher given the schedule, resources, and budget for the project. All of the tasks that ranked a 6 or above were highlighted. Where to `` draw the line '' differs for each project and is based on the available time, money, and resources, typically defined by the product manager.

TABLE 5-4.
Completed Matrix with Highest Priority Tasks Highlighted

Function	User Experience	Business Potential	Technical Feasibility	Overall
Find previous authors/users	2	3	2	7
Become aware of files, reports, artifacts	3	3	1	7
Determine relevancy of information	3	3	1	7
Collect artifacts in project "folder"	1	2	3	6
Contact relevant experts/authors	2	1	2	5
Search shared drives	3	2	2	7
Search local hard-drives	3	2	3	8
Search multiple intranets (cross firewalls)	3	2	1	6
Search email folders	3	2	1	6
Search the internet	2	1	3	6
Search pay sites and groups	2	1	2	5
Search hard copy libraries	1	1	3	5
Share useful artifacts with others	2	3	2	7
Keep some content private	2	1	3	6
Share some content locally (group/dept)	1	2	2	5
Compile selected data	1	3	3	7
Create working docs	1	3	1	5
Distribution 1's	5	5	5	
2's	6	6	6	
3's	6	6	6	

STEP 5: CONDUCT SECOND ROUND OF OBSERVATIONS

The second round of observations uses the same observation methods as in the first round but focuses specifically on the previously determined high priority tasks. Further, greater detail about the higher priority tasks is collected during second-round observations. For example, while the first round identified the key user roles and identified the tasks, the second round focuses on how to optimize the high priority tasks and then design for the new optimized tasks. This means developing a greater understanding of the details of the tasks to ensure that the optimization does not alter the tasks in a way that they become unfamiliar to the users. The following questions should be answered in the second round of observations:

◆ *Triggers.* What causes the user to start the task? These may be alerts, requests, schedules, etc. Of interest, too, are the priorities that users seem to assign these triggers and the reason for these priorities.

◆ *Artifacts.* What objects do users employ in the performance of the task? Anything used in the course of the task is an artifact, such as pencil, eraser, calendar, checklist, sticky note, spreadsheet, report, book, and so on. Frequently, encountered artifacts are important to record in observation notes.

◆ *Outcomes.* What results are users looking to achieve? Each triggered task requires some type of outcome for the user to consider the task completed. Sometimes the trigger defines the desired outcome, other times not. Some outcomes are represented in the form of artifacts, such as a report or a spreadsheet. Other times, it may take a bit of insightful interpolation to define the desired outcome, such as the trends or exceptions found across several different spreadsheets. Such an outcome could be better represented in a single chart or graph.

◆ *Metaphors.* What terms or symbols are used to describe or identify a task? The more common a task, the more likely a set of users have coined a single term to describe the task. These terms help identify how the users' perceive the boundaries (beginning and end points) of a task. For example, when anyone says they are getting dressed for work, we understand this to include several activities, such as selecting clothes, taking off clothing, putting on multiple items of clothing, performing a quality check in the mirror, and so on.

◆ *Handoffs.* Who gets the results and what are they going to do with them? Note what artifacts get transferred between users and between tasks. These are often related to triggers, especially where the artifact represents part of the trigger. This is where it is important to understand the different roles a user or set of users play in the task environment. Even in the case of a single user, they may complete one subtask, generating an artifact that will be used at another time by another task.

Create a Task Flow Diagram

At this point, the observations should yield enough detail for the team to create detailed flows diagrams of the users' tasks. These task-flow diagrams reflect the users' perception of their tasks as opposed to how the technology works. A good task flow should also include all of the artifacts, triggers, issues, and decision points that are integral to the tasks.

Begin by creating an overview task flow composed of major steps or subtasks, then break out each major step into its own task flow. One useful technique is to use different colors to represent different facets of the task flow. For instance: green to indicate steps that the user typically performs, yellow for actions performed by the product, orange for questions and issues, purple for artifacts, pink for triggers and outcomes, and so on.

The task flows capture the general or common tasks and not necessarily the exception tasks. These exceptions can be captured and noted, but try to avoid focusing on the exception and pay more attention to the generalities. The goal is to create flow diagrams for the most common way that users complete the tasks and to almost ignore the edge cases. Focus on the 80 percent, not the 20 percent.

Optimize the Tasks

Once the task flow is complete, review the flow diagrams looking for opportunities to replace user actions with product actions. Using the color codes suggested previously, this step converts *greens to yellows* in the task flows. Given that green indicates user actions and yellow product actions, this means finding ways to make the product do the work that the user now does. Many new product opportunities can be discovered during this process.

KNOWLEDGE MANAGEMENT CASE STUDY: OPTIMIZED TASK FLOW

The second round of observations focused on the 12 tasks that had a composite ranking of 6 or higher in the priority matrix. The analysis of these detailed observations resulted in the task flow diagram shown in Figure 5-2. Reviewing the task flows, the design team was able to identify a number of steps where the user relied on several different tools to repeat the same task, searching different information repositories. Optimizing the tasks aims at reducing the total amount of work the user has to perform in order to achieve their desired outcome. In this case, optimizing the task flow reduced the number of steps users performed and the number of tools they employed. Previous experience by the design team suggested that an integrated search mechanism could search all necessary repositories of information in a single user step instead of the many separate searches currently required.

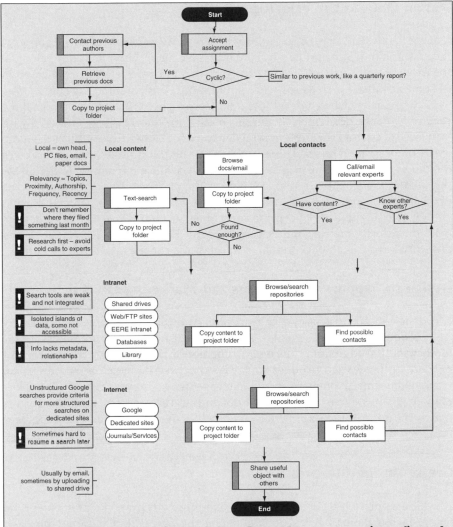

FIGURE 5-2. Example optimized task flow. Rectangles are user actions, diamonds are logic or system actions, ovals are artifacts or objects, and the exclamation marked rectangles are issues to address.

STEP 6: DESIGN TO THE TASKS

Since the task flows represent the user priorities and the priority matrix is based on the tasks balanced by the business objectives, designing for the tasks inherently results in a product that supports the users' needs as well as the business objectives. This also helps discourage scope creep, which occurs with alarming regularity in most product design efforts.

Design Only to the Prioritized Tasks

During the design phase it is important to review how well a feature supports the task at hand. Sometimes a feature may have merit, but it does not belong in the product if it is not required to accomplish a prioritized task. In the creation of some products, such as software applications, it is common for *gold plating* to occur, which is when a developer adds features because doing so is somehow rewarding to the developer. When this occurs, products become cluttered with little needed or used features, making it harder for users to accomplish the more common or general tasks. Though difficult at first, it is imperative to design to the tasks identified by the priority matrix. Also, while it is tempting to design for the what-if or edge cases, remain focused on the prioritized tasks. Otherwise, the product reflects more of a feature-oriented design than a task-oriented design.

Of course, there are exceptions and times when it is reasonable to deviate from the priority matrix, but remain skeptical. Finding new ideas in the design phase that are valuable to users and appropriate for the task *is* the exception.

Design to Support the Business and Marketing Objectives

The design should also support the prioritized business objective. For instance, if the product is being designed to attract new customers, it is not necessary for the new design to maintain design or interaction consistency with the current product. If the current product was really that well designed in the first place, we would not need a new product. The opposite is also true, as well. If the key marketing objective is to strengthen the brand, then reusing specific elements that relate to a brand experience could help drive the design. The important thing is to remain true to the key objectives.

Create Design Blueprints

A typical approach to creating a new building is to draw a set of blueprints before laying the first brick or hammering the first nail. The same approach is successful in product design. Blueprints alleviate much of the misunderstandings that occur in typical product development. Adding blueprints to the requirements specification usually results in a product that more closely resembles what the users need.

Frank Lloyd Wright once said it is far easier to use an eraser on the drafting table than a sledgehammer on the job site. Drafting a set of blueprints allows everyone on the team to express design considerations before going down a particular path, only to find out too late that something is terribly wrong with the design. Good design blueprints help overcome the issue of passing the design task off to more than one product development group, such as industrial designers, engineers, and graphic designers. A coherent set of blueprints ensures that everyone is working from the same conceptual model of the product. Moreover, since not everyone on the various design teams was involved in the observations, the blueprints are a simple method to communicate the results of that research that does not rely on interpretation.

KNOWLEDGE MANAGEMENT CASE STUDY: BLUEPRINTS

The design blueprint shown in Figure 5-3 was created from the detailed task flow previously presented in Figure 5-2, taking into account user interface principles and cognitive psychology.

The design process was straightforward in that the design team walked through each of the optimized task flows and designed the user interfaces to support the actions that the users needed to perform. Each of the rectangles in Figure 5-2 became separate user interface screens. The design was provided to the development team in the form of a design specification that included all of the previously mentioned deliverables and artifacts, Figure 5-1, Figure 5-2, Figure 5-3, and Table 5-4. This project was eventually divided between several development teams, each responsible for a different aspect of the product. The use of several development teams was not anticipated, but the blueprints supported this type of division of labor quite well.

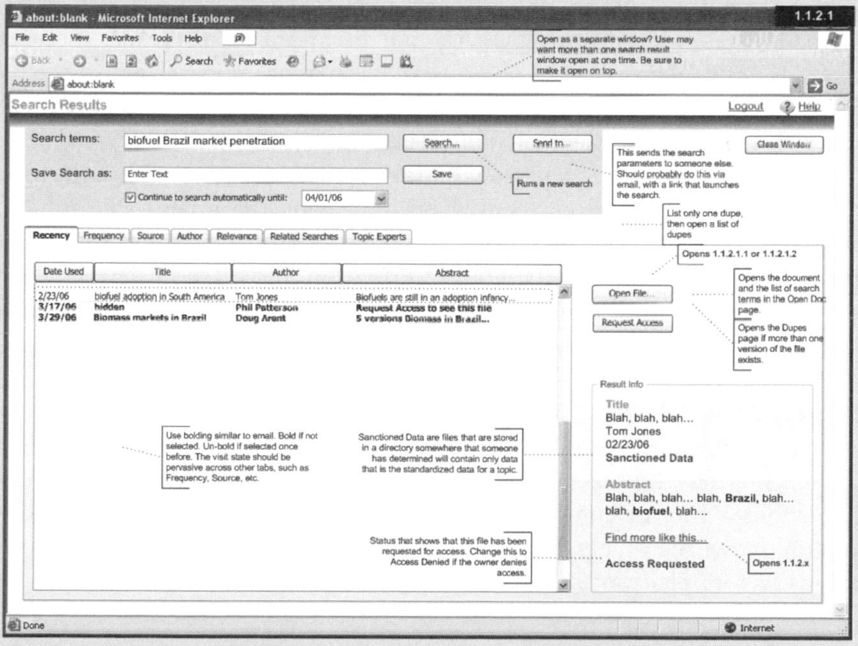

FIGURE 5-3. Sample design blueprint.

COMMON PITFALLS

As with any effort, something can always get in the way and limit the success of a project. The following is not an exhaustive list, but presents mistakes that are easily recognized and avoided or corrected.

BUILD IT AND THEY WILL COME   As attractive as it may be, it is less successful to build a hypothesis (or prototype) and try to prove it with the users than it is to learn what users need by observing them first. As soon as users are shown a prototype, they typically constrain their perceptions of the potentials to the "box" created by the prototype, which tends to give the false impression of what users want or need from a product. Prototypes are useful for verification of capturing tasks in a product, not for innovation.

FEATURES INSTEAD OF TASKS   Consider the users' task from beginning to end and not just a few steps in the middle. Many product features actually increase the effort it takes users to complete their tasks. Remain focused on understanding the tasks and objectives of the user and the product features will naturally follow.

AUTOMATING THE USERS' FRUSTRATIONS   A main objective of good observation research is to look beyond what the users are currently doing and to uncover ways to eliminate or reduce user tasks. Too often products merely automate users' current processes. Do not be afraid to reengineer the users' tasks. An optimized task is an obvious and apparent improvement to the user and readily exposes the value of a product. Look for points of pain with the users and find innovative ways to eliminate that pain as opposed to automating the pain. For instance, early personal information managers, such as the Sharp Wizard and Apple Newton, focused on improving the input of information with mini keyboards or handwriting recognition. However, user observations showed that users infrequently input information, but often retrieve it. Therefore, optimizing the system for information retrieval, which was the focus of the Palm Pilot design, provided a more successful solution.

KEEPING UP WITH THE COMPETITION   One of the most common market research approaches is to include competitive analysis as part of the product requirements development. This is a rather incestuous approach that mires products in a box of competitive muck. Truly innovative design comes not from following the others, but forging a path ahead of the pack. Competitive analysis has its place in product development, but observations will produce the truly innovative ideas.

ANALYSIS PARALYSIS   Success is not absolute, it is relative. A product does not have to be perfect; it just has to be better than the competitors'. That is where prioritization and focusing on the objectives really helps. Much of this work depends on extrapolating generalities from the observations and avoiding the temptation to look for statistically significant findings. If the same information is gained from a half-dozen users and it is rather obvious that the finding is important, it is not necessary to observe 100 more users just to prove the findings are significant.

YOU ARE NOT THE USER   Asking internal staff to play the role of a user will likely result in collecting unrealistic data from unrepresentative users. Moreover, internal staff are biased, or more accurately, affected by their own concepts, perceptions, and environmental factors of what they know about the design potential for the product. It is nearly impossible for them to accurately portray users who are unaware or ignorant of the design potential.

SEEING THE FOREST FOR THE TREES   With practice, it becomes easier to avoid thinking about the technology and trying to force users into a preconceived idea of the product. One way to avoid biasing the observations by the current vision of the product is to have someone unfamiliar with the new product plans be part of the observation team, and to listen to their objective questions and criticisms.

SUMMARY

Organizations employ many methods for determining the design of new products or enhancements to existing products. The product design process that integrates user observations with business objectives was created specifically to develop products that become market winners. The components of the process are not new to product design, but the process does reorient how and when things are done.

Organizations typically have a business objective in mind, which this process codifies into an actionable form that drives product design. Good product design teams already engage users to define the product direction; this process describes how and when to observe users. Many product design approaches result in writing requirements, while this process provides structure to translate user observations to design blueprints. Observational user research always identifies new opportunities—opportunities that the competitors have missed. Aligning those new opportunities with the business and marketing objectives creates winning products.

One key step offered by this process is the priority matrix, which helps align the business objectives and the users' needs. Though not entirely a novel concept, this tool does provide a visible means to understand and prioritize the opportunities measured by the various objectives and constraints. It is not the matrix itself that achieves product success as much as it is the communication and interaction that occurs between all of the key stakeholders during this process. The matrix and the process serve mainly to help organizations focus on the objectives and the users' needs, simultaneously. That focus achieves success.

Many organizations trying this process for the first time were skeptical of the process until they saw the results of the first round of observations. Each of these organizations realized the ah-ha experience that they were addressing the wrong needs of the users, and this process got them on track. This is one product design tool you should certainly consider adding to your toolbox.

REFERENCES

Adams, M., and D. Boike. 2005. *PDMA Foundation CPAS Study reveals new trends: While the "best-rest" gap in NPD widens.* Retrieved 7/30/2005, from *http://www.pdma.org/visions/july04/cpas-highlights.html.*

Cooper, R. G., S. J. Edgett, and E. J. Kleinschmidt. 2004. Benchmarking best NPD practices: Part I, *Research Technology Management* (Vol. 47, pp. 31–43): Industrial Research Institute, Inc.

Davis, A. M. 2005. *Just Enough Requirements Management: Where Software Development Meets Marketing.* New York: Dorset House.

Gallivan, M. J., and M. Keil. 2003. "The User-Developer Communication Process: A Critical Case Study." *Information Systems Journal*, 13 (1): 37–68.

Kazmierczak, E., P. Dart et al. 2000. "Verifying Requirements Through Mathematical Modeling and Animation." *International Journal of Software Engineering & Knowledge Engineering* 10 (2): 251–272.

Nielsen, J., and T. K. Landauer. 1993. *A Mathematical Model of the Finding of Usability Problems.* Paper presented at the ACM/IFIP INTERCHI '93, Amsterdam, The Netherlands, April 24–29.

Market and Technology Attack Teams: Tools and Techniques for Developing the Next Breakthrough Platform Product

Peter A. Koen
Associate Professor, Stevens Institute of Technology

Thomas C. Holcombe
President, THolcombe LLC

Christine A. Gehres
Vice President of Marketing for Engelhard's Appearance and
Performance Technologies Division

INTRODUCTION

Apple Computer's iPod, which now dominates the audio market, is an example of a new platform product that was developed from an idea to initial sales in six months by a dedicated team. In a similar fashion, Hewlett Packard developed the Deskjet printer platform in 1990. Even today the inkjet platform continues to command large market shares of the printer market. Kim and Mauborgne (2005) assert that new platform products or services, in their study of 108 companies, accounted for 38 percent of the total revenue and 61 percent of the profits—in contrast to incremental improvements which account for 62 percent of the revenue, but delivered only 39 percent of the profit. IBM's amazing turnaround from 1993, when it lost $8 billion, to net earnings of $4 billion in 1995 and $6 billion in 1997 was in part due to the company's "...embrace of platform thinking..." (Meyer and Mugge 2001, p. 36). "Product failures in high tech companies frequently can be traced to an incomplete product platform strategy" (McGrath 2002, pg. 54). But how does a company go about developing the next breakthrough platform?

The concepts in this chapter grew out of the frustrations from trying to develop new platforms while simultaneously meeting the demands of the current business that were focused on the next incremental product. Creating a separate business unit and infrastructure for developing new platforms was not viewed as a feasible option, as there are examples of such separated units having difficulty transitioning the new platform to the sustaining business. Another constraint to be overcome is the difficulty of commandeering critical human resources—even for an exciting project—for extended lengths of time. Finally, another objective was to embrace both the risk taking culture of venture capitalists combined with the rigor of Stage-Gate™—but to get to a conclusion quickly. *The objective of this chapter is to describe how to create and use market and technology attack teams to help the firm develop its next platform.* Market and technology attack teams may be thought of as short-term, ad-hoc business development groups that use temporarily assigned resources from the sustaining business units. The overall objective of the market attack team effort is to define a specific set of products or services in a new market where the company can achieve a sustainable competitive advantage and be able to win.

The chapter is targeted both at executives in an organization who are responsible for developing new platforms and at staff who are part of the market attack team. The chapter begins with a brief overview of what others have published, followed by a discussion of the key principles associated with the market attack team. These key principles will be helpful for executives thinking of adopting this approach. The actual market attack team approach is discussed after the principles section. This section serves as a guide to the actual process and should be valuable to both the team who needs to implement the process and the executives who need to guide it. Shaded inserts are included throughout the chapter that describe the market attack team approach from Engelhard's perspective, a firm that has embraced this approach (Engelhard Corporation was acquired by BASF in June 2006). The chapter also briefly discusses a technology attack team that is similar to the market attack team approach, but focuses on the technology hurdles of a project where the market is well known. Key learnings from Engelhard's experience are discussed in the concluding section.

PROLOGUE: WHY WAS ENGELHARD INTERESTED IN MARKET ATTACK TEAMS?

Engelhard Corporation is a surface and materials science company that develops technologies to help customers improve their products and processes. A Fortune 500 company, Engelhard is a world-leading provider of technologies for environmental, process, appearance, and performance applications. Although Engelhard had been a very successful company, the company wanted to improve its track record on new growth projects. Once a project/product was defined, the company had the required project management skills to take the project to commercialization. However, the real challenge was deciding which projects to pursue, especially those that required entry into market spaces that were new and unfamiliar to Engelhard. Typically, projects in new market areas are initiated because of a strong champion that has the skill and courage to convince senior management to invest in the project. The problem with this approach was that in some cases, decisions to proceed were not based on solid market research but rather, on the persuasiveness of the champion. As a result, there had been several cases where projects were stopped because the market potential was determined to be less than satisfactory, but only after significant investment in the project had been made. With this as background, Engelhard's new business development group, Ventures, set out to explore the market attack team approach.

WHAT'S PUBLISHED?

There are only a few articles in the literature that describe how companies develop the next generation platform. Meyer and Mugge (2001) in their article on platform innovation indicated that each business unit would have a "...subsystem development group... responsible for current and next generation ..." platform development (p. 37). This appears to be the approach used by IBM at the time the article was written, even though details of how a new platform is actually developed are not described in the article.

A more popular approach is embodied in the *Harvard Business Review* case written by Dean Whitney (1997). This article describes how P&G established a separate business development group for developing entire "...new products and categories of products based on the sectors' core competencies..." However, the head of this new business group indicated that it had made a "...major error at the outset..." by making the business development group "...too corporate and too high level." He indicated that he should have established more sense of ownership in the projects by the sustaining business units. There are many anecdotal stories where such separated business units were unsuccessful in ever transitioning their projects to the sustaining businesses, since the sustaining business was neither involved in the project definition nor in the subsequent formulation of the business solution. A recent book by Govindarajan and Trimble (2005) indicates how new business opportunities should be managed and organized, but doesn't discuss how they originate. One of the ways to correct this problem is through the use of market attack teams.

PRINCIPLES

Typically the market attack team is focused on a market segment with which the company has some familiarity. Nevertheless, this chapter also will describe approaches for how companies develop knowledge for entirely new markets in which they have little experience. Five key principles associated with the market and technology attack team approach are discussed in this section.

Dedicated Multifunctional Team

It is essential to have the key team members dedicated to spend at least 80 percent of their time on the project during the entire three-month time period. Tersesa Amabile (2002) showed in her study of 22 project teams from seven companies that creative thinking can occur under extreme time pressure, but only when people "...can focus on one activity for a significant part of the day..." In contrast, this same study also showed that creative thinking is unlikely to occur when people experience a highly fragmented day. Second, it is important to have domain knowledge relevant to the problem. This involves both market and technology domain knowledge.

Typically, the market attack team includes at least one market or businessperson and one to two technology employees. Projects that involve markets in multiple countries should also include additional market and/or business development people from those countries. Projects that appear to have a high likelihood of an acquisition should include someone from the acquisition and mergers group. In addition, it is valuable to include someone from the financial group. Though the team members from the acquisition and merger groups as well as finance are not dedicated full time to the effort, they serve as support for the team and attend many of the team meetings. Most market attack teams consist of three to five team members who spend at least 80 percent of their time on the project.

One other recommendation is also to include *inventors* from the company who are already recognized as being successful innovators. Invention in most companies resides in just a few unique individuals. Narin and Briztman (1995) in their study of Xerox, AT&T, Fuji, and Matsushita Electric over an eight-year period, from 1984 to 1991, found that the top 1 percent of the inventors are 5 to 10 times more productive as the average inventor and that the top 10 percent of the inventors are three to six times as productive as the average inventor. This same result has been shown in other studies and there are many anecdotal experiences supporting this contention in other companies. Many companies also assess the personality profile of the participants—such as a Meyers Briggs test—in order to assure creative diversity in the team (Leonard and Swap 1999) when determining the team make-up.

The key features that allow the market attack team approach to be successful are choosing the right team with the right domain knowledge, the right mix of marketing and technical skills, and ensuring that the key members of the team can dedicate at least 80 percent of their time to the project.

Effective Team Leaders

The best teams have leaders with the following three attributes:

1. *Previous profit and loss responsibility.* Typically, these individuals have an intuitive feel of the key elements of what will make a profitable business.

2. *Strong leadership skills.* The compressed time frame, combined with the high energy level of the people chosen for this effort, often stresses the individuals and tends to accentuate team members' weaknesses. A team leader who can deal with this environment is essential to the success of the project.

3. *Credibility at the senior management level.* Senior management needs to have confidence in the effort since the resources committed to these market attack team efforts typically are significant. The team often will be going into areas that are completely new to the company. Further, the team may deal with sacred-cow issues, such as, "We tried that channel before and it didn't work—so why are we trying it again?" "The market is dominated by a major competitor—so why bother?" "It's a commodity market—we can never make any money." A leader with senior management credibility typically will be given more latitude to explore these new areas without being continually second guessed, micro-managed or prematurely shut down.

Large Revenue Potential

The market attack team approach is relatively expensive, both in people and monetary resources. In addition, our experience indicates that only 50 percent proceed past the three-month time frame, and only 50 percent of the approved projects are successful in the marketplace.

Engelhard conducts market attack teams on projects that appear at their outset to have at least a $50 million revenue potential during the first few years of sales. This represents an arbitrary threshold that would be expected to vary by company size and expectation. The hurdle rate should represent both a real and "big enough" number to be able to gain senior management attention so that they are willing to assign resources to the project. In some cases market attack team projects have been done on smaller revenue opportunities for shorter amounts of time with fewer dedicated resources.

Short Time Frame

There are many unknowns at the start of a market attack team. A market that initially looked attractive might soon be found to require a large investment in order to establish a channel, might have intellectual property or regulatory constraints, might require capabilities and competencies that are unattainable by the company, or might present other barriers to entry. As a result, establishing a time limit of a maximum of three months for the market attack team effort is strongly recommended.

Similar market attack team efforts exist in other companies—though many extend for six months or longer. However, three months seems to be the maximum time for the market attack team effort in order to make this process sustainable over time within a company. Essential to the success of the project is choosing the right team with adequate domain knowledge of both the market and technology. The market attack team requires essentially full-time commitment of the key people, and these people are usually committed to other projects. Experience shows that one *can* get these key people for three months—*but* not a day longer.

Getting key people for any time longer than three months seems to be impossible. The three-month time period for studying unknown markets is very short and therefore represents another challenge, since it is impossible to meet with many of the desired customers and market experts in order to gather all of the information required.

Senior Management Involvement and Prompt Decision Making

At the end of the three-month process, senior management must make a definitive go/no-go decision. Only 50 percent of the market attack teams are expected to be funded beyond the three-month period. In addition, the majority of the team members are doing this effort as a temporary assignment. Thus, it is critical that the market attack team effort have a definitive ending. It is important to quickly come to a conclusion without continuing to waste resources. Senior management engagement with the project and commitment to the time frame assures that the market attack team effort will conclude at the end of three months.

In projects of this type, there are always many unanswered questions, and there is never enough time. As such the project could easily be redirected to do more market research and competitor analysis. It is often far easier for the senior management team to ask for more information than to make a decision to kill or proceed with the next step of the project. However, redirection to gather more information defeats the goal of the market attack team by requiring more people and monetary resources than originally planned.

Senior management involvement with the team, which usually occurs only at the decision meeting, will usually result in project redirection, since the team will not have had a chance to voice the key issues. Senior managers must be engaged with the project early enough so that all of their major issues are resolved, giving them the information so they can make a decision at the end of three months. Senior management's buy-in to make a definitive decision at the end of three months will alter their behavior so that they make time to work with the team and understand the project deliverables during the course of the market attack team effort. In this way, senior managers can effect a course correction and, if necessary, direct the team to do additional or different analysis early enough to affect the team's recommendations. This can happen only if they are engaged with the project during the market attack team effort.

MARKET ATTACK TEAM

A schematic of the market attack team approach is indicated in Figure 6-1 and consists of four distinct phases. An overall schedule of the major activities and events are indicated in Figure 6-2.

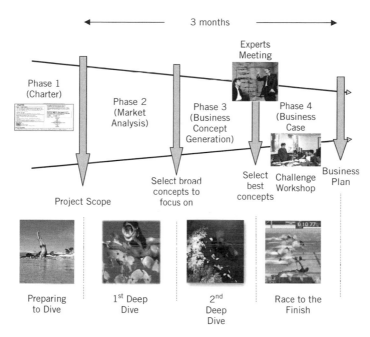

FIGURE 6-1. Schematic of the market attack team approach.

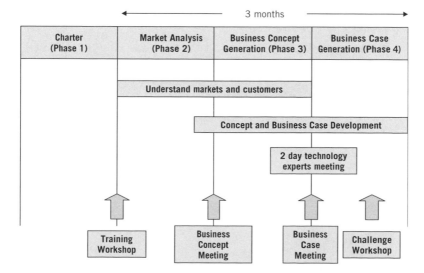

FIGURE 6-2. Generic schedule for the market attack team.

ADOPTION OF THE FIRST MARKET ATTACK TEAM EFFORT AT ENGELHARD

At Engelhard, new ideas, processes, or programs are not successfully adopted without an effective enrollment process, which aligns key management with the idea. The new-business development group at Engelhard, called Ventures, felt that market attack teams were worth pursuing. Ventures is a group, separate from the sustaining businesses divisions, charged with developing new business opportunities.

Since the market attack team effort would involve people from the divisions in order to gain access to critical domain knowledge skills it was important to obtain alignment to pilot a market attack team from these sustaining businesses. This was done through a series of formal and informal discussions over a three- to five-month period, which included presentations on the market attack team effort, a review of potential projects, the funding necessary to support such an effort, a review of past efforts to develop new business opportunities and buy-in from the sustaining division presidents to allow key people from their organizations to be part of the effort. A small team was formed with members from Ventures and one of Engelhard's business segments to decide on the particular pilot project. A charter for the project was drafted (an example of a charter is shown in Figure 6-3) and presented to senior management for approval. The first market attack team was launched shortly thereafter.

Scope: Project Mission
- Identify opportunities globally for Engelhard to win in personal care based on competencies in materials and surface science (including appropriate acquired capabilities)
- "To win" means to add $75-150 million of revenues in three years
- Outside the scope:
  - o Color-based materials
  - o Commodity businesses
  - o Business selling to end users (consumers)
  - o Equipment or service-based businesses

Timing
- Kick-off: July 10-11, 2003
  Senior management review: October 17, 2003

Deliverable
- Comprehensive business plan that defines best way for Engelhard to "win" in personal care market and recommendations on "go-no go" decisions

Assumptions
- Engelhard is willing to acquire company with core competency in organic materials
- Suitable acquisition(s) will be available within timeframe
- Engelhard is prepared to do acquisitions in this market if appropriate

Resources: Core Team
- Team sponsor (50%)
- Team leader (70%)
- Lead technologist (50%)
- PC technical support (40%)
- Process owner (35%)
- Patent support (30%)
- PC marketing support (20%)

Resources: Financial

Market studies	$70,000
Travel & misc.	40,000
Technical testing	20,000
Consultants (business & technical)	50,000
Competitive intelligence	20,000
Total	$200,000

Preliminary Opportunity Analysis
(Assuming appropriate acquisition(s))

FIGURE 6-3. Charter for personal care market attack team.

The market attack team approach is broken up into four distinct phases.

Phase 1: Charter

This initial phase defines the scope of the project and represents an agreement between the market attack team and senior management. The charter is typically a one-page document discussing the project scope, market size, project risks, and people and resources needed to accomplish the project. Preliminary competitor and intellectual property assessments are also sometimes included. The charter represents the key communication vehicle between the team and senior management. Achieving clarity around the project and expected deliverables is essential prior to the start of the project. The charter is usually developed by the team leader and, in most cases, requires several meetings with senior management before arriving at a completed document.

In some cases, the team also is involved in the development of the charter. While this is desirable, it represents a trade-off between gaining team consensus and utilizing the team's resources prior to the start of the project. Obtaining full-time or nearly full-time support from key team members with the correct domain knowledge is always an issue. Often the negotiation involves an agreed-to commitment of time not to exceed the three-month period between the kick-off meeting and the senior management decision meeting. Involving team members in the development of the charter is sometimes seen as violating the negotiated time allocation agreement.

EXAMPLE: ENGELHARD'S PERSONAL CARE MARKET ATTACK TEAM

Engelhard Corporation was a global supplier of effect pigments to the cosmetics and personal care market and wished to add between $75 million and $150 million of sales within three years. At the time of the market attack team launch, the global retail value of personal care products was between $180 billion and $200 billion. The market for raw materials sold to personal care was $12 billion. This included approximately $5.2 billion in specialty raw materials ($2.5 billion without surfactants and fragrances). The annual growth rate of 3 to 15 percent depended on the product category and varied by region. The market was very fragmented. The largest product categories were rheology modifiers (>$300 million) and silicones (>$300 million). The largest suppliers globally were BASF, ISP, and Dow Corning, with $140 to $280 million each in sales to the personal care industry. Adding incremental sales of $75 million to $150 million to Engelhard's business would make it one of the top suppliers of specialty raw materials to this industry.

The charter for Personal Care Market Attack Team is shown in Figure 6-3. The charter includes the following items:

◆ *Project mission:* Identifies what is within and outside of the project scope. This should provide a crystal-clear explanation of the project.

◆ *Timing:* Both the initial kick-off meeting and the senior management review meeting are set prior to the start of the effort.

◆ *Deliverable:* A detailed business plan.

◆ *Team and financial resources:* These detail the actual people and monies needed to support the market attack team effort.

◆ *Risk spider diagram:* The risk spider diagram assesses the project risk along four dimensions: technical and commercial success and strategic fit and leverage. A scale for potential reward is also included since the potential financial return is usually quite difficult to determine. Anchored scales for determining these values are given in Table 6-1 and are similar to those developed by Cooper, Edget, and Kleinschmidt (2001). The overall objective of the risk spider diagram is twofold. First it is used as a communication tool between team members and between the team and senior management to discuss the most critical areas of project risk. Second, it is used as a portfolio tool as companies begin to adopt the market attack team approach. The information on the risk chart can be loaded into a portfolio Excel chart and used to balance one potential project against another. The charter, along with the risk chart, is used as a mechanism to sort between various market attack teams waiting for resources and approval.

EXAMPLE: MARKET ATTACK TEAM MAKE-UP

A multi-functional team that included technology, market, process and patent support was assigned to the project. Team member commitment to the project was generally in the 40 to 70 percent area, as indicated in Figure 6-3. While 80 percent commitment from key team members is highly desirable, the team was able to effectively achieve great results with somewhat less time commitment but effective teamwork. Nevertheless, key team members became overloaded when preparing for the workshops and developing the final business case.

Phase 2: Market Analysis

This phase, which lasts for approximately one month, is focused on understanding the market defined by the charter. The phase begins with a two-day workshop, led by the team leader, that is focused on understanding the market and unmet customer needs. The workshop is broken into four parts:

1. *Introduction.* The morning of the first day's objective is to understand the project, the market attack team approach, and the expected deliverables.
2. *Understanding the current market knowledge.* The afternoon of the first day and morning of the second day are focused on understanding the perceived market knowledge, competitor and intellectual property space. The team often has limited knowledge in markets that are yet to be explored. However, the objective at this first meeting is to ascertain what the market attributes are through the current knowledge that exists either within or external to the company. This is typically handled by inviting people in the company, but external to the market attack team, to be present at this initial meeting. In addition, the team will purchase market studies when they are available to supplement their knowledge. Finally, a number of market attack teams have hired external market consultants who help fill the knowledge gap at this first meeting.

TABLE 6-1.
Anchored Scales Used to Construct the Spider Diagram Shown in the Charter Document (Figure 6-3)

Probability of Technical Success

Factor	1–3	3–6	6–9	>9
Technical Gap	Large gulf between current practice and objective; must invent new science	Order of magnitude change proposed	Step change short of "order of magnitude"	Incremental improvement; more engineering focus
Program Complexity	Difficult to define; many hurdles	Easy to define; many hurdles	A challenge; but "do-able"	Widely practiced in the company
Technology Skill Base	Technology new to the company; almost no skills	Some R&D experience, but probably insufficient	Selectively practiced in the company	Widely practiced in the company
External Technology	Technology does not exist	Technology exists, but have no idea where to find it	Technology exists and is available—but Company has never used it.	We know where to get it and have integrated technology before
Manufacturing Capability	Manufacturing process unknown to us or anyone else	Manufacturing known to others; but unknown to us	Minor modifications to existing technology	Manufacturing technology known and capacity (internal or external) is available

Probability of Commercial Success

Factor	1–3	3–6	6–9	>9
Market Need	Extensive market development required; no apparent need	Need must be highlighted to customers; product tailoring required	Clear relationship between product and need; one-for-one substitution of competitor's product	Product immediately responsive to customer need; direct substitute for existing company product
Market Maturity	Declining	Mature/embryonic	Modest growth	Rapid growth

(continued overleaf)

TABLE 6-1.
(*continued*)

Factor	1–3	3–6	6–9	>9
Competitive Intensity Channels to Market	High Channels required are new to world	Moderate/high New channels required	Moderate/Low Need to modify existing channels to accommodate platform	Low Existing channels in place and can be used
Brand	No brand or image awareness/new to market	No brand awareness, but can leverage company name and image	Company has brand recognition, but is not leading in the market segment	Company is industry leader and has leading brand image
Regulatory/Social Political Impact	Negative	Neutral	Somewhat Favorable	Positive impact for high-profile issues
Raw Material Supply	No known suppliers	Single supplier, but no commitment to supply	Single source, stable contact	Multiple suppliers with acceptable pricing, easily negotiated

Strategic Leverage

Factor	1–3	3–6	6–9	>9
Proprietary Position	Easily copied	Protected but not a deterrent	Solidly protected with trade secrets, patents	Position protected upstream and down stream through a combination of patents, trade secrets, raw material access and so on
Platform for Growth	Dead end/one of a kind	Other opportunities for business extension	Potential for diversification	Opens up new technical and commercial fields
Durability (Technical and Market)	No distinctive advantage; quickly "leapfrogged"	May get a few "good" years	Moderate life cycle (4–6 years) but little opportunity for incremental improvement	Long life cycle with opportunity for incremental improvements

Strategic Fit

Factor	1–3	3–6	7–9	>9
Synergy with other Operations within company	Limited to single channel	Can be applied to more than one channel	Can be applied to multiple channels at company	Could be applied to all channels at company
Congruence	Only peripheral fit with business strategies	Modest fit, but not key element to strategy	Good fit with key element of strategy	Strong fit with several key elements of strategy
Impact	Minimal impact; no noticeable harm if platform is not done	Moderate competitive, financial impact	Significant impact, difficult to recover if platform is not done	Business unit future depends on the platforms

Potential Reward

Factor	1–3	3–6	7–9	>9
Absolute Contribution to Profitability (5 year cumulative cash flow from commercial start-up)	< $2 Million	$3–6 Million	$7–9 Million	> $10 Million
Market Size	< $10 Million	$30–$70 Million	$70–$90 Million	> $100 Million
Market Growth	>5%	5–10%	10–15%	>20%
Potential Market Share	>5%	5–10%	10–15%	>20%

Two key charts are developed during this phase to help guide the team. The first is a market segmentation map. The market segmentation map developed by the Engelhard team for the personal care market segment is shown in Figure 6-4. A product/process map was also developed for some of the key segments. A product/process map for the delivery systems segment of the personal care market is shown in Figure 6-5. The purpose of these maps is to capture the overall knowledge of the group at the very beginning of the project. These maps are continually refined and edited as the project proceeds.

The objective of the mapping process is not the map itself but the discussions that transpire in creating the maps. Most teams have found that the creation of the segmentation and product/process maps are an important ongoing activity of Phase 2. In almost all cases the maps were dramatically changed after the Phase 2 visits.

3. *Identifying target segments and customers for Phase 2 visits.* Once the maps are completed, they are used to identify market segments that are targeted for customer visits. This is accomplished in the afternoon of the second day. The challenge for the team during this first phase is to spend significant time with the customers in segments that *appear* to be attractive. To help narrow the field down to potentially attractive segments, the personal care team identified criteria for a winning strategy. Each of the customer/consumer factors are shown on the left-hand side, the relative degree they are satisfied by the various competitors in the middle and the relative importance of the different factors to both the customer and consumer in the right column. Three

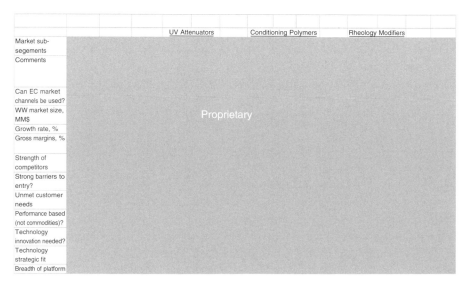

FIGURE 6-4. Personal care market segmentation map.

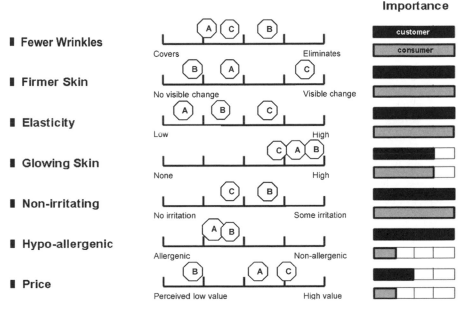

FIGURE 6-5. Product/Process map for anti-aging.

ENGELHARD WINNING CRITERIA FOR THE PERSONAL CARE SEGMENT

The market attack team selected the following criteria for a winning strategy in personal care:

◆ Avoid commodities
◆ Good strategic fit (involves core competencies in surface and materials science and leverages current operations)
◆ Large market (>$100 million)
◆ Strong market growth (>5 percent per year)
◆ Materials suppliers achieve good margins (>percent gross profit, >percent EBIT)
◆ Sustainable competitive advantage
◆ Technology based
◆ Can capture significant market share
◆ Opportunities for new growth
◆ Good cultural fit
◆ Offers broad platforms

seperate competitors are shown as designed by "A", "B" and "C". For proprietory reasons the data is artificial.

The team is cautioned to avoid thinking of solutions during this phase and is encouraged to spend a significant amount of time in the market in order to become intimately familiar with it. This focus on the market space, as opposed to trying to identify solutions, is consistent with approaches followed by innovative companies (Koen et al. 2001).

A major focus of this phase is to gather first-hand knowledge of the target markets through detailed interviews of customers, lead users, and industry experts. Typically, between 25 and 50 interviews are completed by the end of the program, with most of them being done during Phase 2. Before initiating the interviews, the team develops introductory slides about its own company and prepares an interview guide with typical questions[1]. Each interview is conducted by a minimum of two team members, one with a technical background and one with a business/marketing focus. During the interview, each member tries to record in detail everything that is discussed. Within 24 hours after each interview, the notes are e-mailed to all members of the team so the entire team moves up the learning curve as the interview process proceeds. The actual interview guidelines and sanitized results from some of the interviews for the personal care segment market attack team effort are shown in inserts at the end of this section.

4. *Setting the stage for Business Concept Generation* The team, at the conclusion of Phase 2, identifies rough business concepts and market segments in which they believe the company can win. Assignments for developing each part of the final business case are also made during this Phase. An outline of a typical business plan is shown in Table 6-2.

EXAMPLE: INTERVIEW GUIDELINES FOR PERSONAL CARE GROWTH TEAM

Engelhard company description and a discussion of the purpose of the visit. For example the introduction might sound like this:

We are a leading supplier of optical effect pigments to the Cosmetic and Personal Care market. We are looking at expanding further beyond our current business area. Most of our customers feel that we bring strengths to this market, and our expertise in Materials Science and Surface Chemistry has helped us develop a good reputation for innovation in the market place. We would like to find ways of helping your company/the industry to meet the challenges of the future.

The following are a list of guideline questions that the team prepared:

◆ Could you briefly describe your company's role in the personal care markets?

◆ What are the top-five unmet needs that you see in each of the following personal care market segments? (segment 1, segment 2, etc.)

◆ What do you see as the major problems and challenges over the next five years in each segment?

---

[1] See also ToolBook 2, Chapter 7, "Obtaining the Voice of the Customer, for further information, on how to conduct interviews of this type.

◆ Which are the most difficult to satisfy, and why?

◆ What is your strategy for developing new products and choosing new raw materials? (e.g., lowest cost, differentiated products, ease of formulation, natural sourcing, etc.)

◆ What criteria do you use for raw material selection?

◆ What current suppliers are the strongest, and why? What are their weaknesses?

◆ How do you think Engelhard could bring real value to your organization outside of our current business area?

Approximately 90 percent of the month was spent on interviews. As the interviews progressed, the team members begin to understand the players, markets and technologies. About halfway through the process, the team identified important patterns and fine-tuned the questions to better highlight critical information.

EXAMPLE: INTERVIEW RESULTS FOR PERSONAL GROWTH TEAM

Example from Company A: (In What Market Areas is Your Company Involved?)

Skin care and colors are our main focus. At this time, the focus is on color products, which we hope will pull us out of our current financial difficulties. We are also interested in adding skin treatments that we can combine with our color products. Sunscreen formulation is a big area of interest. We are interested in developing formulations that use more physical and less chemical ingredients, although we don't want to use ``tons'' of TiO_2. Small particles seem to offer the best value. There are also issues with odor from chemical sunscreens (octyl methoxycinnamate has a malodor at higher concentrations).

Example from Company B: (Do You See Pharmaceuticals Being A Big New Trend in Personal Care?)

No, I don't. There is really no evidence that this trend is occurring. Also, there is a concern about the benefit/risk ratio. What you do see are companies trying to take concentrated forms of actives (such as botanical extracts) and increasing their performance so that they begin to act like drugs. You see this in vitamins, too. You have to be careful about the claims or the inferred benefits from a material's use. A few years ago, many pharmaceutical companies thought their medicines were going to make their way into personal care products, so they started buying up personal care companies. But it didn't happen and the pharmaceutical companies had to turn around and divest themselves of the acquisitions. Neutraceuticals, on the other hand, are real and growing.

Example from Company C: (What are The Market Trends and Customer Needs in Preservatives?)

Preservatives are not a good business to go into. Most companies are global and are looking for formulas they can use globally. If you were only selling in the United States, it would be much easier. In the United States, if you prove a product is safe to use, then you can use it in anything. Many other countries, however, have a positive list of approved materials, and getting other materials approved is extremely difficult. Japan is the worst. Europe is bad and getting worse. The cost of developing new preservatives is very high, so nobody is doing it. Most companies are looking at combinations of existing preservatives. Very few preservatives are good enough to use by themselves. So companies blend actives together to get the desired breadth of conditions needed.

TABLE 6-2.
Outline of a `` Typical Business Plan. ''

Section	Description	# of pages
I	Executive Summary	1
II	Scope of the Project	1
III	Market and Customer Definition	2
IV	Competitor Analysis	1
V	Regulatory Issues	1
VI	Why This Project Will Win	1
VII	Market Entry Strategy	1
VIII	Intellectual Property Analysis	1
IX	Technology and Development Strategy	2
X	Manufacturing and Operations Strategy	1
XI	Project Plan	1
XII	Financials	1
XIII	Risk Assessment and Reduction.	2
	Total	16
Appendix A	Details of Key Competitors	

Phase 3: Business Concept Generation

This phase begins with a one- to two-day team meeting that is focused on identifying the most attractive segments that have inherent wins for the company. Business concepts identified should ideally have:

◆ Large unmet customer need and a winning value proposition
◆ Sustainable competitive advantage
◆ Large and growing market
◆ High margins

EXAMPLE: PHASE 3 ACTIVITIES

This phase began with a one- to two-day meeting, where the team members brought in a consultant with industry expertise and debated different business concepts presented by different team members. As the lively discussions proceeded, we began to develop a consensus of support for a couple of competing business concepts. Each team member was then assigned different sections of the business plan and prepared rough drafts around these promising concepts. In the process of preparing and reviewing these drafts, it became clear that more details were required in different segments to make the draft business cases more complete and to properly compare the competing concepts. The team then conducted another round of interviews to fill in the missing information. They also organized a *technology experts* meeting and discussed each business concept with them to obtain their technical feedback and advice. All of this information set the stage for Phase 4.

This phase has the team once again out in the marketplace—but now focusing on the particular segments that could be a win for the company.

Although Phase 2 was mainly focused on understanding unmet customer and consumer needs, this phase is focused on identifying solutions that will provide a win to the company in specific targeted segments. The team also focuses its efforts on identifying the competitive and intellectual property landscape. Although detailed patent analyses are not expected, a broad overview of the patent space is evaluated to obtain a general understanding of potential infringement and freedom to practice issues.

The team also determines if a technology experts meeting is needed and who should attend the challenge workshop to be held in Phase 4. In order to get the appropriate people, it is important to identify them early to ensure availability and subsequently invite the attendees to these meetings. The focus of the technology experts meeting is to identify the technology hurdles that are anticipated with the project. Market attack teams often will identify potential solutions that require technology capabilities well beyond those that exist in the company. The objective of the technology experts meeting is to identify the key technology barriers. The details of the technology experts meeting is discussed in the *technology attack team* part of this chapter, since it is always part of that effort.

EXAMPLE: ENGELHARD PHASE 3 ANALYSIS

The team identified delivery systems as the focus of the Phase 3 efforts. This category was relatively small, but represented significant upside potential when actives were included. It was one of the fastest-growing categories in personal care, with some of the highest *value-add* potential.

Delivery systems are technologies that enable the protective encapsulation of a sensitive and unstable active until it is applied to the skin or until it is needed. Delivery systems can help combine products from different categories. The team believed that Engelhard could become the world's leader in delivery systems and sell them in combination with skin appearance and performance products. Delivery systems grow about 10 percent per year, fueled by the demand for new actives and skin performance materials that need some modification. Since most players in delivery systems are regional, Engelhard could add additional growth through globalization. The challenge for the team in Phase 3 was to identify a specific winning strategy for Engelhard.

Phase 4: Business Case Generation

This last phase is focused on building the business case with specific and detailed recommendations to be presented to senior management. The team meets again for a one- to two-day meeting to work on the business case—see Table 6-2. A challenge workshop is then held before the business case is presented to senior management for a decision.

The purpose of the challenge workshop is to critically evaluate the business case that the team has developed. Ideally, total attendance should not exceed 20 people so that the meeting can be interactive and does not turn into a symposium. The attendees should consist of the team, and internal as well as external experts.

Bringing in external experts often creates some controversy due to confidentiality requirements, despite the fact that external experts attending the challenge workshop are required to sign a confidentiality agreement which assigns the intellectual property rights back to the company. Having external experts at the meeting is valuable, however, because they often demonstrate more latitude in their critique since they do not need to be politically correct in their recommendations. In addition, the external experts often have more specific and specialized expertise than exists at the company. Selected members of the senior management team also are asked to participate. This allows them to become more familiar with the business case. However, the division president or chairperson of the executive committee is not invited because there is a concern that his or her opinions will unduly effect the discussions.

The challenge workshop, again led by the team leader, is conducted over a two-day period beginning in the afternoon of the first day. During the first day the team presents its business case—typically in a PowerPoint presentation. A dinner is held for all of the attendees at the end of the first day. This allows the attendees to socialize and further discuss the business case informally. In addition, the evening break allows the attendees to have more time to reflect on the business case, although in some cases companies have held the entire challenge workshop in one day.

The following morning the attendees are divided into three or four "challenge groups." A leader is appointed to each group. The leader is someone external to the team and ideally one of the external experts invited to attend the meeting. The market attack team members are distributed throughout the challenge workshop teams in order to provide clarifying information, but are cautioned against advocating their own or the team's position. Each group is charged with the following:

- ◆ Developing a whole new concept
- ◆ Building on the existing concept
- ◆ Recommending that the concept is not feasible

After about three to four hours, each team presents its conclusions, followed by a general discussion of the findings. The challenge workshop ends in the early afternoon of the second day.

There is no attempt made by the market attack team at the challenge workshop to weave the conclusions together or to pick a "winner." The conclusions from the challenge workshop allow the team increased clarity around what concept the team feels comfortable with recommending. It identifies areas that will need more thought and team discussions before the business case is finalized.

In one case, the team completely changed its recommendation after the workshop, recommending that the project be stopped because certain hurdles became clear. As another example, the team responsible for the case discussed in this chapter questioned whether one acquisition, rather than the two being recommended, could meet the growth requirements set forth in the charter. The team evaluated whether the difference could be made through internal growth. Overall, the challenge workshop enhances the clarity of the conclusions.

EXAMPLE: ENGELHARD'S PHASE 4 BUSINESS CASE

Early in the process, it became obvious to the team that the aggressive growth target for this project could not be accomplished in the required time frame by internal means and with existing Engelhard competencies. The team had to carefully consider the fit of a product category in the personal care area with the Engelhard strategy, not just from a technology fit that would help future growth, but from the perspective of sustainability of a competitive advantage that would initially come from the outside via an acquisition and/or licensing. The market for active ingredients and delivery systems was relatively fragmented. It included many small-sized players with varying capabilities and strengths.

Success is based on know-how and science, as well as closeness to the customers. Going through the winning criteria, as well as customer feedback during the interviews, allowed the team to make a better decision. The product categories of actives and delivery systems were selected as entry points into the personal care market outside of colors. The team recommended that Engelhard acquire two of the larger delivery system companies (Collaborative Labs in E. Setauket, NY and Coletica in Lyon, France) with competencies that are complementary and that have significant skin care actives in their line.

During initial implementation of this approach, market attack teams sometimes consider the challenge workshop to be superfluous. The teams believe that they are essentially done with the business case, having spent almost two and a half months studying the market. *Without exception, companies who have participated in the market attack team approach consider the challenge workshop to be one of the most valuable parts of the process.* Often times teams develop a group think perspective, which the workshop overcomes. This is a nonthreatening environment in which they are able to fully discuss their business case.

After the challenge workshop the team begins final preparation of the business case for the executive decision team. The make-up of the executive decision team depends on the project and organizational structure of the company. When market attack teams are done within a division, the executive decision team typically consists of the division president and many of his or her direct reports, including the VP of R&D, VP of marketing, and the chief financial officer. When the projects are cross-divisional, there may be presidents and their senior management from each division.

This is a decision meeting! *The senior management team is required to make a decision within one hour of the team's presentation.* At this meeting, the market attack team ends. This is important, since a clear ending date allows the team in the beginning to recruit key team members throughout

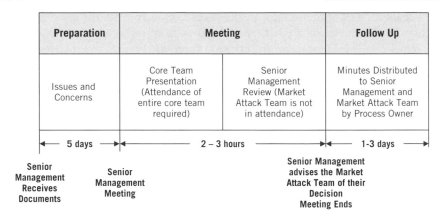

FIGURE 6-6. **Generic schedules for the senior management decision meeting.**

the corporation, while maintaining a commitment to only have them involved with the project for three months. The actual date of the meeting is set prior to the start of the market team and is included in the team's charter.

In order to help assure that this is a decision meeting, the business case is sent out to the senior management team five business days before the meeting. The overall logistics of the meeting are shown in Figure 6-6. Teams are also responsible for getting tentative approval for the specific people required to take the project to the next milestone. It is not enough to say that so many full-time equivalents are needed. The business case must state the specific people needed to continue this project. This is usually a key discussion topic that occurs during the senior management meeting. Teams are encouraged to have this issue resolved prior to the senior management meeting.

During the five-day interval prior to the senior management meeting, each member of the executive team meets with selected members of the market attack team for one-on-one discussions. Market attack teams find that these one-on-one discussions are also a critical part of this process. These meetings allow the executive team to fully understand the business case and allow the team to make course corrections prior to the senior management decision meeting. Additionally, the best teams maintain some communication with key members of the senior management team throughout the entire market attack team effort.

The actual presentation at the senior management meeting is focused around the recommendations and key discussion areas that transpired during the previous five days. A typical presentation outline is shown in Table 6-3. In many cases, the team modifies its initial recommendations as a result of the discussions that occurred during the intervening five days. The format indicated in Figure 6-6 indicates that the team is asked to leave the room during the senior management discussion. This is done so the senior management team can have a more confidential discussion about the recommendations. In some cases, teams become upset when the executive team modifies the recommendations,

TABLE 6-3.

Outline of the Executive Team Presentation. The Objective of the Presentation Should be Decision Oriented and Should NOT be a Summary of the Business Plan

---

Slide 1. *Title Page:* Project Name and names of the team members.

Slide 2. *Decisions Required:* This slide should be focused on what specific decisions are required from the executive team to proceed.

Slide 3. *Customer Map:* The slide is an overview of why this project will meet customer needs and beat competitors.

Slide 4. *Product Features:* This slide is an overview of product features, critical dates and product costs.

Slide 5. *Why will the company win:* This slide explains the key product features and strategies which will allow the company to win.

Slide 6. *Technology and Development Strategy:* This slide explains the keys areas which will be addressed at the next stage.

Slide 7. *Project Overview Chart:* The overall project is shown with key review dates indicated. The expected tolerances for the product release dates are also shown.

Slide 8. *Team:* This slide indicates the percent of time required for all the people necessary to take this project to the next milestone along with an estimate for project completion. Specific names and time requirements for the people not on the market attack team, but who also are required to take the project forward also should be included.

Slide 9. *Financial Analysis:* Highlight the key financials.

Slide 10. *Risk Assessment:* Present the risk score as well as highlighting critical risks and indicate how they will be reduced.

Slide 11. *Executive Team Questions:* Discuss any issues which remain unresolved.

Slide 12. *Decisions Required:* This is the same as slide 2 and should remain projected during the executive team discussions and subsequent Q&A.

---

and the discussions become more difficult when the market attack team is in the room. However, in some cases critical information that will help the decision is missing during these discussions—so having the team there may actually help the decision. In the end, this is a cultural issue that needs to fit with the company.

The senior management team may accept or modify the team recommendations. The decision to go back and do more analysis is not an option for the senior management team. In the majority of cases, the senior management team, especially in the initial market attack team efforts, modifies the team's recommendations. A typical example would be when a team recommends both additional market efforts to better identify the size of the market, combined with a technology effort to solve particular high-risk issues in order to assure a viable product. In many cases, the senior management team will fund only the technology part to limit the company's exposure and overall resource commitment and will postpone the more in-depth market analysis. The analysis done during the market attack team project is often more than adequate to justify the expenditures on the technology.

> **EXAMPLE: ENGELHARD'S PERSONAL CARE MARKET ATTACK TEAM EPILOGUE**
>
> Following approvals by senior management, Engelhard formed a new team to implement the market attack team's recommended business plan. The new team consisted of a full-time leader (director of business development, personal care), a marketing resource, a technical resource, as well as support from the legal, intellectual property, and mergers and acquisition departments. The two acquisition candidates were Collaborative Labs in E. Setauket, New York, and Coletica in Lyon, France. On July 30 2, 2004, Engelhard announced that it had acquired The Collaborative Group, Ltd. (including its wholly owned subsidiary, Collaborative Laboratories, Inc.), and on March 17, 2005, Engelhard announced completion of its acquisition of a majority stake in Coletica, S.A.
>
> In a public announcement Engelhard stated that the acquisition `` ... further strengthens Engelhard's position as a leading global supplier of materials technology to the cosmetic and personal care industries. It expands the company's existing capabilities in the growing market for skin-care materials used in such applications as anti-wrinkle creams, sun protectants, moisturizers and materials that improve the overall complexion of the skin. ''

TECHNOLOGY ATTACK TEAM

The objective of the technology attack team is to focus on the technology hurdles of a project where the market is well known. This is in contrast to a market attack team, which focuses on markets that are not well understood by the company. The technology attack team follows a similar four-phase process. The first phase is the same for both projects. However, there are some differences the remaining phases, as discussed in this section. Table 6-4 summarizes the differences and similarities between the technology and market attack teams. The last three phases of the technology attack team are outlined next, with the differences noted.

Phase 2: Market Analysis

Phase 2 of the project begins with a market analysis. Presumably, the team has a deep understanding of the market for which it is developing or needs to develop the technology. Detailed segmentation (Figure 6-4) and product/process maps (Figure 6-5) are developed as was done in the market attack team approach. In addition, a technology performance table, like the one shown in Table 6-5, is also developed (Ajamian and Koen 2004).

The technology performance table links specific key customer needs to specific product specifications. These maps and tables are developed during the first two-day kick-off meeting of the technology attack team. *In almost all cases the development of these maps and technology performance tables creates considerable discussion with the realization that additional market information is needed to define the technology goals.* As a result, the first

TABLE 6-4.
This Table Highlights the Major Similarities and Differences Between the
Market and Technology Attack Teams

Phases	Market Attack Team	Technology Attack Team
1. Charter	same	
2. Market Analysis	Broad market analysis. (The major activity is a deep dive into the market arena.)	Assessment of market knowledge and development of the technology performance table – see Table 6-7. (Activities include a focused market assessment of unknown areas and a deep dive into technology expertise throughout the world.)
3. Business Case Generation	Identification of market focus and initial generation of a business concept.	This phase begins with a technology experts meeting to gain consensus on the technology hurdles. The major activities during this phase are technology experiments in order to better understand the hurdles and limits.
4. Business Case Generation	Preparation of the business case. The challenge workshop has a predominate focus on the market hurdles. The challenge workshop often includes some of the market consultants from Phase 2.	Preparation of the business case, but where the challenge workshop is focused on the technology hurdles. The challenge workshop includes many of the technology experts from Phase 3.

month of the technology attack team effort is focused on a deep dive into the marketplace to better understand the user requirements. In many ways, this is similar to the "deep dive" done in Phase 3 of the market attack team effort, but is much more selective on the market area that the project is focused on. *All of the market issues should be known by the end of this phase.* In addition, the team identifies the key technology experts in order to better understand the technology hurdles and who should be invited to participate in the technology experts meeting, which is part of the Phase 3 effort.

Phase 3: Business Concept Generation

This phase is focused on identifying potential technology solutions. Phase 3 usually begins with a technology experts meeting, which is focused on forming a consensus view of the hurdles involved. The technology experts meeting is a one- to one-and-a-half-day meeting that includes the technology attack team

TABLE 6-5.

Example of a Technology Performance Table for the Next Generation Office Copier Capable of Reproduction Speeds of 25 Copies Per Minute. For this Example Assume that Current Copiers can only Make 10 Copies Per Minute. Ultimately the Goal Would be to Develop a Copier Capable of 200 Copies Per Minute. The Table Specifies the Overall Performance Criteria, Technology Feasibility Points and the Confidence Level for each of the Desired Characteristics. More Details Concerning the Use of this Table can be Found in the PDMA *Toolbook2*, Chapter 11, "Technology Stage Gate" (Ajamian and Koen, 2004)

Market Need	Desirable Performance	Technology Performance Criteria	Technology Feasibility Point	Confidence Level	Ultimate Performance Criteria	Confidence Level
	Significantly more copies per minute	High evaporation rate for faster ink drying	30 microsecond drying	50%	2 microsecond drying	<30%
	Significantly more copies per minute	Infrared Sensitive ink for faster drying	Infrared absorbency by test 41X >2.5%	<50%	Infrared absorbency by test 41X >75%	<30
Higher Productivity		High speed paper transport	25 feet per second	70%	195 feet per second	<30%
	No paper jams	Less static generation to reduce jamming	<50 volts at 50% relative humidity by test #21	50%	<2.1 volts at 50% relative humidity by text #21	<30%
Environmentally friendly	No waiting	No warm up time	1 second preheat	50%	0.1 second preheat	<30%
	No dangerous fumes	No environmentally restricted solvents	0.0 ppm hydrocarbons	30%	0.0 ppm hydrocarbons	30%
Lower Operating Costs	Longer wearing brushes	Maintains stiffness	<20% stiffness loss after 1 year	90%	<10% stiffness loss after 1 year	50%

and key technology experts identified on a worldwide basis. Key technology individuals from the senior management team are sometimes invited as well.

Most of the technology experts are academics, since technology experts working at competitor companies are excluded due to conflicting interests. In many cases, the technology experts are individuals working within the company but in other divisions. All external technology experts are required to sign a confidentiality agreement that assigns intellectual property rights to the company. In the majority of cases this is not a problem. External technology experts unwilling to sign such an agreement are not included in the meeting. Typically, the meeting will have two to three technology experts.

The best way to identify technology experts involves talking to individual gurus, scanning the literature, searching the Internet, consulting with in-house colleagues, and reviewing the inventors listed on patents. Attending a technology meeting where technology experts gather is often ideal. One such example is the American Chemical Society meeting. However, the timing of the technology attack team effort typically does not allow for such fortuitous events to occur. After several weeks, a pattern usually emerges where multiple sources converge on a few technology experts. Telephone calls, or in some cases individual visits, will allow the team to review the project with the identified technology experts in order to validate that they can provide value.

Although each technology experts meeting is different, they generally fall into two types. The first type is a comparison of a technology pathway with those of competitors. For example, a company was comparing the relative merits of integrated circuit cleaning solutions against its competitors. The second type involves understanding the technology hurdles utilizing Technology Performance Tables. An example is shown in Table 6-5, where a company is trying to develop the next-generation office copy machine. The first type is competitor focused, while the second is customer focused.

Table 6-7A shows the agenda for the first type of technology experts meeting—a technology pathway agenda. The first day is concentrated on sharing of information. Initial discussions on the subject are presented by each of the technology experts. The technology approach of the leading competitors, its limitations, and the competitors' intellectual property position are then reviewed. The final presentation is the company's technical approach, its limitations, and intellectual property position. This last presentation would include why team members expect their approach to be superior to that of the competition. The remaining part of the meeting is a general discussion, followed with dinner, which is held in a private dining room so that confidential discussions may continue. The meeting then continues on the following day. Each of the consultants is asked to prepare a 30-minute presentation for the following morning. Consultants are asked to prepare their presentations separately to prevent *groupthink*. Overall recommendations are then developed after the presentations.

The second type of technology experts meeting (see the agenda in Table 6-7B) begins with a review of the Technology Performance Table (see Table 6-5), the technologies used to meet them, and both the technology

TABLE 6-6.
Anchored Scales for the Technology Confidence Levels Used in Table 6-5

Level	Overall Expression	Influencing Variables	Information Sources
<30%	Uncertain	Totally uncontrollable, many unknown variables and unpredictable experimental results	Instinct and intuition. Belief of the technology team and few if any experiments
30–50%	Possible	More uncontrollable than controllable, some unknown variables and low predictability of experimental results	Experience in a few analogous areas, some preliminary experiments
50–70%	Probable	Few not controllable, few unknowns and moderate predictability of experimental results	Extensive experience, theoretical and experimental foundation combined with broad internal input
70–90%	Highly Probable	Controllable, most variables are known and understood and the experimental results are predictable	Preliminary database, independent confirmation with broad mult-functional internal input.
>90%	Certainty	Totally controllable, variables are known and understood and the experimental results have been reproduced.	Large database and familiarity, multi-independent confirmation with broad multi-functional external input.

feasibility points and ultimate performance criteria. The technology experts then separately vote on their confidence levels. A discussion is then conducted to gain consensus around the confidence level. The meeting topic then moves to a discussion of the areas of alternate technologies and solutions that may increase the lowest confidence level. Each row of the table that is below the 70 percent confidence level is discussed. The experts are asked to make their 30-minute presentations on their recommendations the following morning, with overall recommendations discussed at the conclusion of the meeting.

The experts' meeting usually prompts the need for additional experiments to better understand the hurdles, to resolve issues around the confidence level when there are wide discrepancies, to learn about alternate solutions, and to try different technologies that might accomplish the same goal. The remaining time in Phase 3 is spent on these issues. An example would be a technology attack team effort focused on identifying an adhesive that would allow a device to stay attached to wet, hairy, and irregular skin surfaces for several days without causing degradation in fragile elderly patients. The experts identified this as the key hurdle for the project to continue and decided that the team did not adequately address it during the experts' meeting. The team ended up doing a number of experiments with different adhesives in order to determine

TABLE 6-7.
Generic Agendas for Technology Experts Meeting

TABLE 7A—Technology Pathway Agenda

Agenda Item	Description	Allotted Time
1. Introduction (Day 1)	Introduction and discussion of the objectives of the meeting	30 minutes
2. Expert Perspective (Day 1)	Current and Future View of Technology Solutions (30 minute presentations by each of the technology experts)	90 minutes
3. Competitor I (Day 1)	Presentation of Technology Approach Used by Competitor I	1 hour
4. Competitor II (Day 1)	Presentation of Technology Approach Used by Competitor II	1 hour
5. Competitor III (Day 1)	Presentation of Technology Approach Used by Competitor III	1 hour
6. Company Approach (Day 1)	Presentation of Technology Approach Hypothesized by Company	1 hour
7. Open Discussion (Day 1)	Discussion	Remaining Time
8. Expert Perspective (Day 2)	Separate recommendations by each of the consultants	90 minutes
9. Overall Recommendations (Day 2)	Discussion of Final Recommendations	Remaining Time

TABLE 7B—Technology Performance Agenda

Agenda Item	Description	Allotted Time
1. Introduction (Day 1)	Introduction and discussion of the objectives of the meeting	30 minutes
2. Discussion of the Technology Performance Tables (Day 1)	Detailed discussion of the confidence levels detailed in the Technology performance table (see Table 6-7).	2–3 hours
3. Discussion of area with the lowest confidence levels (Day 1)	Discussion of experiments which may be performed to test the assumptions of the confidence level	2–3 hours
4. Discussion of area with next lowest confidence level (Day 1)	Discussion of experiments which may be performed to test the assumptions of the confidence level	2–3 hours
5. Expert Perspective (Day 2)	Separate recommendations by each of the consultants	90 minutes
6. Overall Recommendations (Day 2)	Discussion of Final Recommendations	Remaining Time

the technology risks and clinical experiments that must be discussed in the business case for continuing the project.

Phase 4: Business Case Generation

Much like the market attack team, the technology attack team will develop a business case that will be presented to senior management for a decision. A challenge workshop is also held—though in this case it is often attended by many of the technology experts who worked with the team in Phase 3.

ENGELHARD LESSONS LEARNED

Visible senior management commitment to the project is a must. This empowers the team. Without it, team members will be more inclined to do their regular job rather than work on the attack team, which involves a significant time commitment and often a chaotic challenge.

The team must have a clear charter with project scope, mission, timing, and deliverables. The chartering process is a critical first step. Before senior management will agree to charter a market attack team, it must first be convinced that the opportunity is worthy of significant effort. Typically, a single champion or a small team must first do enough work to develop a case for action and sell management on the idea that the project should be pursued further. Once that is complete, multiple discussions are held with senior management to clearly define the scope of the project, timing, deliverables, and so on.

A dedicated group effort is essential. One of the greatest challenges in launching a market attack team is defining the team membership. It's always a challenge to extract key commercial and technical people from operating businesses for three months. A dedicated team of high-caliber people is a must. If you don't have this, don't charter the market attack team.

Frustration, confusion, and lack of clarity are all hallmarks of a team kick-off. It's OK; this is normal. However, a strong team leader is vital for getting the team through the rough spots.

Interviewing multiple lead users, customers, and external and in-house experts is critical. Through these interviews, a clear understanding of the market can be developed.

It is important to keep an open mind when first studying the market/opportunity, because with just a little bit of information, the tendency is to jump to `` the answer'' or try to make the market input fit a preconceived idea.

The challenge workshop improves the team's output. Having others review the plan improves the strategy and content as well as clarifies the message. Having a clear concise message with key recommendations and resource request is critical to gaining senior management support.

Market attack teams generate comprehensive facts that allow for confident, effective decision making.

CONCLUSIONS

The market attack team, as well as its cousin, the technology attack team, is a four-phase process for allowing companies to develop new platform projects in markets and or technologies unfamiliar to the company. The market attack team approach typically challenges an organization. Entirely new concepts, sometimes even larger than the original concept, often emerge at the end of the market attack team effort, even though 50 percent of the market attack team efforts are not continued after the three months. These new concepts form the basis for a subsequent market attack team. The benefits of this approach are that real actionable recommendations with an understanding of the risk/reward profile are developed. Rich information on emerging and future needs is typically uncovered.

WHY HAVE MARKET ATTACK TEAMS TAKEN HOLD, AND HOW HAVE THEY BEEN SUSTAINED AT ENGELHARD

By the winter of 2006, Engelhard had completed seven market attack teams and one technology attack team. Senior management had found that the detailed market analysis, competitive assessment, the products and technology, and the growth potential, along with the team's assessment/recommendations, allowed Engelhard to more effectively make decisions about entering new markets.

Put another way, the teams came back with such extensive market knowledge and understanding that senior management was confident in making decisions. About half of the team efforts have resulted in senior management approving continued efforts that are ongoing today; the other half led to fast abandonment of the concepts that were explored by the teams.

Ownership of the market attack process was retained by Engelhard's business development group. In almost every case, the market attack team leader was a business development manager from this group. This assured that a reasonably consistent process was followed as well as consistent output.

However, this approach is quite challenging. Assigning full-time or almost full-time people to a market attack team *always* represents a hurdle. The key people in organizations who you would like to participate on the market attack team are always fully booked. Efficiently finding and gaining access to customers, lead users, and experts always represents a challenge for the team during the three-month effort. In addition, the Phase 2 effort seems quite chaotic, with lots of visits to many customers—all of whom at times seem to be leading the team down multiple paths. Translating all of this information into a coherent story presents a continuing challenge.

Teams could always use more time. Our experience indicates that the team gains enough market knowledge during the three-month interval to make an informed recommendation, however. It is difficult to predict the pathway from start to finish. In the majority of the cases, the preconceived notions of where the final business case would end up differed from what was initially envisioned.

Further, the process stretches people. Everyone on the team is expected to perform at his or her highest level. The weaknesses of mediocre performers become highlighted very early in the process. In some cases, the team com-position changes because the weak team members begin to bring down the overall effort.

Ultimately, the success of the market attack team effort relies on the people chosen to be on the team and their ability to focus significant amounts of time, energy and passion on a particular project for a three-month interval of time. The approaches outlined in this chapter provides a methodology for guiding and accelerating the thought process of the team and its company into entirely new platform projects, which in many companies never happens.

REFERENCES

Ajamian, G., and P.A. Koen. 2002. Technology Stage Gate: A Structured Process for Managing High Risk, New Technology Projects, In P. Belliveau, A. Griffin, and S. Somermeyer, eds. *PDMA Toolbook for New Product Development*. New York: John Wiley and Sons, 267–295.

Amabile, T., C. Hadley, and S. T. Kramer. 2002. "Creativity under the Gun," *Harvard Business Review* August: 3–11.

Cooper, R. G., S. J. Edgett, and E. J. Kleinschmidt. 2001. *Portfolio Management for New Products*, 2nd ed. Cambridge, MA: Perseus Publishing.

Govindarajan, V., and C. Trimble. 2005. *10 Rules for Strategic Innovators*. Boston: Harvard Business School Press 2005.

Kim, W. Chan, and Mauborgne, Renée. 2005. *Blue Ocean Strategy*, Boston: Harvard Business School Press.

Koen, P.A., G. Ajamian, R. Burkart, A. Clamen, J. Davidson, R. D'Amoe, C. Elkins, K. Herald, M. Incorvia, A. Johnson, R. Karol, R. Seibert, A. Slavejkov and Wagner. 2001. "New Concept Development Model: Providing Clarity and a Com-mon Language to the 'Fuzzy Front End' of Innovation," *Research — Technology Management*, 44 2 (March–April): 46–55.

McGrath, M.E., *Product Strategy for High-Technology Companies* 2nd ed. McGraw-Hill, NY 2001.

Meyer, M. H., and L. Leonard. 1997. *The Power of Product Platforms*. New York: The Free Press.

Meyer, M. H., and P. C. Mugge. 2001. "Make Platform Innovation Drive Enterprise Growth." *Research Technology Management*, 44 (1): 25–39.

Narin, Francis, and Anthony Breitzman. "Inventive Productivity," *Research Policy* 24: 507–519. Whitney, Dean. 1997. "Corporate New Ventures at Proctor & Gamble." Harvard Business School, Case 9-897-088.

7

Segmenting Your Market so You Can Successfully Position Your New Product

Brian Ottum
President, Ottum Research & Consulting

If you want to be successful in archery, you must do four things well. First, you must be able to see the target clearly. Second, you must aim carefully. Third, you must understand how to release correctly. And finally, you must actually hit the bull's-eye. Successful new product development requires these exact same steps. This four-step process is called *STUP:*

Step 1: Segmentation—What customers are out there; which could be my target?

Step 2. Targeting—Who am I aiming for?

Step 3: Understanding—What do they need, and think of current products?

Step 4: Positioning—What do I want them to think of *my* product?

Figure 7-1 shows the four steps. The arrow signifies time, because the steps can take months to complete. The results of each step are critical to the success of subsequent steps, so overlapping is not possible.

All four steps are very important for new products. If you have not thought through the segmentation structure of the entire market, you may pick a tiny segment or one that is shrinking. If you don't consciously design for a specific customer segment, then your new product might try to be all things to all people (and beloved by none). If you don't take the time to deeply understand your chosen segment, your product might not truly meet their needs. Finally, if you don't think about positioning, you might offer an undifferentiated, "me too" product that is just like competitors' products.

This chapter will take you through a process by which you can identify key segments in your market, enabling you to develop a compelling positioning for your new product.

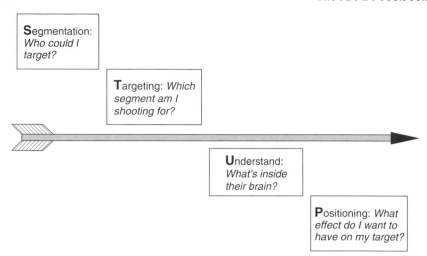

FIGURE 7-1. The four step targeting process: STUP.

AUTOMOTIVE EXAMPLE

A simplified view of the recent automotive market demonstrates the STUP process in action throughout this chapter. In this simplified auto market, you will see how five customer segments are identified (this is the S in STUP). Then you will observe how successful manufacturers have targeted these segments (T). An example survey will be used to show how you can understand (U) the driving needs of a particular segment. Finally, we will use insights into current perceptions to create a unique and compelling positioning (P) for a new product.

STUP is a task for the very early stages of new product development. Segmentation, targeting, and understanding provide the insights upon which the development stage is based. Positioning can happen in parallel to the development stage, while the new product or service is taking shape. The rest of this chapter demonstrates how to perform each part of the STUP process.

SEGMENTATION

Market segmentation is the act of dividing potential customers into groups. The groups (segments) can be defined using many criteria, but it is critical that the customers within a group share similarities with others in the same group. Target marketing is successful because specific products can be designed and marketed to specific market segments. This is better than a one-size-fits-all approach.

There is definitely a hierarchy from *crude* to *powerful* segmentation schemes. Most companies start with the conventional demographic segmentation. This uses obvious defining characteristics to clump customers into segments. In consumer products, the defining characteristics can be age, gender, geography, education, or income. Business-to-business firms might use geography, size of the firm, NAICS industry code[1] or number of employees. All segmentation schemes must contain demographics to help *find* customers, but they are insufficient to fully describe a segment by themselves. This is because two people can have the same demographics (like male, in his forties, Caucasian, and middle income), but buy strikingly different products.

At the other end of the spectrum, the most insightful segmentation is based on customer needs or the benefits that customers seek. It is very powerful to identify a group of customers who are all motivated by the same need. They are very likely to react similarly to a particular new product. Unfortunately, needs-based segmentation is difficult. Because the data do not exist in published databases, *you* have to uncover the diverse set of needs

TABLE 7-1.
Ways to Segment a Market

Segmentation Scheme	Example Segments	Pro's	Con's
Demographic	◆ Men age 18-35, ◆ Southern U.S. households, ◆ Firms over $2 million in sales	Simple, already have the data, can easily find them	A very blunt instrument for predicting reaction to your new product
Product usage and other behaviors	◆ Frequent beer drinkers, ◆ Crime drama watchers, ◆ Firms with complex buying processes	Data are concrete, some already in-house, rather easy to gather	Two customers may do same thing for very different reasons
Attitudes toward products	◆ Health conscious, ◆ Price sensitive, ◆ Design-driven	Shows why people do the things they do	Often non-actionable alone because you cannot find a given segment
Needs	◆ Need intensive PC graphics, ◆ Need a mortgage but have no documentation, ◆ Need heavy support and training after the sale	The best explanation of purchase behavior	Often non-actionable alone because you cannot find a given segment

[1] North American Industry Classification System (NAICS) replaces the Standard Industrial Classification (SIC) and groups economic activities into 20 sectors and 1,170 industries in the United States version. It was developed to provide common industry definitions for Canada, Mexico, and the United States to facilitate analyses of the economies of the three countries.

customers have with respect to your particular product. Then, you have to survey customers and develop the segmentation scheme. All too often, an elegant but useless needs-based segmentation is developed because the company failed to include any demographic criteria in the scheme—and therefore cannot tie known customer data to the descriptive segments. Table 7-1 provides a number of different potential segmentation criteria.

Data for Segmenting a Market

The first step in segmenting your market is to identify the disparate needs that customers have. You may already have this information. If not, some of the qualitative research techniques illustrated in Chapter 1 of the *PDMA Tool-Book 1 for New Product Development* and Chapters 7 and 8 of *ToolBook 2* can be used. The second step is to field a quantitative survey with customers. Writing the survey is the key step. The survey should contain the following subjects, in the following order:

1. Importance of various needs
2. Behaviors (which illustrate underlying needs)
3. Products purchased (which illustrate underlying needs)
4. Demographics

A key challenge is keeping the survey short, preferably 10 minutes or less to complete. Figure 7-2 provides a sample survey that segments the market for automobiles. It is roughly 10 minutes long. The issues raised in the survey match the automotive example used throughout this chapter. The shaded text in Figure 7-2 provides explanations for why each section is included.

Note that many different types of questions are asked: needs, products purchased, related behaviors, and demographics. All of these help in the creation of the segments. The core need assessment questions use a 1 to 10 importance scale. This is a simple way to present the needs for rating. However, if the needs list gets long, the bank of questions gets very tedious for respondents and the quality of the data can suffer. Therefore, it is important to test the fewest possible distinct and different needs (with no overlapping needs).

A frequent problem with using the 1 to 10 importance scale for questions is that many respondents will rate everything at the top of the scale. A valuable tip is to always state the needs in a strong positive tone to dissuade the use of the top of the scale. Note how this was done in the Figure 7-2 survey (*extremely* important used instead of just very important). Another way to accurately measure the relative importance of various needs is to use sophisticated conjoint questions.

Data Collection Methods

Once the segmentation survey is designed, there are many ways to collect the data. Table 7-2 illustrates the pros and cons of each method.

The key to data collection success, whatever the method, is to work from a good respondent list. In the consumer world, many vendors have well-maintained consumer databases and panels. In the business-to-business world, your customer list must be current and accurate. Usually, data collection firms ask for lists containing about 10 times as many names as the ending sample size of the survey.

It is important to plan for a sufficient sample size. There is no hard and fast rule here, except that the smallest segment should not have less than 30 respondents (the standard statistics hurdle). As a general rule, sample size must be higher if the market being segmented is large and highly diverse. For example, the market for car buyers in the United States is huge and highly diverse. By contrast, the commercial market for semi-tractors (Class 8 trucks) is more modest and more homogenous. A final sample size of roughly 1,000 would be needed to fully segment the myriad of U.S. car buyers; while a sample size of about 200 should be sufficient to segment the companies who

---

Thank you for agreeing to take our survey. Your responses will be used to help design exciting new cars in the future. Your responses will remain anonymous and confidential. This survey will take approximately 10 minutes to complete.

These are the needs assessment questions, core to the segmentation analysis.
For the following questions, please think about buying (or leasing) your next vehicle. When you are shopping for your next vehicle, please rate the importance of the various features in your decision.

	Not Important								Extremely Important	
Very high gas mileage	1	2	3	4	5	6	7	8	9	10
Room for many passengers	1	2	3	4	5	6	7	8	9	10
Room for lots of cargo	1	2	3	4	5	6	7	8	9	10
Very safe in a crash	1	2	3	4	5	6	7	8	9	10
Very quick acceleration	1	2	3	4	5	6	7	8	9	10
Very tight steering	1	2	3	4	5	6	7	8	9	10
Never needs repair	1	2	3	4	5	6	7	8	9	10
Lowest price	1	2	3	4	5	6	7	8	9	10

Here are concrete behaviors that illustrate underlying needs.
What do you *frequently* use your vehicle for? Please check the boxes below:

[ ] Haul lots of passengers (one way) [ ] Drive more than 20 miles to work
[ ] Haul lots of cargo [ ] Weekend trips (more than 4/year)
[ ] Taking children to sports and activities [ ] Drive mainly freeways
[ ] Tow a boat or camper [ ] Drive mainly suburban streets
[ ] Long trips (over 500 miles total) [ ] Drive mainly unpaved roads

These behaviors help us find the various segments when it comes time to target them.
What do you do when shopping for a vehicle to purchase or lease? Please check the boxes below:

[ ] Talk to friends or relatives [ ] Read brochures
[ ] Visit dealerships [ ] Watch vehicle-related TV shows
[ ] Read *Consumer Reports* [ ] Look at vehicles on the road
[ ] Research vehicles on the Web [ ] Look at vehicles in parking lots

FIGURE 7-2. Sample automotive segmentation survey.

What you drive today is a good proxy for what your needs are: cargo room, dependability, safety, etc.

What is the brand, model and year of the vehicle you drive most often?
Brand _____ [written in for paper survey, pull-down menu for Web survey]
Model _____
Year _____

What is the brand, model and year of another vehicle owned by your household, if any?
Brand _____
Model _____
Year _____

Needs change over time, so it's a good idea to take a forward look.

If you were to buy or lease another vehicle, what type would it be? (please choose as many types as you might consider)
[ ] Sport Utility Vehicle (like Ford Explorer or Honda Pilot)
[ ] Sedan (like Honda Accord or Toyota Camry)
[ ] Truck (like Chevy Silverado or Ford F-150)
[ ] Minivan/van (like Mazda MPV or Nissan Quest)
[ ] Performance car (like Mini Cooper or Ford Mustang)
[ ] Wagon (like Subaru Outback or Ford Freestyle)
[ ] Luxury car (like Cadillac CTS or Lexus LS430)
[ ] Hybrid (like Toyota Prius or Honda Accord)

Demographics often correlate with needs, helping us find the various segments.

Please tell us about your household:

How many licensed drivers in your household? ____ [written in for paper survey, pull-down menu for Web survey]
Your age? ____
Your gender? [ ] Female   [ ] Male
What is your marital status? [ ] Single   [ ] Married   [ ]
Divorced/Widowed/Separated
How many children do you have living with you at home? ____
What are the ages of children living with you at home? ___ ___ ___ ___ ___ ___
Do you work outside of the home? [ ] No   [ ] Yes, part time   [ ] Yes, full time
What is your household's total annual income? (remember, your responses are anonymous and confidential)

[ ] less than $40,000
[ ] $40,000 to $69,999
[ ] $70,000 to $99,999
[ ] $100,000 to $149,000
[ ] $150,000 or more

FIGURE 7-2. (continued)

buy semi-tractors. The special situation is business-to-business segmentation, when the entire universe of customers is small (often less than a hundred). In this case, finding two or three segments based on a survey is still valuable, as long as a substantial fraction (at least a third) of the universe is included in the survey.

TABLE 7-2.
Data Collection Methods

Method	Pro	Con
In-Person	High quality data, good for complex surveys	Expensive and logistically difficult
Mail	Low cost, anyone can participate	Low response rates make you wonder about those who do respond
Phone	Fast, live interviewer ensures quality	High rejection rate due to telemarketing, cannot be long or complex
Fax	Fast, good for business respondents	Have to alert respondent prior to arrival, decreasing use of fax machines
Web	Very fast, inexpensive if you have email addresses or a popular Web site	Some respondents may not have Web access, confidentiality concerns

Once the segmentation survey data are collected, sophisticated analysis is needed to identify the segments. This is a two-step process:

1. Look for underlying simple themes among the many needs questions asked in the survey. To do this, you will use the statistical tool called *factor analysis*.
2. Find groups of respondents who tended to have similar needs. For this, you will be using the statistical tool called *cluster analysis*.

The first step is *data reduction* in order to reduce a large number of survey questions down to a small number of underlying themes. Factor analysis is used to find the underlying theme by identifying the questions that tend to get answered in similar fashion. The second step is using cluster analysis to find underlying segments by identifying the respondents that tended have the same need themes. Factor analysis clumps questions, while cluster analysis clumps people.

Reducing the Data to a Manageable Size

Let's illustrate this two-step factor analysis process using a popular statistical analysis program, SPSS.[2] Imagine that data have been collected from 1,000 drivers using the hypothetical auto segmentation survey shown in Figure 7-2.

[2] Other popular statistical analysis programs are MiniTab, STATISTICA, SAS, and S-PLUS. Base SPSS software for a single user is about $1,700. MiniTab is $1,200. STATISTICA is $1,000. SAS and S-PLUS are very large enterprise programs annually licensed and used by large organizations.

The SPSS data set would have a row for each driver's answers. Each column would contain the answers to a single survey question. So the data set would look much like a Microsoft Excel spreadsheet, with 1,000 rows (people who responded) and over 50 columns (their answers for each question).

As stated earlier, the most powerful segmentation is based on customer needs. In the survey, there are two sets of questions that inform us about needs. There are the eight core needs assessment questions (with the 10-point scales). Immediately following are 10 behaviors, which also reveal needs. So we'll include all 18 questions in our two-step process.

The first step is to conduct an SPSS factor analysis of the 18 questions by clicking on the "Analyze" menu at the top, and then selecting "Data Reduction," and finally, "Factor." This opens a menu box that allows the highlighting of the 18 questions from the left-hand list. Click on the right arrow, and the 18 questions are now variables in the factor analysis. The other choices at the bottom of the menu box can usually be left at their defaults, except for two. Click on "Rotation" and select "Varimax." This allows us to see the most distinct and different themes (the technical reason is to get orthogonal dimensions). Click on "Options" and select "Suppress absolute values less than..." Then enter something in the range of 0.2 to 0.3. This simplifies the output, allowing a focus on only the most meaningful results. Finally, you can click "OK" in the Factor Analysis menu box, and the program will very quickly produce the output shown in Table 7-3.

TABLE 7-3.
Factor Analysis Results

Question	Factor 1	Factor 2	Factor 3	Factor 4
Very high gas mileage	−0.6		0.8	
Room for many passengers	0.7	−0.7		
Room for lots of cargo	0.6	−0.4		
Very safe in a crash	0.4		0.3	0.8
Very quick acceleration		0.8	−0.3	
Very tight steering		0.7	−0.4	
Never needs repair			0.7	
Lowest price		−0.4	0.6	
Haul lots of passengers	0.8	−0.5		
Haul lots of cargo	0.7			
Taking children to sports and activities	0.7		0.4	0.3
Tow a boat or camper	0.5	−0.9		
Long trips (over 500 miles total)	0.5			
Drive more than 20 miles to work (one way)			0.6	0.4	−0.3
Weekend trips (more than 4/year)	0.6			
Drive mainly freeways		0.3		
Drive mainly suburban streets		0.4	0.4	
Drive mainly unpaved roads	0.3			−0.3

Table 7-3 is the *factor-loading table*. It shows how well the answers to specific questions correlate with the underlying theme (whose statistical name is factor). The higher the loading, the more the question defines the underlying theme. Negative loadings mean that people who tended to answer highly on the question feel very low on the underlying theme (or vice versa). The factor loadings in Table 7-3 show us that four distinct and separate themes (factors) are emerging from the data. Some entries are missing, because anything less than 0.3 is suppressed.

The program does not name the factors/themes. This is done by observing which questions had the highest loadings (in absolute value). In the automotive example, Factor 1 is defined by "Room for many passengers," "Room for lots of cargo," "Haul lots of passengers," "Haul lots of cargo," "Taking children to sports and activities" and "Weekend trips (more than four a year)." So the name "Need to haul lots of people and stuff" can be given to factor one. The other two factors are named using the same procedure. Factor two is defined mainly by the "Very quick acceleration," "Very tight steering," and NOT "Tow a boat" or "Room for many passengers" questions. So a good name for factor two is "Need great performance." Factor three can be called "Need practical transportation" because it is high for "Very high gas mileage," "Never needs repair," and "Lowest price." Factor four can be called "Demand safety" because it is, by far, the highest on "Very safe in a crash." Table 7-4 shows the four names.

Once some clear underlying themes have been successfully identified, it's time to run the factor analysis procedure again (Analyze > Data Reduction > Factor). The previous selections are still there, including the 18 variables. This time, select "Scores... ," followed by "Save as variables," and finally "Continue." The same output will be repeated, but this time the program will add four additional columns to the data set (one column for each of the four factors). Looking down the rows, you could see that the numbers range from roughly −2 up to +2 (they are standardized so they average 0 and have a standard deviation of 1). The larger negative numbers mean that that particular respondent *did not* have that need. The larger positive numbers mean that that particular respondent *did* have the need. Numbers near zero mean the respondent has just typical need. So now there is an estimate of the degree to which each respondent has the four underlying needs. Label the new factor columns in the data with their descriptive names.

TABLE 7-4.
Example Factors

	Factor 1	Factor 2	Factor 3	Factor 4
Name of Factor	"Need to haul lots of people and stuff"	"Need good performance"	"Need practical transportation"	"Demand safety"

Finding Similar Respondents

The second step in segmentation is to cluster the respondents to find those respondents who share the same needs. There are several statistical tools available, including k-means and hierarchical clustering, regression, discriminant analysis, and automatic interaction detection (AID). The technical discussion of each method is beyond the scope of this chapter. However, the k-means clustering method is popular, so an SPSS example is used here.

To cluster the respondents, the first step in SPSS is to pull down the "Analyze" menu, select "Classify," and then "K-Means Cluster." Scroll down the list of variables at the left to select the final four variables created in the factor analysis, which are your need themes. Press the right arrow button to place these four as variables in the clustering analysis. Finally, the number of clusters to create needs to be specified. An iterative approach works best. First, select the largest number of clusters that might be feasible, given the size of the data. Segmentation schemes containing more than about six individual segments are rare unless the data set runs into the thousands. However, a sample of 200 companies who buy semi-tractors might be big enough to find only three segments. Given our example of 1,000 car drivers, a good first try is six clusters. The results of this first clustering run are shown in Table 7-5.

Some respondents (50) did not answer all of the questions in this analysis, so they are "missing" and not included. Note that cluster #5 contains only 11 respondents. For statistical reasons, the desired minimum is about 30. So one can reset the number of clusters to five and rerun the analysis, giving the results shown in Table 7-6.

TABLE 7-5.
Number of Cases in each Cluster

Cluster		
	1	114
	2	272
	3	140
	4	238
	5	11
	6	175
Valid		950
Missing		50

TABLE 7-6.
Number of Cases in each Cluster

Cluster		
	1	114
	2	275
	3	142
	4	238
	5	181
Valid		950
Missing		50

TABLE 7-7.
Final Cluster Centers

	Cluster				
	1	2	3	4	5
Need to haul lots of people and stuff	1.12	.28	−1.04	−.25	.10
Need good performance	.55	.25	1.59	.95	−.46
Need practical transportation	−.62	1.22	−1.48	.83	.20
Demand safety	.49	.32	−.29	−.36	.97

TABLE 7-8.
Statistical Clustering Results

Cluster #1: "**Experience Seekers**" – highest score of any cluster for "need to haul lots of people and stuff," also rather high for "performance" and "safety" (114/950 = 12% of all drivers).

Cluster #2: "**Practical and Pragmatic**" – these drivers are the highest on the "need practical transportation" but also have all the other needs as well (275/950 = 29% of all drivers).

Cluster #3: "**Will Pay for Performance**" – these drivers have, by far, the greatest need for performance, and they don't require "practical" transportation which is low priced and high mileage. They are also least concerned about safety. (142/950 = 15% of all drivers).

Cluster #4: "**Affordable Performance**" – these drivers are not quite as willing as the previous cluster to trade off low price for performance (238/950 = 25% of all drivers).

Cluster #5: "**Safety Conscious**" – these drivers are highly focused on safety 181/950 = 19% of all drivers).

Now there are a sufficient number of respondents in each cluster to fully define a segment. The key output table is shown in Table 7-7. The numbers are the degree to which the average person in each cluster feels he or she has the particular need. Higher numbers mean a greater need.

The clusters can be interpreted and named just like the factors before them. These clusters are the market segments, shown in Table 7-8.

TARGETING

Targeting is picking the market segment that will be the focus of your new product efforts. The most effective targeting requires a careful matching of company with customer. The following questions are pertinent on the company side:

◆ What is the overall corporate strategy?
◆ What are the company's strengths and weaknesses?
◆ What is the goal of the new product program?

On the customer side, the following questions need to be answered:

♦ Which segments are growing?
♦ Which segments have the most buying power?
♦ Which segments have the greatest need?

Once these questions are answered, then the number-one targeting question can be answered:

♦ Which segment(s) *need* what I will offer?

In our ongoing auto example, five segments of U.S. drivers have been identified, based on what they need in a car. In order to further profile each segment, let's add some demographic and behavioral results from the segmentation survey. This is done by looking at how each of the five segments differs from each other demographically. Figure 7-2 shows that the demographic questions were asked at the end of the survey. The answers to these demographic questions are critical if we are to ever find the five segments in the larger world beyond just those in survey.

Table 7-9 can be used to show how auto manufacturers have targeted these customers for their specific vehicles:

Ford Explorer Sport Utility Vehicle—Experience seekers are the obvious target market because those people haul lots of people and stuff, and say they planning on buying a SUV for their next vehicle. These households are the most likely to have children and have higher incomes.

Toyota Camry and Honda Accord—These two sedans offer good performance, good ability to carry passengers, good styling, good gas mileage, good reliability, good safety, and reasonable pricing. The Practical and Pragmatic segment is the target market for these two vehicles, because these customers value practical transportation but also demand safety and a moderate ability to haul people and stuff. The Practical and Pragmatic profile is skewed strongly female, has children at home, and is most likely to be a *Consumer Reports* reader.

BMW Sedans—These are high-performance vehicles that cost significantly more than the average. The Will Pay for Performance segment is the obvious target market. This segment is very likely to be shopping at the dealership as opposed to researching new vehicles on the web, and very likely to be male (and not have children at home).

Pontiac Sedans—These are sporty vehicles that are competitively priced. They are well matched to the Affordable Performance segment, because the segment has the lowest median household income (yet they still value performance). Pontiac can find these shoppers on the Web.

Volvo Station Wagons—The Safety Conscious segment has helped the Volvo brand be successful by buying these extremely safe vehicles. Volvo can find these shoppers on the Web. They are primarily women.

TABLE 7-9.
Segment Profiles

	Segment				
	Experience Seekers	Practical and Pragmatic	Will Pay for Performance	Affordable Performance	Safety Conscious
Need to haul lots of people and stuff	1.12	.28	−1.04	−.25	.10
Need good performance	.55	−.25	1.59	.95	−.46
Need practical transportation	−.62	1.22	−1.48	.83	.20
Demand safety	.49	.32	−.29	−.36	.97
Type of vehicle to be bought/leased next:	SUV SUV	Sedan or Hybrid	Luxury car	Performance car	Sedan
Read Consumer Reports when shopping	5%	28%	6%	11%	12%
Visit dealerships when shopping	50%	24%	76%	46%	29%
Research vehicles on the web	25%	35%	10%	55%	50%
Male/Female split	50/50	35/65	75/25	65/35	35/65
Median age	40	49	42	33	40
Children at home	80%	60%	30%	20%	50%
Median income	$70,000	$60,000	$85,000	$35,000	$60,000

Figure 7-3 graphically shows now the cars above target the various segments. It should now be clear that when a company does a good job of segmenting its market, the most promising target segments are immediately obvious. This is why so much of this chapter was devoted to the details of segmentation.

UNDERSTANDING

Once a company has decided on its target segment(s), it is not yet ready to develop a positioning statement. This is because it is usually lacking answers to two critical questions:

1. What are my target customers' needs, in their own words?
2. What do customers think of current products in the market?

The typical segmentation survey is usually too superficial to uncover answers to these two questions. Therefore, after making a targeting decision, the company needs to study its chosen segment(s).

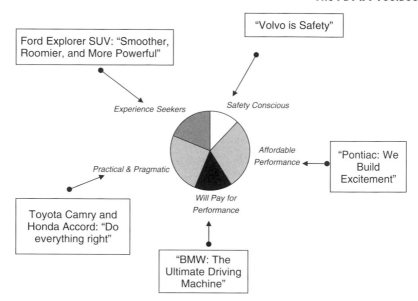

FIGURE 7-3. Targeting in the auto industry.

The first goal of the understanding phase is to understand the needs (met and unmet, stated and latent, current and future) of each target segment. The company will use this information to develop a positioning statement that resonates with target customers. The positioning statement needs to be specific, personal, and compelling enough to stimulate sales.

In order to understand target customer needs, the company must conduct focused and in-depth qualitative research. The tools to do this are covered in Chapter 1 of *The PDMA Toolbook 1 for New Product Development*, and Chapters 7 and 8 of *The PDMA Toolbook 2 for New Product Development*. All of the tools allow insight into the hopes, fears and language of the target customer. Only after immersion into the target customer's world can new product developers competently create a positioning statement.

The second goal of the understanding phase is to understand the current structure of the market. This is not the structure as the company sees it. This is the structure as the customer sees it. This understanding is critical, so that the resulting positioning builds on current perceptions and leverages current perceived strengths.

The perceived structure of the market can be superficially uncovered using the same in-depth qualitative research already explained to find the unmet needs. However, good understanding requires quantitative data. Another survey is necessary—a current perceptions survey. The deliverable from such a survey is hard data showing how customers view products currently in the market. The results can be used to conduct a *TOWS* situational analysis: to identify threats (T) and opportunities (O) in the external environment, assessing your current product's weaknesses (W) and the strengths (S) of your competitors' products.

Let's illustrate the market structure survey by continuing with our automotive example. As you may recall, five segments were found. Let's assume that a foreign automaker, new to the U.S. market, uses the segmentation to target the Affordable Performance segment. Also assume that the automaker did in-depth interviews to understand the Affordable Performance drivers. Over the course of the interviews, it would uncover about 20 to 35 attributes that are important in a new car and the buying experience. Let's make our example simpler by considering only seven:

1. "The car is an extension of my personality."
2. "The car makes a statement about me."
3. "The car has edgy styling."
4. "The car looks fast."
5. "The car has a quick 0 to 60 time."
6. "I can afford the payments."
7. "I don't have to haggle over the price."

Now the goal is to uncover the structure of the car market, from the perspective of the Affordable Performance driver. A copy of the hypothetical survey is in Figure 7-4.

Note that only the Affordable Performance segment of drivers is included in this survey. Only the questions that discriminate the most are used, based on the segmentation survey. The questions that identify the Affordable Performance segment best are age (young), gender (mostly male), high importance of performance, and high importance of low price. Respondents who don't fit the profile are terminated from the survey.

Table 7-10 shows the hypothetical results from the perceptions survey (1 = does not apply, 10 = applies perfectly). Under each vehicle are the average ratings it receives for each attribute. Note that the Tiburon is fantastic for "I can afford the payments," while the Beetle is poor for "The car has a quick 0 to 60 time."

Conventional bar charts of these results can be used to understand how target Affordable Performance drivers perceive the affordable performance car market. However, there are a lot of data here. What is needed is a concise, one-page graphical summary of the market structure.

The statistical tool for boiling down a lot of data to one chart is the perceptual map, shown in Figure 7-5. A perceptual map will boil down our seven questions into a simple two-dimensional chart. There are many different types of perceptual maps, but they all seek to do this *boiling down*. All use some variation of the multidimensional scaling method. For this example, the multidimensional unfolding procedure within SPSS will be used, which attempts to find a common quantitative scale that allows the visual examination of the relationships between the set of cars and the set of attributes.

The first step in the production of the perceptual map within SPSS is to input Table 7-10. Then pull down "Analyze" and select the "Scale" option

Thank you for agreeing to take our survey. Your responses will be used to help design exciting new cars in the future. Your responses will remain anonymous and confidential. This survey will take approximately 10 minutes to complete.

Quickly identify those "Affordable Performance" drivers who are very likely to be young males, while terminating everyone else.

How old are you? ____

__Male   __Female

Find drivers who don't need room for passengers, but want acceleration, steering, low price. Terminate the rest from the survey.

For the following questions, please think about buying (or leasing) your next vehicle. When you are shopping for your next vehicle, please rate the importance of the various features in your decision.

	Not Important								Extremely Important	
Room for manypassengers	1	2	3	4	5	6	7	8	9	10
Quick acceleration	1	2	3	4	5	6	7	8	9	10
Tight steering	1	2	3	4	5	6	7	8	9	10
Low price	1	2	3	4	5	6	7	8	9	10

From here on, only target "Affordable Performance" drivers should be completing the survey.

For the following questions, please rate how well each statement applies to each car. Use the pull-down menu to select anything from "doesn't apply at all," to "applies perfectly"

	Hyundai Tiburon	Honda Civic	Mini Cooper	Volkswagon Beetle	Ford Mustang

"The car is an extension of my personality"
"The car makes a statement about me"
"The car has edgy styling"
"The car looks fast"
"The car has a quick 0-60 time"
"I can afford the payments"
"I don't have to haggle over the price"

Thank you for taking the time to complete our survey.

FIGURE 7-4. Sample automotive perceptions survey.

and then "Multidimensional Unfolding." The resulting "Multidimensional Unfolding" dialog box has many options. Select the five cars in the left-hand list and click them to the right into the "Proximities" box. Select the labeled attributes (first column in Table 7-10) and click it to the right into the "rows" box (this puts the labels on the resulting perceptual map). Then under "Model" select the "Similarities" option and "Smooth." This is because the cars need to be analyzed based on their similarities rather than their differences. Finally, click "OK" and the perceptual map in Figure 7-5 will be produced.

The five cars are spatially located throughout the map. The closer together they are, the more alike they are in the mind of the Affordable Performance

TABLE 7-10.
Perceptions Survey Results

Attribute	Hyundai Tiburon	Honda Civic	Mini Cooper	Volkswagen Beetle	Ford Mustang
"The car is an extension of my personality"	7.1	6.0	8.9	8.4	7.5
"The car makes a statement about me"	5.0	5.5	8.4	8.0	8.9
"The car has edgy styling"	6.0	4.5	5.9	5.0	6.0
"The car looks fast"	7.8	5.5	4.5	4.5	9.5
"The car has a quick 0–60 time"	4.5	5.9	3.5	2.5	9.9
"I can afford the payments"	9.8	8.2	7.6	4.9	4.0
"I don't have to haggle over the price"	6.5	3.9	6.5	7.5	2.9

driver. The seven attributes are represented as circles. They are also spread out throughout the map. The closer together they are, the more they tend to represent similar features in the mind of the driver. Most importantly, the closer a car appears to an attribute, the more that car embodies that attribute.

Here's what can be learned from the map:

1. The Hyundai Tiburon and Honda Civic are perceived as similar cars. They are located in close proximity to each other on the map. The reason they are located well away from the others is that they both do very well on "I can afford the payments."
2. The Volkswagen Beetle and the Mini Cooper are perceived as similar cars. They are located in close to each other on the map. They both seem to do well for "The car is an extension of my personality." The target segment also seems to think that they "don't have to haggle over the price" of these two vehicles. However, because these two cars are located well away, they are not perceived to have "a quick 0 to 60 time" or "look fast."
3. The Ford Mustang sits out by itself. It seems to be the best for performance ("0 to 60" and "looks fast"). It does poorly for "no haggle."

Look back at the table of values and you can confirm these observations.

The foreign manufacturer contemplating entering the U.S. sporty car market, targeting the "Affordable Performance" segment, now knows what its target customer thinks of current competition. But there's even more to learn from this perceptual map. There appear to be three "holes" in the market where current performance cars don't exist (Figure 7-6). These represent opportunities for new product positioning.

FIGURE 7-5. Perceptual map of sporty cars.

Opportunity 1 is a car that would combine the Mustang's performance with the Mini or Beetle's personality. It would have better styling. It would also be a bit more affordable. This car offers a good mix of styling, performance and price (it's near the center).

Opportunity 2 is a car that has a low price, like the Tiburon and Civic (but perhaps not quite as low). The car wouldn't require haggling at the dealership. The car would not have to have extremely quick acceleration. This opportunity is a less expensive Mini Cooper.

Opportunity 3 is a car combines the Mustang's performance and "looks fast" appearance with the Civic's affordability.

In our automotive example, the new foreign manufacturer is ready to develop a positioning for its new sporty car, targeted to the Affordable Performance segment.

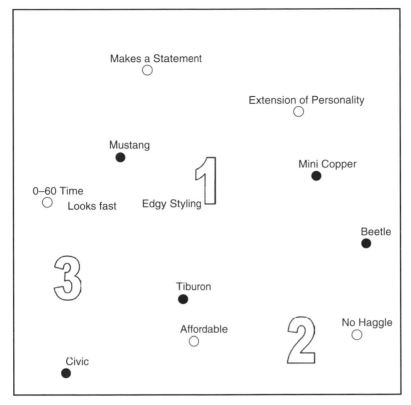

FIGURE 7-6. Perceptual map of sporty cars with market ``holes''.

POSITIONING

The positioning of a new product is the impression the company wants to create within the mind of the customer. It is a succinct statement of the nature and benefits of the new product. The positioning statement is a communication between the company and the customer. It needs to satisfy both the needs of the company and the customer. The company needs the positioning to be an integral part of its corporate and brand strategy. But the customers won't care about the new product unless the positioning resonates with them personally.

In our automotive example, one can see that the overall Toyota and Honda brands have positioned themselves as the "dependable" alternative. The Pontiac brand is "exciting" for drivers. Volvo is "safe." These are overall brand positionings. The positioning for a new product is much more specific. For example, the reborn 2006 Dodge Charger was positioned as an affordable, stylish sedan with muscle car roots.

One thing about positioning is that it's all relative. All products are positioned against each other—faster, cheaper, better, and so on. That's why the perceptual map is such a useful tool for understanding customer-perceived positioning. It shows where products fit in against each other.

Four factors are important in positioning a new product:

1. You must create and clearly communicate the positioning. Leave nothing open to interpretation.

2. A positioning must be based on product factors that are important to the customer. You position a new product by trumpeting features that customers care about. These important factors should be uncovered using qualitative research during the understanding phase.

3. Research[3] has shown that new products must be unique and superior in order to be successful. The positioning should proclaim the uniqueness and superiority.

4. Holes in the perceptual map can point to opportunities for new product positions. But, of course, the new product must live up to the promises.

A good place to start when writing the positioning statement for your new product is to use the following template.

> For [target segment], the [new product] is
>
> [positioning claim] because [single most important support].

Going back to our automotive example, let's imagine the foreign automaker wishes to leverage the "2" opportunity in the Figure 7-6 perceptual map. Assume that it has developed a new car that matches the "2" position: a two-door coupe, moderate performance, great styling and a low preset price. Its positioning statement could be the following.

> *For the driver who seeks affordable performance, the new Galaxy is the best value in stylish coupes because it offers low "no haggle" pricing.*

WHEN TO STUP?

Segmentation, targeting and understanding are critical early steps in the new product development process. The answers provided by these steps are required *before* developing a successful new product. Figure 7-7 shows how the STUP process lines up with the conventional Stage-Gate™ process often used in NPD. Positioning is best developed simultaneously with the development of the new product or service.

Keys to STUP success:

◆ Segment your market based on what they need.

◆ Target a group of customers that need what you could offer.

---

[3] Cooper, Robert G. 2001. *Winning at New Products*. 2nd edition. New York: Perseus Books.

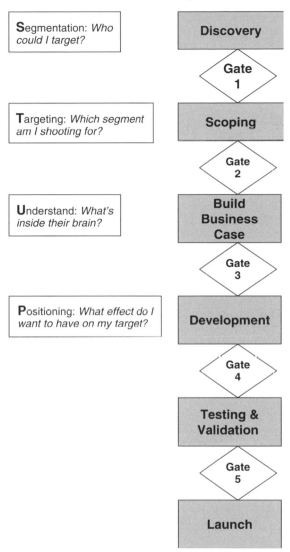

FIGURE 7-7. STUP within the stage-gate™ process.

- ◆ Deeply understand the hopes/fears/perceptions/needs of your target market.
- ◆ Combine what you know about the target market and what your new product offers to create a unique and powerful positioning.

Common STUP pitfalls to avoid:

- ◆ Segmentation based on just demographics
- ◆ Trying to find a target for your new product after you've already developed the product

◆ Using superficial techniques (like a couple of focus groups) to understand your target market

◆ Assuming your new product's position in the market is obvious, and therefore not articulating a positioning

CONCLUSION

It is important not to skip any step in the STUP process. Segmentation is required in order to understand the differences among your customers. Deliberate Targeting helps you keep a specific customer in mind when designing and positioning a new product. Once a target is chosen, it's time to gain a deeper Understanding of this critical customer segment. Only after these three steps are complete can new product developers successfully Position their new product. Two of the most valuable tools along the way are statistical clustering to uncover segments and perceptual mapping.

REFERENCES

Boike, Douglas G., Ben Bonifant, and Tony Siesland. 2005. "Market Analysis and Segmentation for New Products." In K. Kahn, ed. *The PDMA Handbook of New Product Development*. 2nd ed. Hoboken, NJ: John Wiley and Sons.

Cohen, Steven H. 1996. "Tools for Quantitative Market Research." In M. Rosenau, ed. *The PDMA Handbook of New Product Development*. 1st ed. New York: John Wiley and Sons.

Crawford, C. Merle. 1985. "A New Positioning Typology." *The Journal of Product Innovation Management* 2, 4: 243–253.

Davis, Robert E. 1996. "Market Analysis and Segmentation Issues for New Consumer Products." In M. Rosenau, ed. *The PDMA Handbook of New Product Development*. 1st ed. New York: John Wiley and Sons.

Green, Paul E., and Abba M. Krieger. 1991. "Product Design Strategies for Target-Market Positioning." *The Journal of Product Innovation Management* 8, 3: 189–202.

Lilien, Gary L., and Arvind Rangaswamy. 2003. *Marketing Engineering: Computer-Assisted Marketing Analysis and Planning*. Upper Saddle River, NJ: Prentice-Hall.

Moore, William L., and Edgar A. Pessemier. 1993. *Product Planning and Management: Designing and Delivering Value*. New York: McGraw-Hill.

Ottum, Brian D. 2005. *Quantitative Market Research*. In K. Kahn, ed. "The PDMA Handbook of New Product Development." 2nd ed. Hoboken, NJ: John Wiley and Sons.

Weinstein, Art. 2004 *Handbook of Market Segmentation: Strategic Targeting for Business and Technology Firms*. 3rd ed.. Binghamton, New York: Hayworth Press.

Belliview, Paul, Abbie Griffin, and Stephen Somermeyer. 2002. *The PDMA Toolbook 1 for New Product Development*. New York: John Wiley and Sons, Inc.

Belliview, Paul, Abbie Griffin, and Stephen Somermeyer. 2004. *The PDMA Toolbook 2 for New Product Development*. New York: John Wiley and Sons, Inc.

Wilson, Edith. 1996. "Market Analysis and Segmentation Issues for New Business-to-Business Products." In M. Rosenau, ed. *The PDMA Handbook of New Product Development*. 1st ed. New York: John Wiley and Sons.

WRC Research Systems, Inc. 2005. *BrandMap Software*. Downers Grove, IL: WRC Research Systems, Inc.

8

Giving Your Product the Right Name
Do It Yourself While Avoiding the Pitfalls

Leland D. Shaeffer
Managing Director, PLM Associates

James S. Twerdahl
Managing Director, James S. Twerdahl & Associates

This chapter will step through the process of naming a new product. It will address the don'ts as well as the dos of researching and choosing the name. Depending on the company and the product, this might be a relatively simple and straightforward activity that is done in-house, or an elaborate process involving outside specialists. This is written for the product manager or marketing manager who plans to lead the project using in-house resources, although an understanding of the process is useful to those engaging outside specialists or to anyone involved in the product-naming process.

The chapter starts by discussing the various categories of names you are likely to encounter and then launches into the actual naming process. It focuses on the creative process and the typical activities you will want to do yourself, and it suggests when you *might* consider engaging outside resources and specifies when you *should* use outside expertise. Resources are listed in the body of the chapter and summarized at the end.

The don'ts (pitfalls) of product naming are summarized in Table 8-1 for those who want a quick takeaway. The example at the end of the chapter provides a case study of an in-house naming project.

CATEGORIES OF NAMES

Before embarking on the actual naming process, it is useful to understand the various categories of names that may exist at your company. This will provide

TABLE 8-1.
A Summary of Product Naming Pitfalls

Failing To Anticipate The Future Uses of the Name: Names are often chosen because they are cute, humorous, or expedient when the person responsible is running out of time. These names may be unsuitable when the company expands and wants to use the equity in the product name for product extensions or related product lines, or when the company decides to sell internationally. Fiera became a controversial name when Toyota began selling it in Puerto Rico, where it meant "ugly old woman".

Not allowing sufficient lead-time for the process: It is all-too-common for product naming to be left until the last minute, as part of the market launch. Product naming can be a lengthy process, particularly if the product will be sold internationally and must be researched in several countries. While much of the naming process can be time-compressed, some of it cannot (at least not without significantly increasing risk). The naming process should be started as soon as the initial product positioning is determined, and it should be performed concurrently with product development.

Failure to allocate sufficient resources: A good naming process takes work and effort. Often, the person in charge of naming (e.g., the product manager) has many other responsibilities, and he/she fails to set aside enough time to perform the job properly.

Failure to identify the decision makers in advance: Product naming can be an emotional endeavor, everyone has an opinion, and more people than you might expect consider themselves to be stakeholders. If you do not get agreement up front on who the decision makers are—and are not—you run the risk of later stage problems and restarts.

Deciding on a name that is comfortable: The best names may be provocative and controversial at first. (*Yahoo!* was coined before the Internet became highly commercial. Imagine the difficulty that name would have today making it through a corporate screening process.) Conversely, a name that is comfortable may be safe but uninspiring and not memorable (the Graphics 100—a PC graphics card, fell into this category. The PC itself could have, but for the marketing might of IBM behind it.) Pick the name that best achieves the marketing objective and give it time to grow on you.

Using Too Many Decision Makers: Too many decision makers bog you down, lengthen the process and produce a decision that is safe but not necessarily the best. The naming decision should not be based on a democracy or consensus. Use a small group who understand the objectives of a name and who are willing to stretch out of their comfort zone. Of course, the CEO needs to participate or specify who will act on his/her behalf.

Picking an early favorite and running with it: If you become attached to a particular name, it can be tempting to focus the attention on that name and short circuit much of the testing. One risk is that the name might become disqualified late in the process, and you might need to start over if you have not developed viable alternatives along the way (particularly painful when close to the launch date). Another risk is that you consciously or unconsciously "sell" the name during the process and fail to get objective feedback from the people who matter most—customers, channel members and other partners from the target markets.

Cutting corners on the trademark attorney: The availability of on-line, "do-it-yourself" resources increase the temptation to conduct the comprehensive name search and registration process in-house to avoid outside legal fees. A qualified trademark attorney can navigate through the intricacies of the process and deal with the nuances that would not be obvious to a layperson. The downside risks can be great and certainly not worth the savings on the attorney's fees.

(continued overleaf)

TABLE 8-1.
(continued)

Failure to identify negative meanings in other cultures: The formal translation of a
  word or phrase may appear good, yet there may be slang or cult meanings that are
  negative and/or inappropriate for the message you are trying to get across. You
  should check with people in the country or ethnic group, who are current on the
  culture, not just translators in the United States who may be out of touch with
  localized trends. The Ford Pinto flopped in Brazil, where the name was slang in
  Brazilian Portuguese for small penis. Pajaro (a Mitsubishi car model, since renamed)
  is slang in Spanish for "masturbator". Additional examples appear in Table 2.
Failure to keep trademark registrations current: Once you register your trademark,
  you need to take steps to ensure you don't lose it. Observe usage rules, which vary
  by country, and keep your address current with the various registration offices so
  you receive their notices.

---

a context for your product name, set boundaries on what you should and
should not do, and help avoid confusion.

Product Name

The *product name* is the actual name assigned to your product and the
focus of this chapter. Depending on the breadth of the product line, a product may have its own name, or it may be referred to as a model number
within the line. Examples of product names are Motorola's RAZR (when
it was first introduced), Explorer (Ford) and Secret (Procter & Gamble). Of
course, if a product is successful, it usually will spawn numerous extensions
and derivatives; the original product's name then becomes the name of the
product line. Apple's iPod and Procter & Gamble's Swiffer are examples of
highly successful products whose names are now associated with extensive
product lines.

A highly complex product line may have an even more extensive hierarchy of names and models. In the example shown in Figure 8-1, a consumer
ceramic glaze company had over 1,600 products that sold under three separate brands. The main brand was organized into product *categories*, which
were subdivided into product *lines*. Within the product lines were the individual products, which themselves had further designations based on size and
package type.

If the naming and model scheme is not already established, it is important
to plan this before you actually start the naming process. Think of how the
product line might evolve over time and consider, for example, what would
happen if you added five to six brand extensions or derivative products to
the line. You don't want to spend time developing a name for a product that
should be assigned a model number, nor do you want to assign a number
when a name is most appropriate. It is much easier to do the naming with an
established naming/model architecture than to retrofit a plan after you have
named several products.

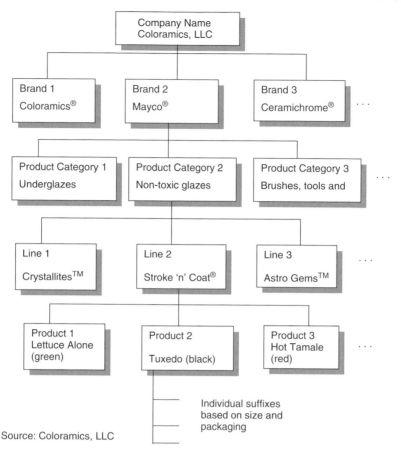

Source: Coloramics, LLC

FIGURE 8-1. An example of a complex hierarchy involving over 1600 products. It is important to establish the scheme for your situation before starting the naming process.

Model Numbers/Names

The *model name and number are* the unique identifiers for a particular product within a product family or product line. The product name applies to a range of products and gives the product line a common identity. Examples of model numbers include BMW 740i, Microsoft Outlook 2003, Levi's 501 Jeans. Dell sells desktop computers under the "Dimension" name that has several model numbers, depending on price and performance. Models may also be designated by a secondary name, such as iPod Shuffle, iPod nano, iPod, iPod U2.

Brand Name

The name of the brand under which the product is marketed is called the *brand name*. It may or may not be the same as the product name, and often there is a brand associated with a product that itself might be a brand. Examples of branded products include Symantec's Norton products (e.g., AntiVirus, AntiSpam. "AntiVirus" is the name of the product, "Norton" is the primary brand and "Symantec" is an overarching brand, also the company name). "Coca Cola" is the company name, the corporate brand, a product line, and a product itself. Often, the name of a successful product becomes a brand as a result of the success.

While there is considerable overlap in the processes for developing a product name and a product brand, the stakes are higher for the latter so the process is usually more elaborate. If you are starting the naming process with the intention of naming a brand, not just an individual product, you should plan accordingly and be more willing to engage outside expertise.

Company Name

Although the *company name* is often a brand in its own right, used to position and strengthen the product in the market, many companies prefer to keep their name out of the limelight and rely on product line brands instead. Procter & Gamble is the leading example. It has 21 brands, each having sales in excess of $1 billion.

It is useful to understand these distinctions and relationships, since the brand name and identity may place constraints on your product name, and you will want to determine whether your product needs a new name or simply a model number.

THE IMPORTANCE OF A NAME

Does a name really matter?

In 1969 while speaking at a small scientific conference, Sir Roger Penrose, a Cambridge physicist, announced his discovery of what he called a "gravitationally totally collapsed object." The world yawned.

Months later, he changed his description to a "Black Hole" and the news of his discovery raced around the world. Today, the term Black Hole is a part of the world's working vocabulary.

—lexicon-branding.com

John Wayne was a popular actor who played many heroic parts. Consider what might have happened if he had kept his given name of Marion Morrison.

A good product name will strengthen brand equity, provide distinction and differentiation in a sea of clutter, create an image for the product, and help position the product in the marketplace. In short, it can be an important part of the marketing mix if chosen with care. Although a good name will not eliminate the need for public relations, advertising, and other forms of marketing, it will help leverage the dollars spent in these areas. Examples of product line names with high market value that are not also company names include Walkman, iPod, Pentium, and Crest.

A problem associated with underestimating the importance of the name can arise when a name is chosen that is cute or expedient, without considering the long-range implications. This occurs most frequently in emerging companies, who are moving fast and, in many industrial and high-tech organizations, focused more on product features and attributes than on the softer elements of the product. If the product becomes a big seller, the name will gain recognition and equity in the marketplace. Often, it becomes a candidate for a brand name, the product may spawn extensions or related product lines, and/or the company may begin selling it internationally. A name chosen without proper forethought may be unsuitable as a brand name, too limiting to accommodate line extensions, and inappropriate in certain foreign markets. The company then cannot take full advantage of the equity that the successful product provided to its name. One enthusiastic company named a new high-tech product *Killer*, since it was going to kill the competition. Needless to say, that name did not endure.

WHO SHOULD DO THE PRODUCT NAMING?

There are many approaches to product naming, and many people/organizations who can lead the process. Before embarking, it is important to determine where the responsibility will lie. The following are four alternative approaches:

1. Specialized naming organizations
2. Marketing firms
3. Individual consultants
4. Naming the product in-house

We can look at these approaches in more detail to determine which methods are appropriate for various situations.

Specialized Naming Organizations

There are firms that specialize in developing names for products and brands. Most use formal processes that begin with setting objectives based on analysis of the product, the marketplace, competition, company culture and company/product positioning. They will perform extensive research and testing based on linguistics and phonetic associations, as well as the traditional customer research (described later). Many have worldwide facilities or partnerships with similar firms in international markets who can perform extensive testing in most countries of interest. Most can also do extensive research worldwide to determine if a particular name has already been taken.

Engaging a naming specialist is helpful if not essential when the stakes are high—typically the case at large multinational corporations whose product will sell hundreds of millions or billions of dollars. This is also likely to be the most expensive and time-consuming approach, and it is likely to be overkill—or at least beyond the resource constraints—for most mid-sized or smaller companies.

Often, a company that uses naming specialists has in-house resources that are experienced in managing the outside organizations and with the naming process overall. As a product manager or marketing manager, you may or may not directly engage or manage the firm, although you will be part of the naming team.

Marketing Firms

Many outside marketing firms that specialize in other areas (e.g., marketing research, branding, advertising, direct mail, public relations, general marketing, or a combination of these) will be happy to assist you with naming, either as a subspecialty or an extension of their other services.

Relative to a naming specialist, they are most likely to be less expensive and faster. In many cases, you can use the firm with which your company or division is already working, which has the additional advantage that it already has knowledge of your other products, the markets you are serving, competition, your brand, and your company culture. If you personally are already working with the firm and have an existing relationship, it further reduces the learning curve. However, relative to the specialists, the general-purpose firms will have less expertise and experience at the naming process. Expertise does not guarantee better results, of course, but it does increase the probability.

There are certainly advantages of using a marketing firm relative to doing the naming in-house. The firm is likely to have experience in idea generation and a process for testing ideas with customers and prospects, so it can facilitate the process and provide expertise in an area that many companies—particularly smaller ones—simply do not possess. The firm may be multinational or it may have relationships with counterparts in countries of interest, so it can manage the process of globalization. Outsourcing these activities will reduce the workload on your in-house resources that are already likely to be stretched. The primary disadvantage, of course, is higher cost. If you do hire an outside marketing firm, you should still follow the disciplines outlined in this chapter to make sure that you maintain control of the final outcome.

Individual Consultants

There are many individual consultants who can help with various stages of the process. Some are experienced at facilitating the brainstorming and idea generation stage; others can assist with selection and testing. These consultants typically become extensions of an in-house naming team and will be mentioned during the discussion of the individual process stages.

Managing the Naming In-house

This is a viable approach in many situations when the risks are low, and it may be the only alternative for small companies on a tight budget. In these cases, the responsibility for product naming typically falls on the product manager or marketing manager, and often most of the actual work will be done in-house. The remainder of this chapter is written from this perspective.

THE PROCESS OF PRODUCT NAMING

This section of the chapter discusses the steps in the naming process, which is summarized in Figure 8-2. The major activities in naming a product are:

- ◆ Identify and engage the decision makers and other stakeholders
- ◆ Do the preparation
- ◆ Develop the initial (long) list
- ◆ Select the short list
- ◆ Pick the name/register the trademark
- ◆ Protect the trademark

Identify and Engage the Decision Makers and Other Stakeholders

Knowing in advance who must agree to the product name (and getting agreement on who does not have veto power) will save considerable headaches and potentially painful resets at later stages of the process. The product name can be a highly political and emotional topic, particularly since nearly everyone will have an opinion, many will consider themselves to be a stakeholder, and there rarely will be a consensus. Also, since product naming may not be a routine activity with clearly established formal and informal decision makers, there is ample opportunity for redirection by people outside the normal chain-of-command who have strong feelings and organizational clout. This redirection often occurs late in the process.

It is important to identify the individual who will have the ultimate decision in naming the product. *Ultimate* might mean the actual decision maker, the person who picks the people who should participate in the decision, or the arbitrator if there is a deadlock among the designated decision makers. At a small company, this is usually the president; at a mid-sized company it might be the president or the vice president of marketing. If the latter, it is best for the president to tell you directly.

Identify and Engage the Decision Makers and Other Stakeholders
- Who is the ultimate decision maker?
- Who else must buy into the decision?
- Who should be on the product naming team?
- Who else should be kept in the loop?

Do The Preparation
- Gather the information that will provide the context for identifying and selecting the names (Figure 8-3)
- Ensure agreement among the participants

Develop the Initial (Long) List
- Generate ideas from as many sources as practical; e.g.,
  - Brainstorming sessions
  - Employee Naming Contests
  - Dictionaries and Thesauruses
- Target 20 to 40 viable names

Select the Short List
- Narrow the list using quick and dirty initial testing:
  - Basic "good name/bad name" filters (Figures 8-4 and 8-5)
  - Customer preference
  - Availability
  - International, as appropriate
  - Foreign language suitability
- Use logotype renderings, if these already exist
- Target a short list of 2 to 5 names

Pick the Name/Register the Trademark
- Hire a trademark attorney
- Perform comprehensive versions of the above quick and dirty tests on the shortlist candidates
- Select a final name
- Develop the graphic look and feel
- Register the trademark
- Pray that there are no changes

Protect the Trademark
- "Use it or lose it"
- Keep registrations up-to-date
- Follow use guidelines

FIGURE 8-2. A summary of the naming process.

The ultimate decision maker usually will have several people s/he will want to participate in the decision, either for buy-in or because they are considered a trusted resource. This might include many of the vice presidents or immediate staff. You may add others to the list, but be very selective, since the larger the list of decision makers, the longer and more difficult the process becomes.

For subjective decisions such as naming, an otherwise decisive president might consult with family members or might want to sleep on the decision, so allow extra time for this in the process.

Depending on the company culture, it is often prudent to publish in advance the naming process you intend to pursue and the decision makers involved. Doing so will enable you to resolve, early on, any objections or questions from those who think they should be among the decision makers. You may, of course, have many more people involved in the brainstorming, nomination, and testing process, but everyone should be clear who will participate and who will actually decide.

The naming decision should not be a popularity contest or a democracy, since most will vote for a name that is comfortable but not necessarily the most memorable. Often a truly great name is controversial at first and it takes time to get used to it—consider "Xerox" or "Google." For that reason, the ultimate decision should rest with a small group of people who are familiar with the naming objectives and who are willing to look beyond the conventional.

As with other business decisions, it is advisable to give the decision makers both the reasons you chose the name(s) you did, but also the reasons for rejecting others.

Do the Preparation

There is considerable homework to be done before the initial brainstorming session, since you need to have the proper context for the name. The product plan and marketing strategy will dictate critical factors such as whether you need a new product name or simply a new model number within a named product family, how the product will be positioned in the marketplace, whether and where the product will be sold internationally, and how much is at stake. It is also very important that you have agreement among team members and executives about the critical elements of the product and how it will be marketed.

Figure 8-3 provides a set of questions that you should have answered and for which there should be consensus among the naming team. If there is a complete marketing plan for the product, many of the questions will already have been answered. If not, it is doubly important that they all be addressed. Later on, you will use this information to ensure the name is consistent with other aspects of the product, and it might even provide useful clues for generating the actual name. Important elements of the preparation include:

- ◆ Ensuring alignment of the product name with the other elements of the marketing mix, which in turn should be aligned with the company mission.
- ◆ Ensuring that the name is appropriate for the product's target customers, including channel members who may have different perceptions than the target end customers.

☐ What is your company's mission or statement of purpose, its vision and its statement of values?
    ☐ Does the product's division or group have its own mission statement? If so, include it.

☐ Who are the target customers for the product?
    ☐ Channel partners.   ☐ End consumers or users.

☐ How would a buyer or consumer describe the product in his/her own terms?

☐ What makes the product unique? How is it differentiated from others in your company and from those of your competition?

☐ Why would a buyer or consumer buy the product? What pain does it ease, problem it solves, satisfaction it gives?

☐ What is the product hierarchy and where within that hierarchy does the name fit?

☐ List other product names in your department, division and company.
    ☐ Is it important that the product name have any association to any other product or family of products?
    ☐ Is it important to avoid any connection to other products

☐ Who are the competitors and what are the names of their products that are or may be competitors for this product?

☐ Should the name strongly differentiate your product from others? If not, why not?

☐ What is the product positioning desired for this product? (High tech, low cost, status, image, high quality, value, etc.)

☐ Describe how the end user will actually use the product.

☐ Describe how the product will be merchandised by the company and any channel members.

☐ Are there marketing campaigns or themes planned into which this product will be marketed?

☐ Describe the overall objectives and goals for this product.
    ☐ What are the sales and the market share goals of this product?

☐ Should the name connote any particular feeling or emotion to the user/buyer? Are there any emotions or feelings that should be avoided?

☐ Will the product ever be marketed internationally? Will it be marketed to the Black or Hispanic market in the U.S. or any other ethnic markets?

☐ How will the pricing for your product compare with others of the company's and with those of competition?

☐ What principal media will be used to promote the product?
    ☐ Print   ☐ Video   ☐ Audio   ☐ Multi-media

☐ Many products are given nicknames by other company personnel, distribution channel partners and even by end users. Do similar products have nicknames, and if so what are they? Is it likely that whatever name you give the product will be shortened by people in actual use?

FIGURE 8-3. Pre-naming questionnaire/checklist. This information will provide a context for brainstorming and selecting a product name. It is also important to have agreement amoung the participants.

◆ Seeing the product through your customer's eyes. It is always hard to put yourself in your customers' shoes, but remember for whom the product name is ultimately intended.

◆ Determining the uniqueness and compelling value proposition(s) for the product. These will provide important clues for a good product name (and names to avoid).

◆ As noted earlier, determining where on the current naming hierarchy the new name will be. If there is no naming scheme currently in place, establish one.

◆ Reviewing your existing and soon-to-be-announced products to identify any associations that should me made and that should be avoided.

◆ Analyzing the competition to identify strengths that could be counteracted, weaknesses that could be exploited, and names that you don't want to appear to be imitating.

◆ Determining if you want to strongly differentiate your product from your competition, or if you want to ride on their coattails (as long as you are not infringing on trade names and trademarks).

◆ Understanding your product positioning in order to choose a name that reinforces it.

◆ Describing how the product will actually be used, both the intended use and possible other ways it could be used. A name might suggest the usage, and it certainly should not have a negative impact.

◆ Determining how the product will be advertised, promoted and merchandised by the company and any channel members. While these plans may be preliminary at this stage, you should have ideas based on other products. The name should be consistent.

◆ Reviewing marketing plans and themes that already exist or are planned for other products into which this product might be integrated.

◆ Agreeing on the sales forecast. While this will be a rough number at this stage, perhaps even a range, the magnitude of sales will indicate the stakes involved.

◆ Identifying any feelings or emotions that the name should connote to the buyer, and any feelings/emotions that should be avoided. Many people consider the products they use to be an extension of their personality. This often has a big impact on buying behavior, especially for products that might otherwise be hard to differentiate.

◆ Determining the international markets and ethnic groups into which the product will be marketed. You will need to ensure that the product name is available in the international markets you target, and that it conveys the desired message to ethnic groups.

◆ Reviewing the product's price point against those of other related products from your company and from your competition. Knowing where the price falls within the spectrum can have a significant influence on the name.

◆ Determining the principal type of media that will be used to promote the product (e.g., print, video, audio, multimedia). You should have a good idea based on your other products. A name may be interpreted differently when seen in print, heard aurally and pronounced orally.

◆ Reviewing any nicknames given to your other products by your employees (particularly those who are externally focused), distribution partners and end users. This will give you insights into whether and how a new product name may be given a nickname. The nickname should convey a message consistent with that of the full name.

◆ Last, but certainly not least, is *establishing what (if any) objectives might exist for the product name.* It may be important for the name to help position the product, perhaps it should suggest the product usage, and sometimes you want it to convey a feeling or emotion. The types of names used for other products can provide clues, of course, but you can use the opportunity to break from tradition if past names have been rather mundane. The marketing plan may suggest an objective. If no clear objective exists, leave this undefined for now since that will give you more creative headroom.

You will be using this information later, when selecting short-list names from the initial long list.

It is important to have agreement on this information before you start, so you do not go in circles because people are operating with different assumptions. How you gain this agreement will depend on the process used to generate the information. If it resulted from a discussion among the stakeholders, it usually is sufficient to document the results and circulate the write-up for review. Of course, if any of the participants had strong positions that are not reflected in the final output, and this was not resolved during the meeting itself, it is helpful to discuss the reasons with them separately. If you generated the information on your own, using existing marketing plans and other information, you should call a meeting to vet it with the stakeholders These meetings may seem to be overkill, but the effort spent gaining agreement at this stage can reduce thrashing and frustration later.

Develop the Initial List

The initial list is the *long list*, as contrasted to the *short list* that you will use later when the cost for managing each name is considerably higher. The objective here is to identify as many candidates as reasonably possible, without worrying about how good a particular candidate might be. Any name might later inspire another name that turns out to be the winner.

There are several sources from which a name might emerge:

BRAINSTORMING A common source of generating ideas is one or more *brainstorming* sessions. These can done informally as part of a team meeting, but results are much better when you structure a meeting for the purpose, allow ample time to let the creative juices flow, and use an outside facilitator. A facilitator trained in brainstorming, especially brainstorming product names, will help navigate through the intricacies of the process, and s/he will be viewed as neutral and objective.

Doing the initial homework and following a disciplined approach actually helps foster creativity, contrary to the common belief that structure is inhibiting. Understanding the objectives and constraints helps concentrate efforts with less wasted time and frustration. Sometimes people come up with a great name, but it is not appropriate for the situation or it has too many trademark issues, for example. By narrowing the focus, the energy is better spent. For that reason, it is important to review the results of the prenaming questionnaire (Figure 8-3) with the team beforehand.

A more thorough approach to brainstorming involves using several smaller teams, which increases the likelihood that more creative directions will be explored. You can also give brainstorm teams assignments that do not relate directly to the product. For example, if a possible objective of the name is to convey high performance, the team could brainstorm ideas for high performance in another product category—many of the resulting names could apply.

EMPLOYEE NAMING CONTEST It is very common in small companies to solicit input from employees, sometimes with prizes for people who submit the best, most creative, or most humorous nomination. Some consider this to be a waste of time, but it can be fun, it enables everyone to feel part of the process, and it is another source of potential names. However, the naming contest does not always produce the final name, so do not rely on that exclusively, and be careful to specify that the resulting names will be combined with those from other sources. If there are prizes to stimulate input, be sure to mention how the judging will be done and whether the prize for best submission will be awarded even if it does not become the finalist.

DICTIONARIES AND THESAURUSES You will want the most complete sources available. The Oxford English Dictionary, with approximately 450,000 words, is a start. In it, you can find many root words and derivatives, which will greatly expand your search capability over a standard dictionary. If you have access to dictionaries of foreign languages such as Greek, French, German, and Italian, these can further expand your search area. (These should be cross-referenced to English if you are unfamiliar with the language.) A great source, if you can find it, is a compendium of Latin words.

You can use any one of these sources, plus any others you can think of, but the richest set of results will come from a combination of all three. Often, the brainstorming and employee-naming contest produces a wealth of creative ideas, which can be expanded using information in the dictionaries and thesauruses.

ADDITIONAL SOURCE CONSIDERATIONS   It is also helpful during the generation phase to keep in mind the various types of names that could apply to a product, since addressing these can lead to new ideas. Types of product names include the following:

- *Personal:* This name is taken from a real person, often the founder or inventor of the product. Examples include: Lamborghini, Hewlett-Packard, Edison, Toyota, Ben & Jerry's. This can be an advantage if the person has a famous name or he/she is well known and charismatic. The disadvantage is that the person's name may say nothing about the product, and the name may already be in use for a similar product. It is also hard to register a person's name, since the U.S. Patent and Trademark Office (PTO, the U.S. registry) generally does not want to prevent people from using their own name in their business. Although many companies and some products use personal names, most were derived years ago and today there are fewer cases going forward (particularly for products).

- *Descriptive:* This describes what the product is. Examples include PowerBook (Apple), ExecuStay, and Residence Inn (Marriott). The advantage of a descriptive name is that it communicates information about the product. The disadvantage is that it cannot be protected if it is too generic, and it might be too narrow if the product is successful and spawns a brand. A descriptive name, if too generic, also runs the risk of seeming flat and undistinguished.

- *Functional:* This describes what a product does or suggests the experience it may give the user. Examples include Norton Internet Security, Dell Photo All-in-One Printer, and TurboTax. These potentially have the same advantages and disadvantages of a descriptive name, although you may find the functional description to be less constrictive.

- *Emotional:* These names suggest a feeling, emotion or image that you want the customer to associate with the name. It may have nothing to do with the actual product. Examples include Malibu, Pampers, Cougar, Lynx, Secret, Escapade (hotels), Eclipse (private jets) and Zoom (data communications, airlines). Notice that some of these are geographical locations, animals and other common English words that have strong image associations. These associations may also include important attributes such as color, smell, and texture.

- *Invented:* This is a word that has no direct meaning in the English language, although it may be a variant of a word that does. Examples of invented words with no clear meaning include Kodak, Zytel, Kofax, and Viiv. Examples of derivatives include Acura, Visteon, Inspiron, Encarta, and Pentium.

- *Numbers, Initials and Acronyms:* Panasonic relies heavily on model number, such as TH-65XVS30U, a 65″ plasma TV. Cadillac offers the "SRX". Lucas has the "THX" sound standard. IBM originated the PC. (PC is short for *personal computer*, so it could also be considered a descriptive name.) For future planning, where possible leave gaps in series of numbers so that if you later develop a product whose features fall between two existing products the numbering system will have logic.

◆ *Puns and humorous names:* These are fun to generate but may not convey the image you want for your product, since humor is context-specific (i.e., it will be limiting going forward and may not be humorous to the ethnic groups and other cultures you are trying to reach). These are unlikely to translate well into other languages. If a takeoff on an existing trade name, there may also be infringement problems ("Dogiva" was a clever rearrangement of the Godiva name for chocolates. Cute and memorable—until Godiva Chocolatier, Inc. sued and won.) Generally, you should avoid puns and humorous names unless your product will have a very limited market, which is generally the case for local businesses, or unless it clearly reinforces your brand positioning (as is the case with Ben and Jerry's ice cream flavors). Examples of puns and humorous names include Tex's Chain Saw Manicure (a tree and yard care service), The Come On Inn (a bed and breakfast) and TEA-ReX (a line of premium teas).

At the early stages of generating names, you should use these categories to help expand your imagination, not to narrow the search.

A consideration when generating and screening names is the common associations of letters and sounds. Words with consonants are considered more masculine and suggest hardness and sharp angles. Vowels are softer and more feminine. "X" implies high-tech or extreme, but proceed with caution, since the letter is currently overused.

A good target for the long list is 20 to 40 viable names after you have weeded out the obviously bad and inappropriate candidates. More is generally better, but you need to balance quantity with the time and effort of generating the names.

Selecting the Short List

The next step is narrowing the long list down to a short list that will be rigorously tested and researched. As a first step, pass the candidates through the rough screens listed in Figure 8-4 (characteristics of a good name) and Figure 8-5 (naming blunders and things to avoid). This can quickly eliminate some of the candidates and make the testing process easier.

There are several types of tests you should perform in narrowing down the list. These can be conducted serially, if time permits, or concurrently. The testing and research at this stage will be quick and dirty and intended to produce a short list of two to five candidates that will be scrutinized more carefully. You may have a clear favorite at this stage, but it is helpful to have at least one and ideally several backups, in case the favorite later fails an

☐ Simplicity

☐ Memorability

☐ Ease of spelling

☐ Ease of pronouncing

☐ Ease of understanding

☐ Evokes positive images

☐ Does not have hidden meanings

☐ Translates well into other cultures

☐ Does not have negative meanings in other languages

☐ Can be graphically represented appropriately

☐ Fits with the corporate mission and brand strategy

☐ Complements other product names

☐ Does not conflict with other product names

FIGURE 8-4. Checklist of criteria for a good product name.

☐ Names with cult meanings

☐ Names based on what may be short-term fads

☐ International faux pas

☐ Offensive to religious or ethnic groups

☐ Names with two or more differing connotations.

FIGURE 8-5. Checklist of naming blunders and things to avoid.

important test (e.g., it may already be taken, it may not be appropriate for another culture). The greater the stakes, the more backups you should have. Five is not a magic number for the upper limit, but the research on the short list can become expensive and time consuming, which places practical limitations on the number.

The objective of the testing at this stage is to weed out names that would likely fail the more rigorous testing you will perform on the short list of candidates, so do not worry about being overly thorough. You want to invest a little bit of time at this stage to avoid significantly more time and cost downstream.

◆ *Testing against the marketing context and objectives:* You should evaluate the candidates using the criteria established during the preparation phase to eliminate any clear losers and marginal candidates.

◆ *Testing for availability:* At this stage, you will want to perform some quick tests to determine whether the name is available. The final candidates will be subjected to more rigorous testing, of course, but the objective here is to eliminate names that are clearly off limits. Testing at this stage includes:

◆ Enter the name into search engines and see how others may be using your proposed name. You can also type the name directly in the browser as a URL (e.g., *http://www.namebeingtested.com*, _____.*biz*, _____.*net*, _____.*org*, _____.*edu*, _____.*tv*) to see what comes up.

◆ Search on the names on the Patent and Trademark Office Web site (*www.uspto.gov/*) and search under "trademarks" (*not* the general search field that appears on the home page). Another handy source is "Name Protect" (*www.nameprotect.com*).

◆ Repeat these tests using misspellings, abbreviations, hyphenations, potential nicknames, and alternative spellings. Just because a similar spelling is in use does not automatically mean the name is off limits, but you will need to proceed with your eyes open, since you may have increased risk of infringement.

◆ Check the potential domain name derived from the product name candidate at *www.register.com*. You will want to avoid any names that are confusingly similar to others or where others have some negative connotation that could rub off on your product. If you have a very successful product, you may also want it to have its own Web site in addition to your company Web site. For example, Tide and Crest have their own sites independent of the Procter & Gamble site.

◆ If you plan to market internationally, you can search the trademark databases in the countries of interest. You can locate these databases through a search engine; a good place to start is typing in "Trademark Search {*Name of Country*}"

◆ An additional test that can be performed at this stage is contacting the Secretary of State's office in your home state to see if anyone has registered your name as a company name. You should also check Delaware and Nevada, since those are popular states in which to incorporate a business due to their favorable laws and taxes. Although this test will be conducted later, on the short list candidates, weeding out names at this stage will reduce cost later. You can also access *www.trademark.com*, a fee-based service that provides access to state trademark databases.

Remember that a name may be in use but not registered as a trademark. The acid test for ownership is first use, not first registration. The first user may be a local business in another state that does not have a Web site and that has not registered its name. Their presence will prevent you from registering your trademark: the U.S. Patent and Trademark Office will deny your registration if you try. You can always take your chances with an unregistered trademark and hope the business does not come after you, a calculated risk if the other business is small and unlikely to take action, but can be costly if you assume wrong.

Another consideration is that usage of a name in one category may not preclude you from using it if your product falls into a noncompeting category. While there are no clear rules and *noncompeting* can be a subjective interpretation, there are internationally recognized classes of products for trademark purposes that can provide some guidance. You can find these on the World Intellectual Property Organization Web site at *www.wipo.int/treaties/en/classification/nice/index.html*. (You can also navigate there from *www.wipo.int*. There are 45 classifications; class 38, for example, is "Telecommunications.") Ultimately, this can become a discussion with a trademark attorney having experience in the subtleties.

◆ *Testing in foreign languages:* If you plan to market in other countries, you should determine meanings in foreign languages. You can enter the name into a site providing translation services (e.g., *www.free-translation.com*, *world.altavista.com*), translate it into a foreign language, and translate it back. If the results are out of bounds, you can do further investigation or reject the name. If you have distributors or other affiliates in other countries, you may wish to ask them for their reactions to the name and how they think it would play in their country. You can't do this in every country, but a preliminary indication may save grief later on. Table 8-2 shows examples of why this step is important.

◆ *Testing for customer preference:* The objective of customer testing is to try the names out on your customers, channel members, partners, and other stakeholders. The streamlined version of the testing that occurs at this stage involves company employees (including those who are emotionally detached from the product) and easily accessible customers and channel partners. Types of testing include, but are not limited to:

  ◆ Providing the subjects the proposed name, together with a brief product/market context, and getting them to describe what the name evokes. Without describing the context, you may ask a subject

TABLE 8-2.
Examples of Names That Translated Poorly into Certain Languages or Had Unintended Slang Meanings

Product Name	Language	Meaning
Toyota MR2	French	MR2 = Merde
Chevrolet Nova	Spanish	No Go
Buick LaCrosse	Quebecois	masturbation
Mercedes GST	Canadian	Goods and Services Tax (Gouge & Screw)
Ikea Gutvik bed	Swedish	Good F***
Perdue "It takes a tough man to make a tender chicken"	Spanish	"It takes a sexually aroused man to make a chicken affectionate."

to make up a story based on the name—the longer and more interesting the story, the richer the imagery that the name evokes and the more memorable it is likely to be.

◆ Showing pairs of names and asking the subject to choose their preference, then continuing to substitute a new candidate for the less favored of the pair until they are all reviewed.

◆ Handing the complete list to the subjects and ask them to rank the list.

When asking for a reaction, you will get more information by asking the subject why s/he has the particular reaction, as well as capturing the what.

You may also wish to survey with and without graphic treatments for the name. *Without it* gives you information on the name itself, but with the graphic treatment you get a more complete reaction. If there is an existing graphic theme or template that your product name will use, you should have your graphic artist render the name accordingly and then conduct the preference testing. If there are few restrictions in putting graphics around the name, it will become a major project in its own right and is best done later, after the final name is selected.

The goal of this research is to be directionally correct, not statistically significant, since you are doing the naming in-house. (If statistical significance were important—as it may be when the stakes are high—you would undoubtedly have made the decision to use outside resources to perform the naming in the first place.) A good sample size to target is in the range of 15 to 30, preferably toward the high end. If you experience a wide divergence in the responses, you should increase the number. The research methodology can be informal, but you should be careful to avoid any personal biases when you present the information.

The objective in narrowing the list down to two to five names is to identify the showstoppers and weed out marginal candidates, and then perform the more extensive (and expensive and time-consuming) testing on a smaller, more manageable list. The stakeholders should, of course be involved in the decision regarding which names make the short list. Hopefully, you have been keeping them involved, or at least informed during the initial testing, so the decision should be fairly straightforward at this point. Although the voting process can range from a formal discussion to sending out an e-mail asking for a response, it is important that each stakeholder be given an opportunity to have a say. (The effectiveness of e-mail depends on your company's culture. The risk of e-mail is that someone doesn't get to it in time, or it gets overlooked.)

Depending on how much time can be devoted to this activity and how closely the stakeholders have been involved during the research phase, narrowing the long list down to a short list can take several days (if it is a high priority) or several weeks. Regardless of how much the time for research is compressed, be sure to allow sufficient time for stakeholder input.

Pick the Name/Register the Trademark

There is more testing and research that needs to be done before picking the final name, much of it a more in-depth version of the types of tests performed in the previous step using additional expertise where appropriate

ENGAGE YOUR TRADEMARK ATTORNEY Skilled attorneys will perform comprehensive research on the availability of your name, including a state-by-state search, and they will work with you on class designations, alternate spellings, and the registration process. A comprehensive domestic search may cost $250 to $350 per name. International searches cost more. Since the cost is proportional to the number of countries, you can limit the cost if you conduct the search only for those countries in which you plan to do business. Your attorney can also advise you regarding any economies of dealing through the European Union, which as of this writing is in a state of transition regarding trademarks. To learn the latest, you can search on "EU Trademarks" and probe the results; for a quick overview, go to Wikipedia.

There are a growing number of online services that promise to help you conduct your own comprehensive search and even register the trademark. Although it is tempting to use these and eliminate outside attorney fees, this course of action is riskier. A good attorney can navigate through the search and registration process, which can be quite complex, and will deal with subtleties and nuances that would escape the layman. The consequences of a mistake can be very costly when it results in an infringement lawsuit or provides an opportunity for someone to poach your improperly protected name. A savings of a few thousand dollars does not justify the downside risk.

The comprehensive search is essential, and it is the most likely cause of a name disqualification at this stage. If time allows, you can do the comprehensive searches first and perform the next steps only with names that you are confident are available.

CONDUCT IN-DEPTH CUSTOMER RESEARCH Based on the preliminary customer research conducted during the initial screen, you may be satisfied with your choices. When the stakes are high, however, you will probably want to do more extensive testing on existing customers, prospective customers, channel members, and targeted ethnic groups. If you do this research yourself, be careful not to sell the name that you have decided is your favorite—you want to get objective feedback. Better yet, you should consider engaging an outside firm—perhaps the marketing agency with whom you are already working or a specialist—that has experience in testing and the resources to execute.

DEVELOP THE GRAPHIC LOOK AND FEEL Work with your graphic artists to develop graphical representations of your name candidates. Since this can be a complex process in its own right, you probably will want to save this step until you have your final name selected, unless there is an existing format that you must use. However, in certain situations the graphical representation might

TABLE 8-3.
Examples of Graphical Treatments of Names. Note
How Much More Distinctive the Name Becomes
When Rendered in the Logotype, Even in Black
and White. Color and Graphic Symbols Further
Increase the Distinctiveness.

Basic Name	Logotype Rendering
Nestle	**Nestlé**®
Hertz	*Hertz*
IBM	IBM
American Icon Vodka	AMERICAN ICON VODKA
Disneyland Resort	Disneyland.
Google	Google
Intel	intel.

provide a tiebreaker between two leading candidates. Table 8-3 illustrates
how a logotype increases the memorability of the name; a full graphical
representation with color and symbols would become even more distinctive.
Of course, you will want to choose a name that is powerful in its own right
and not rely on graphics to turn a mediocre name into a winner.

PRAY THAT THERE ARE NO CHANGES   It is possible that someone could be
working on the same name as you are concurrently, and neither party would
recognize it until deep into the process. Hopefully, you will beat them to it,
but if not, be prepared to step back and start over (at least with an alternate
short-list candidate).

Again, the testing of the shortlist candidates should be more extensive
than the quick and dirty testing at the previous stage. How extensive depends
on the expectations for the product and the amount of risk you are willing to
take. Regardless, you should not cut corners on the comprehensive search and
registration process. For this stage, you should allow several weeks and prepare
your stakeholders accordingly. It may take less if everything goes smoothly
(and you can devote the time to the customer research), but you must allow for
resets in the event the favorite name is taken. You also can buy insurance that

you will be more likely to meet your deadline by simultaneously conducting the search on several prospective names at once.

The final step, of course, is actually registering the trademark. As with the trademark search, it is advisable to engage a trademark attorney, who will have expertise in the nuances of the process. If you want to register yourself, the U.S. Patent and Trademark Office (*www.uspto.gov*) and the World Intellectual Property Organization (*www.wipo.int/madrid/en/*) provide the means to do so. There are also independent full-service sites as such as *www.tmcenter.com*, *www.marcaria.com* and *www.legalzoom.com*.

You do not need to register your trademark domestically, since you will have ownership due to usage, but doing so gives you additional legal clout and signals to others that you are serious about defending your trademark. You can use the trademark symbol (™) whether or not you register; you can only use the registration mark (®) if you are indeed registered. An important note—you must submit a "Statement of Use" that includes proof of your trademark's "use in commerce" order to complete the registration. "Use in commerce" means all commerce that U.S. Congress may lawfully regulate (e.g., interstate or international) and it must be bona fide (i.e., not a superficial act performed simply to reserve rights for future use). *Proof* includes a sworn statement that the mark is in commerce, a listing of the date of first use anywhere and first use in commerce, and a specimen showing use of the mark in commerce. Alternatively, you can file an *Intent to Use* application, which protects your rights to the trademark and gives you six months in which to file a *Statement of Use*. The Intent to Use may be extended six months at a time for up to three years. (Further information on this is available at *www.uspto.gov*. and, of course, from your trademark attorney.)

One optional but potentially valuable step is to register your product name as a domain name—if the product becomes highly successful, you eventually will want to redirect that URL to your company site or even establish a separate Web site. As examples, try *www.ipod.com* and *www.crest.com*.

Protect the Trademark

Congratulations when you have made it this far and have registered your name as a trademark! The last step is to ensure that you do not lose the trademark due to lack of use. In the United States, you must file a *Declaration of Continued Use* or *Excusable Non-use* between the fifth and sixth year after registration. In other countries, the maximum length of time a trademark can go unused is typically three to five years, after which another party can successfully file a claim of abandonment (assume three years to be conservative).

This could become an issue if you do a progressive international rollout. It could also present a problem if you decide not to release the product that is currently under development and use the name instead for a product that will be brought out later (resets do happen as a result of competitive introductions or longer-than-anticipated development of new technology). Also, be sure to keep your address current at the various registry offices so you don't miss any notices they send to you. Your trademark attorney can give you specific advice regarding length of time and what constitutes usage in the various countries, and can manage at least part of the part of the process if you maintain an ongoing relationship.

For more information on lack of use provisions and how to prove use —and some compelling evidence why you might not want to manage this yourself—turn to *http://www.ecta.org/position_papers/Trademark%20Use% 20in%20Opposition%20survey.doc#INTRODUCTION*. Some of the information may be out of date. But then, so is much of the information you will find online for specific countries, since laws and their interpretation are changing rapidly as the economy goes global.

One other consideration is to use—and make sure others use—the trademark as an adjective or modifier, not as a noun or verb. This will prevent the trademark from becoming so widely used that it becomes a generic description, at which time you will lose your trademark protection. Examples of products that were once trademarks include xerox, aspirin, thermos, and escalator. It is a nice problem to have, since it means that your product has become very popular, but it is a problem nevertheless when you must forfeit rights to a name that you invested in establishing and that has considerable equity. An example of proper usage would be "a Xerox copier." Improper usage would include "make a xerox" or "please xerox this document." While you cannot prevent or police every instance of improper usage, you should use the trademark properly yourself and take reasonable steps to ensure others do as well (e.g., in collateral and media). Your trademark attorney can give you specific guidance.

CONCLUSION

Although naming a product can involve considerable creativity and work, the process itself is relatively straightforward. How much work is involved and how extensively the process is implemented will depend on the stakes involved—naming a brand-new product that is expected to sell in 20 countries and produce several hundred million dollars annual revenue will require considerably more care than naming a product that is an extension of an existing product line that produces several million dollars in specific domestic markets. The dollar cost can be in the low four figures if the process is done mostly in-house and only the legal work is outsourced, or it may be well into six figures if a top naming firm is engaged. The information in this chapter should have given you insight into which direction you wish to pursue, and it provided a practical step-by-step approach if you elect the in-house route.

Table 8-4 summarizes some of the many on-line resources that are available to facilitate the process.

Even a simple naming project can involve a reasonable amount of work, but it is well worth the investment when you consider the downside risks

TABLE 8-4.
A Summary of On-Line Resources. There Are Many Useful Sites on the Internet; This Is a Partial List to Help You Get Started. While Some of the Resources Discuss Brand Names, Not Product Names Per Se, the Information Relevant to Selecting Brand Names Is Also Generally Applicable to Product Names

Activity	Links/comments
Obtain general information from the US Patent and Trademark Office.	*www.uspto.gov/*: contains a wealth of information and resources.
Obtain information on international trademark registration.	*www.wipo.int/madrid/en/* contains general information, resources and registration forms.
Search for existing name usage:	All major web search engines.
Search for existing trademarks.	*www.uspto.gov/*: (U.S Patent and Trademark Office)
	www.nameprotect.com
	www.trademark.com is a fee-based service with access to state databases.
	www.tmcenter.com
	www.legalzoom.com
	Searchable databases of trademarks registered in foreign countries can be found using search engines. Enter "Trademark search name of country" as a starting point.
Access information on trademark registration.	*www.uspto.gov/www.marcaria.com*
Test for foreign language appropriateness.	*www.freetranslation.com* *http://world.altavista.com*
Determine the category into which your product is classified.	*www.wipo.int/treaties/en/classification/nice/index.html*
Review US Trademark Law - Rules of Practice and Federal Statutes.	*www.uspto.gov/web/offices/tac/tmlaw2.html*
Online sites about branding that includes collections of papers and lists of resources.	*www.brandchannel.com* *www.namedevelopment.com/articles*

and upside benefits. The *BusinessWeek/Interbrand* 2005 annual study of the world's most valuable brands indicated that the value of the Microsoft brand was over $60 billion, at least a portion of which is attributable to the name itself. Your brand value is unlikely to reach that level, of course, but it can easily be worth tens or hundreds of millions of dollars. Consider that many mid-size business are valued at approximately two times sales, and of that premium, 60 percent to 80 percent is often attributable to the brand—roughly $70 million for a company with $100 million annual sales and a strong brand.

NOTE: All trademarked names mentioned in this chapter belong to their respective owners.

CASE STUDY: A NAMING CASE STUDY

This is an example of a product-naming project that took place at a small, high-tech company several years ago. The product to be named was a layer of middleware designed to provide standardized, high-level *application programming interfaces* (APIs) for document-processing applications from different vendors. Collectively, these applications would be integrated to form an enterprise-level end-to-end solution. The key value proposition was the ability for system integrators and value-added resellers to construct customer-specific solutions much more easily using well-known `` best of breed '' applications that were proven and easy to sell. The company also planned to develop several of those applications itself, so the name of the product had to be robust enough to extend to a broader product line.

The product manager was responsible for driving the naming project. The core naming committee consisted of the product manager, the marketing communications manager, the director of product development and the company president. They designated a formal extended committee, composed of the executive committee and development team, whose stated role was to provide additional ideas and feedback and whose unstated role was to provide political buy-in. The decision-making process, defined in advance, was that the core committee would agree on a name and recommend it to the extended committee, who then would vote on it and make a case for one of the alternative finalists if necessary. The final name would also be vetted to the all of the company's approximately 200 employees, who could weigh in, although there was never any intention of making the decision based on a popular vote.

The product manager immediately held an employee naming contest with a nominal prize for the best submission and an additional reward if the nomination became the actual name used. Approximately 10 employees responded to the e-mail announcement, which produced approximately 20 names. The product manager supplemented this list by directly soliciting committee members (an additional 7 names) and by doing a search through thesauruses, the Oxford English Dictionary—the full version with 400,000 + words and Latin derivatives, since standard dictionaries turned out to be too limiting—and various English to French/Spanish/Italian/German translation guides she had handy. This exercise produced another 5 words. (It actually produced many more, but the product manager immediately screened out most of them out. For example, one promising name—Imagery—was already taken by a company in the same industry). While the product manager, with hindsight, agrees that one or more brainstorming sessions would have been valuable, it didn't get sufficient priority at that time in the fast-paced environment of the small company.

Of the original 30 + names, most were ruled out quickly. For example, `` Documator '' invoked images of an Arnold Schwarzenegger movie, `` Build-an-App '' was too plain and uninspired, and `` Solution99 '' was potentially catchy but might be confused with a bathroom cleaner. Some of the possibilities were closely related to other candidates that eventually became finalists. For example, `` Ascend '' was on the long list, together with `` Ascent. '' As a verb, `` Ascend '' had the perceived aura of involving more work than the noun `` Ascent, '' which went on to the finals.

The product manager made a first pass at selecting the finalists, which included one-on-one discussions with other core team members, as well as other employees informally selected for their strong opinions and good intuition. The finalists received a quick test for possible infringement from the outside trademark attorney. (This occurred before Internet search engines were available.) While the trademark attorney was conducting the test, she informed the extended committee of the progress and solicited comments.

The leading finalist was *Montage*, chosen because it was a pleasing word and its meaning conveyed the value proposition of the product— `` a composition of individual pictures [applications] forming a blended whole. `` A related finalist, *Collage*, was less favored: While it also was a collection of individual objects, the result was, by definition, incongruous. *Ascent* had appeal because it suggested upward movement and an advancement (in the state of the art), although it was behind *Montage* in the rankings because it was not suggestive of the key product attribute. *ImageMaker* was considered the best of the employee naming contest submissions, and, as such, received a prize and became a finalist. However, it was not considered consistent with the planned evolution of the product line and therefore never was in serious contention.

The core team, then the extended team, agreed on *Montage*. Approximately six weeks after the beginning of the naming project, the product manager was about to call the outside trademark attorney to begin the registration process. Literally on the evening before the planned phone call, a company executive entered a meeting between the president and product manager holding an announcement he had seen in a trade publication: A start-up company had just launched a database called Montage. (It was never determined how this had escaped the initial search, although the name was not yet public—by a matter of days—when the search was conducted.) The product manager, deflated after becoming emotionally invested in that name, argued that the database product was in a different industry and that the risk was further reduced because it involved a start-up that was likely to have too much on its plate to worry about legal challenges. She was quickly overruled by the president and an influential board member, who did not believe the risk of a legal challenge was worth taking. After several days of further deliberations, Ascent emerged as the name.

As the saying goes, all's well that ends well. Ascent went on to become the flagship product of the company, which then underwent a successful IPO.

REFERENCES

Bates, Colin. "How to Name a Brand." *buildingbrands.com*

BBC News. 2003. "Buick 'masturbation' car renamed," *http://news.bbc.co.uk*, October 23.

Bromberg & Sustein, LLP, "Trademark Usage Guidelines." *www.bromsun.com*

Chang, Victoria, and Chip Heath. 2001. *Lexicon Branding*, Stanford University School of Business.

Charmasson, Henri. 2004. *Patents, Copyrights & Trademarks for Dummies*, Hoboken, NJ: John Wiley & Sons.

Keller, Kevin Lane. 2003. *Building, Measuring and Managing Brand Equity*, 2nd ed. Upper Saddle River, NJ: Prentice Hall.

Stallard, Lizzy. 2004. "The Alphabet Soup of Auto Naming," *www.brandchannel.com* (brandpapers section), April.

Twerdahl, James, and Lee Shaeffer. 2005. "Taking the Mystery from Product Naming," *www.brandchannel.com* (brandpapers section).

Williams, Phillip G. 1991. "Naming Your Business (and Its Products and Services)." The P. Gains Co.

9

Using Assumptions-Based Models to Forecast New Product Introduction

Kenneth B. Kahn
Associate Professor of Marketing and Stokely Scholar. The University of Tennessee

Satellite radio was supposed to be big. One forecast produced in the year 2000 was suggesting 36 million satellite radio subscribers by 2007 (McBride 2006). Within just 12 months, forecasts were becoming less optimistic. A revised forecast in 2001 suggested 16 million subscribers by the end of 2006. Five years later actual figures indicated a subscriber base of 11 million subscribers across the two companies of XM Satellite and Sirius Radio. These significantly lower subscriber numbers resulted in much lower than expected revenue. In turn, the lower-than-expected revenue caused substantial financial losses for XM Satellite and Sirius Radio due to the inability to offset large investments made to secure market channel access, sign radio show personalities, and execute marketing promotions (McBride 2006).

Some say that "numbers don't lie." However, this example shows that numbers can be wrong, if not very wrong. Moreover, the numeric forecasts do not explain why they are wrong. What were the assumptions underlying these numbers? Unfortunately, the sources for these numbers do not present details for any underlying assumptions. The fact that assumptions are not given exemplifies that many managers tend to focus just on the numbers and not the assumptions underlying their numeric new product forecasts. In other words, there is an inherent preoccupation with knowing just the numeric new product forecast and overlooking the assumptions that underlie the given number. Knowing the latter is crucial because new product forecasts are characteristically overladen with emotional hype and optimism (Tyebjee 1987). By knowing what the assumptions are, along with the numbers, we can get a sense of transparency to where hype and optimism may be occurring.

This chapter promotes the theme that companies should focus on the assumptions that underlie forecasts of new product introductions. Doing so provides the necessary understanding to thoughtfully, logically, and systematically evaluate, or even challenge, the numeric forecast. Simply saying that a forecast is too high or low cannot be substantiated, nor acted on meaningfully, if the triggers driving the number are not established. Identifying and understanding assumptions are just as, if not more, important to forecasting new product sales as are the numbers themselves.

One rather straightforward class of forecasting techniques designed to identify and systematically lay out assumptions is assumptions-based models. Judgmental in nature, assumptions-based models are flexible techniques that can be employed at different points during the product development process. Assumptions-based models are especially valuable for discerning critical assumptions during launch planning and identifying necessary strategies that will ensure a successful introduction, given these assumptions. This chapter discusses and illustrates how a manager tasked with constructing a new product forecast can develop and use assumptions-based models for effective new product introduction.

Accordingly, the first section of this chapter discusses what assumptions-based models are and applies a generic assumptions-based model framework to the satellite radio context to illustrate the straightforwardness of these models. Next, discussion shows how the assumptions-based model framework can be adapted to more complex business situations and can incorporate more sophisticated analyses such as risk analysis. Launch planning and assumptions management topics are then discussed relative to using assumptions-based models. The chapter closes with a discussion of pitfalls to avoid and guidelines for proper use of assumptions-based models.

ASSUMPTIONS-BASED MODELS

Assumptions-based models attempt to describe the behavior of the relevant market environment by breaking the market down into components (also called *market drivers*). Values for these components are established and forecasts are generated. These values represent assumptions stemming from judgment because events have not yet occurred to prove the established values. Assumptions-based models are characteristically a class of judgmental forecasting techniques, though numeric in outcome. Other names for assumptions-based models include chain models and market models (cf. Latta 1998).

A generic framework of an assumptions-based model begins with an overall potential target market size and uses various market factors to break down the potential target market proportionally. This emphasizes how important it is to carefully articulate, identify, and detail the intended potential target market before beginning the new product forecasting endeavor and prior to specifying other key market drivers. As suggested in Figure 9-1, the potential

FIGURE 9-1. General framework of an assumptions-based model.

target market is proportioned down to the available target market, then to the qualified target market, then to the attainable target market, and finally, to the penetrated target market. In this chapter, we will look at developing a forecast for satellite radio to illustrate this generic framework and to show how this framework could have been employed in early 2000 to calculate a forecast for the satellite radio industry.

Sizing the Potential Target Market

The starting point in applying the generic framework is establishing the overall potential target market size. By definition, the potential target market size would represent the maximum amount of sales possible for a particular product aimed at a given set of buyers within a given period of time. As will be the case for sizing any market factor in the assumptions-based model, a variety of approaches and thinking may be employed to generate a value for the respective market factor. Time, expertise, and other resources will naturally influence which approach(es) can be employed. There is certainly no one definitive approach.

In the case of satellite radio, a less sophisticated approach is employed, where census data on the number of cars in the United States market serve as the potential target market size. Although apparently straightforward, careful thinking is needed regarding which cars and/or car owners should be considered likely candidates for satellite radio. Is satellite radio being aimed at older or newer drivers, old or new vehicles, and so on? U.S. Census figures in early 2000 indicated that there were approximately 212,706,399 passenger cars, vans, pickup trucks, and sport utility vehicles in the United States, and among these, approximately 17,349,933 were new vehicles. For the sake of simplification, all vehicles are deemed likely candidates, because all cars come equipped with at least some standard radio. Thus, the total potential market for satellite radios is 212,706,399 cars.

Sizing the Available Target Market

The next step in developing the satellite radio forecast is to determine what part of the potential target market is the available market. Available market is defined as the set of buyers who are able to gain access to purchase the product. For the present case of satellite radio, the available market is estimated by distribution capability in terms of how much of the market will be able to purchase from each respective company. Other delimiters for moving from potential target market to available target market are possible as well.

It is presumed that XM and Sirius will be aggressive in garnering channel partners, which would include retailers specializing in electronic products and large mass merchandiser retail chains. A 95 percent estimate of U.S. market availability is suggested due to the large market presence of mass merchandisers in the car radio market and the presumption that almost all retailers that sell car radios will offer satellite radio, too. The 95 percent estimate specifically means that satellite radio will be made available for purchase at 95 percent of all locations where car radios are sold. While this number may seem high, higher percentages are possible, such as in the case of grocery products based on point-of-sale data provided by AC Nielsen (*www.acnielsen.com*) and Information Resources, Inc. (*www.infores.com*), where even a 100 percent figure may be possible (100 percent meaning that the product is sold in all possible locations where that product category can be sold). The 95 percent estimate is applied, leaving 202,071,079 vehicles as the available market for satellite radios.

Sizing the Qualified Market

The qualified market represents the proportion of the marketplace able to actually purchase the particular product of interest. Again, various approaches are possible to determine an appropriate figure to input into the assumptions-based model. One approach could be to use benchmark data from analogous market situations: for example, the percent of migration from free television to cable television (pay-for television) during cable television's first years may be applicable. For the sake of simplification, income is presumed to be an important determinant of whether someone will subscribe to satellite radio. It is further presumed that only half of those individuals owning a vehicle will be able to afford / likely pursue satellite radio. 50 percent of vehicles are therefore labeled as part of the qualified market resulting in 101,035,540 vehicles remaining. Market research via surveys and focus groups would be especially useful, and would be a preferable approach to determining and validating the correct percentage constituting that portion of the available market with the qualifications to buy the respective product.

Sizing the Attainable Market

The *attainable market* represents the reasonable share likely to be obtained by a particular company. One approach may be to consider those buyers who are heavy users of their car radios, such as individuals with a 45-minute or longer commute. These people would be more aware of the value that satellite radio may offer. Awareness also can be a function of marketing communications initiated by the satellite radio companies themselves coupled with publicity through various media sources. For the present illustrative case, attainable market is viewed as the proportion of the market who are aware of satellite radio as a consequence of advertising and publicity. Awareness is an important consideration for gauging the attainable market, because those who are aware might have the propensity to purchase satellite radio, while those who are unaware could not make an effort to purchase satellite radio.

One way to gauge attainability could be the level of marketing communications and corresponding percent of awareness generated by such efforts. Some companies have historical evidence suggesting what the level of awareness will derive from a proposed marketing budget and planned set of marketing communications. Based on the heavy marketing communications employed by satellite radio companies, a 30 percent awareness rate is offered; that is, 3 out of 10 people will have seen advertising and publicity about satellite radio and become aware that satellite radio is available for purchase. Applying this proportion, the forecast is now 30,310,662 vehicles.

Sizing the Penetrated Target Market

The last model component represents the penetration rate, which can be interpreted as that proportion of the marketplace intending to buy the new product. One approach could be to look at the historical penetration rate for similar or analogous products and technologies. Another approach could be to apply an estimate of company market share among competitors. Still a third approach is to rely on diffusion theory, which suggests that customers for new product technologies fall into one of five groups: approximately 2.5 percent are innovators (first users), 13.5 percent are early adopters, 34 percent are early majority, another 34 percent are late majority, and the remaining 16 percent of customers are laggards (last users). Applying the third approach, an estimate of 16 percent is offered to suggest that innovators and early adopters will make up immediate sales—these are the individuals who are most apt to purchase the latest technology in a marketplace (cf. Rogers 1995; Moore 1995).

Taking this approach, 16 percent of the attainable market is presumed to be likely candidates for purchasing the emerging satellite radio technology. 16 percent of the attainable market results in a market size of 4,849,706 vehicles. A further assumption is made that there will be only one satellite radio per vehicle so this value represents an annual estimate of the number of satellite radio subscribers.

Other Market Factor Considerations

One might conclude that if this is an annualized estimate, a five year estimate could be determined by multiplying this number by five. Had this product been a grocery product or similar consumable good, then such logic might be appropriate. In the case of satellite radio subscribers, there are three consumer outcomes per year: a consumer keeps the satellite radio subscription for another year (maintain outcome), the consumer cancels the satellite radio subscription (loss outcome), or a new consumer signs up for satellite radio (gain outcome). Discussion on these values would be undertaken and actual figures from each year could be applied to subsequent years. For purposes of illustration, a subscriber growth rate of 15 percent is used coupled with a customer defection rate of 5 percent, resulting in an incremental gain of 10 percent per year over five years. This value is applied and suggests a market forecast of 7,810,500. As one can tell, this figure is far below the 36 million subscriber estimate but is also below the achieved 11 million subscribers. Determining why this number is wrong would focus on the assumptions. Because of the assumptions-based model framework, such focus is possible due to the transparency of assumptions underlying the forecast.

ADAPTING THE ASSUMPTIONS-BASED MODEL FRAMEWORK

The assumptions-based model framework is flexible and can be adapted to multiple market forecasting situations with any number of assumption combinations. Consider the following example illustrating the generation of a meaningful new product forecast for first-year sales of a new computer network security product. The product to be forecast was a newly developed computer host and network-based architecture and software application. The application was specifically designed to meet security requirements, provide comprehensive protection of any networked environment, detect anomalous activity, and dynamically respond to security events. A distinction of the application was enhanced operational efficiency through customizable, centrally administered configuration tools and automated solutions that isolated and mitigated threats. In light of the information technology nature of the product, the designated target market was information technology personnel, including chief technology officers, wishing to better secure their companies' existing computer server networks. Product management was the function charged with developing the forecast.

A half-day cross-functional team meeting was held to discuss the forthcoming product. It included the product manager overseeing the application's development and launch, the director of product management, lead product managers for other major product lines, director of sales, director of sales support, and director of the sales and operations planning process. Note that in the case of this company, all new products and product launches were managed through the product management department, with launch closely coordinated with sales and operations planning.

One specific task assigned to this team by senior management was the development of a forecast on which to plan and gauge launch decisions for the new application. In the course of meeting discussions, important factors for successfully marketing the product were identified, the nature of relationships among these factors were determined within the structure of an assumptions-based model framework and deemed key model components, and assumptions for each model component were specified. The team decided to put forth to management the following assumptions-based model framework: total market size, the intended/marketed use, company market share, buying intent, and company market coverage. Although other model components could have been included, these five model components were viewed as most relevant, and more importantly, each of these specified components could be quantified based on existing data sources.

Market size was predicated on a recent Gartner Group study (*www.gartner. com*) and was supplemented by customer data collected by the product management group. Gartner estimated the value for total market size for new computer security technology at $3 billion. Note that because this figure is in dollars, the basis of measurement becomes dollars; had this value been in units, units sold would have served as the basis.

Intended/marketed use was defined as the percent of the marketplace using the new product technology as core technology. That is, the product technology could be used as the primary security system or as a peripheral or back-up system. The value of 65 percent was determined through interviews with sales management and product management groups in the company, suggesting that two-thirds of prospective customers would be looking for a primary security system.

Company market share in the core technology use segment was estimated from published industry reports noting competitor market shares, supplemented by sales management and product management personnel intuition. Market share was estimated to be 20 percent.

Buying intent was defined as the percent of the market interested and likely ready to migrate to a new core technology. Sales management was the predominant source for this value and indicated that a follow-up using its sales pipeline tool would be used to validate the number. Buying intent was estimated to be 25 percent.

Company market coverage was based on current worldwide sales networks. This value represented the extent of distribution that the company had worldwide. An 80 percent market coverage rate was given, indicating that the company could serve 80 percent of world markets through its existing distribution system.

The assumptions-based models framework is applied by multiplying the values for the five model components. The forecast for the new computer product is thus $78 million, which is calculated by multiplying $3,000,000,000 × 65 percent × 20 percent × 25 percent × 80 percent to equal $78 million. Presuming that the data are valid, the figure of $78 million stands as the forecast for first year sales of the new computer technology application. This figure can be broken down into quarterly estimates by simply dividing by four quarters, or $19,500,000 per quarter, although it is more likely that some degree of seasonality would exist because sales are very seldom uniform throughout the year. First-year sales fluctuations due to channel pipeline fill, ramping-up, and market diffusion effects are also important considerations. Breaking the annual figure into monthly estimates would require similar thinking. Note that the nature for how the annual forecast should be broken down into quarterly and/or monthly estimates would be predicated on further assumptions.

CONSTRUCTING AN ASSUMPTIONS-BASED MODEL

As these examples show, applying the assumptions-based models framework is straightforward. What is not straightforward is data collection and getting consensus on the assumptions and corresponding numbers to input into the framework. Numerous variations for the types of data to be collected and the specific assumptions to be made are possible, all depending on the company and the business. The best way to start is to hold at least one meeting to lay out assumptions and to identify where data can be found and what numbers and assumptions are most relevant. More likely, a series of meetings that include representatives from different parts of the company such as marketing, operations, sales, product management, and research and development will be necessary. Specifically, when dealing with a new product introduction, a meeting of the sales and operations planning team (cf. Wallace 1999) is the logical starting point to determine which assumptions to include in a tailored assumptions-based model.

When constructing an assumptions-based model, one should be particularly cognizant of issues pertaining to validity, precision, and data availability. Validity of assumptions is paramount to successfully developing a reasonable and meaningful new product forecast. An uneducated, wild guess on any particular assumption will almost always result in an erroneous new product forecast because each individual assumption impacts the forecast outcome. If a number of assumptions are not valid, the new product forecast will only be more erroneous.

Precision is another consideration. Slight deviations in any assumption have the potential to significantly change the resulting new product forecast. This is particularly evident when an assumptions-based model is predicated on only a select few assumptions. Such sensitivity to assumption precision highlights the use of range forecasting versus point forecasting. That is, a range around each assumption is determined, versus reliance on one specific number for each assumption. A what-if analysis can then be employed to determine the range of forecast outcomes, depending on the low and high values for each assumption. This exemplifies how assumptions-based models can be used as tools for what-if scenario analysis and sensitivity analyses can establish critical assumptions and examine risk.

Data availability is a third consideration. While determining the assumptions to include in the model is important, determining how to quantify and collect a value for that assumption is just as important. An assumption that cannot be quantified to serve as an input in the assumptions-based model is not useful, nor meaningful. This does not mean that the assumption must be solely objective; managerial intuition may be a necessary element in deriving a value for a given assumption. The latter, though, will require careful thinking about how to systematically collect and quantify the subjective data.

CONDUCTING RISK ANALYSES

Assumptions-based models lend themselves very easily to conducting risk analyses. Risk analyses are conducted by establishing high and low points around each of the given model assumptions, which corresponds to best case (optimistic) and worst case (pessimistic) scenarios on each assumption, respectively.

Risk analyses were applied to the computer network security product previously discussed, where the base case (also referred to as the likely case) was the original $78,000,000 calculation. Product management personnel held discussions with sales management personnel to lay out potential best-case and worst-case scenarios. As shown in Table 9-1, it was determined that Core Use had the potential to run as high as 80 percent, but could be as low as 40 percent. History had indicated that market share regularly fluctuated between 30 percent in good months and 10 percent in difficult months. Buying intent was seen as a variable, falling between 20 percent and 30 percent. And market coverage, which was viewed as the most certain of the given assumptions, had the potential to increase to 95 percent based on distributor growth. Note that best-case and worst-case scenarios were envisioned for four assumptions, excluding target market size. Managers in sales management and product management determined that target market size could be considered a constant. Table 9-1 presents all best-case and worst-case scenarios.

Using these values, a sensitivity analysis was conducted within the assumptions-based model framework. This is done by holding all values constant to the base case, and changing the assumption under consideration to its best-case and worst-case values. The resulting forecasts when the best-case value is used and when the worst-case value is used are recorded and compared to the initial (base case) forecast. For example, the *core use* assumption is scrutinized on its best-case and worst-case scenarios. Holding all assumptions

TABLE 9-1.
An Example of Pessimistic, Likely, and Optimistic Values for Input
Assumptions

Assumption	Base Case	Best Case	Worst Case
Core Use	65%	80%	40%
Market Share	20%	30%	10%
Buying Intent	25%	30%	20%
Coverage	80%	95%	80%

constant but allowing for a *core use* best case of 80 percent results in a new product forecast of $96 million ($3,000,000,000 × 80 percent × 20 percent × 25 percent × 80 percent). Holding all assumptions constant but allowing for a *core use* worst case of 40 percent results in a new product forecast of $48 million ($3,000,000,000 × 40 percent × 20 percent × 25 percent × 80 percent). These values indicate that if *core use* is really as high as 80 percent, then the market is $18 million larger than the base case ($96,000,000 − $78,000,000 = $18,000,000). If the *core use* is really as low as 40 percent, then the market is $30 million smaller than the base case ($48,000,000 − $78,000,000). These differences portray the sensitivity and risk surrounding the *core use* assumption. The same line of thinking and analyses would be applied to the remaining assumptions to assess model sensitivity and risk surrounding these assumptions. Such sensitivity analyses help identify which assumptions should be deemed too uncertain. Refer to Tables 9-2 and 9-3.

The range between the best case and worst case is another metric for evaluation of risk and sensitivity. For example, *core use* with a best-case scenario of $96 million and worst-case scenario of $48 million, indicates a

TABLE 9-2.
An Example of Model Sensitivity with Regards to Financial Outcome

Assumption	Base Case	Best Case	Worst Case
Core Use	$78,000,000	$96,000,000	$48,000,000
Market Share	$78,000,000	$117,000,000	$39,000,000
Buying Intent	$78,000,000	$93,600,000	$62,400,000
Coverage	$78,000,000	$92,625,000	$78,000,000

TABLE 9-3.
An Example of Evaluating Potential Shortfall, Upside, and Total Risk

Assumption	Best Case: $ Above Base Case	Worst Case: $ Below Base Case	$ Range Between Best Case and Worst Case
Core Use	$18,000,000	−$30,000,000	$48,000,000
Market Share	$39,000,000	−$39,000,000	$78,000,000
Buying Intent	$15,600,000	−$15,600,000	$31,200,000
Coverage	$14,625,000	$0	$14,625,000

possible swing of $48 million. As shown in Table 9-3, the assumption of *market share* has the greatest range compared to the other assumptions. This suggests that *market share* has the greatest level of uncertainty. Interestingly, the lowest worst-case value and highest best-case value also correspond to *market share*, highlighting that market share assumptions are crucial components of the new product forecast (refer to Table 9-2).

One tool for visualizing this sensitivity analysis is a *Tornado chart*. A Tornado chart shows the financial values associated with each assumption. Proper use of the Tornado chart would involve first sorting the assumptions from high range to low range, and plotting accordingly to draw a tornado-like picture (cf. Clemen 1996). The Tornado chart simplifies the effort in determining where risk exists; those assumptions with the longest bars in the negative side of the chart would be assumptions deserving close scrutiny. Those assumptions with the greatest range also would be deserving of discussion by the management team. Figure 9-2 portrays the Tornado chart of the data found in Tables 9-2 and 9-3.

A further analysis that can be conducted using the assumptions-based model framework is a business simulation. Specifically, a Monte Carlo simulation can be applied to those assumptions where there are best-case and worst-case values. The simulation would randomly generate a value falling between the best case and worst case on each assumption and would calculate the resultant new product forecast. Running a number of iterations of the simulation would then provide a distribution of outcomes for determining the probability of attaining a given new product forecast.

There are a variety of ways to generate random numbers and conduct a Monte Carlo simulation. One approach is to use the RAND function in Microsoft Excel, which randomly selects values between the pessimistic and optimistic cases based on a uniform distribution between 0 and 1. The formula for using the RAND function to generate an outcome for each assumption

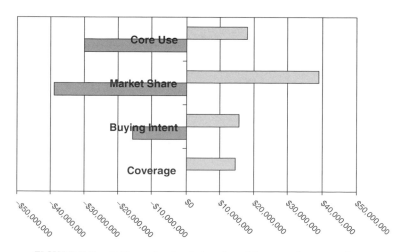

FIGURE 9-2. A Tornado chart showing risk around assumptions.

would be as follows:

$$\text{Assumption Outcome} = [\text{RAND}() \times (\text{Bestcase} - \text{Worstcase})] + \text{Worstcase}$$

For illustrative purposes, presume that the RAND() function generates the random value of .3324 in one iteration of the simulation. This value is used to calculate a core use estimate of 53.296 percent as follows: [.3324 × (80 percent − 40 percent)] + 40 percent. The values of the other assumptions and subsequent simulations runs would be calculated similarly, though with different random numbers generated by the computer program.

Note that use of the Random Number Generator found in Microsoft Excel's Analysis ToolPack is an even simpler approach for generating random numbers and allows for other distributions to be modeled, in addition to the uniform distribution (other computer programs are available to do this as well) (cf. Weida et al. 2001).

As shown in Table 9-4, 1,000 iterations of a Monte Carlo simulation were run using the best-case and worst-case data for each assumption as simulation parameters. Tabulation of the results from the simulation form a distribution around the new product forecast and provide a probabilistic view of the new product forecast. As shown in Figure 9-3, 27 percent of the simulations resulted in a financial value falling between $60 million and $80 million. The cumulative percentages of these results further indicate that 58 percent of the simulations had a financial value of less than $80 million. Referring back to the base-case estimate of $78 million, these results roughly suggest that there is approximately a 42 percent chance (100 percent − 58 percent) of reaching or exceeding this base-case estimate. This therefore forces the issue of whether management is comfortable with the probability in attaining this number. In short, is management comfortable with a 42 percent chance of success? If no, discussions would focus on the level of risk that management is comfortable

TABLE 9-4.
Sample of Monte Carlo Simulation Data

Simulation Run	Core Use	Market Share	Buying Intent	Coverage	Financial Outcome
1	68%	21%	22%	85%	$79,658,870
2	57%	25%	26%	94%	$104,365,372
3	66%	24%	21%	84%	$83,327,863
4	69%	29%	28%	83%	$138,039,675
5	74%	17%	27%	88%	$88,769,425
6	60%	18%	29%	92%	$87,420,466
7	53%	26%	29%	90%	$109,708,715
8	53%	22%	24%	85%	$72,116,748
9	47%	16%	30%	81%	$53,424,659
10	59%	23%	23%	86%	$81,347,268
...	...	...	...	...	...
1000	71%	30%	20%	89%	$115,924,185

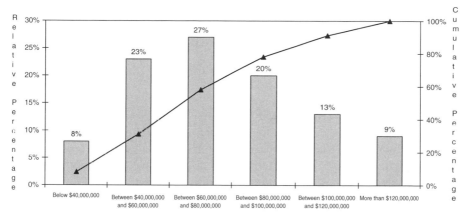

FIGURE 9-3. Distribution of financial outcomes from a Monte Carlo simulation.

with and/or what assumptions need to change and what the values need to be in order to make this initiative have a higher chance of attainment.

LAUNCH PLANNING

Assumptions-based models can be quite useful in the course of launch planning. For instance, following a risk analysis, a list of assumptions deemed critical can be constructed and further evaluated in preparation for the new product introduction. Subsequent discussion would then focus on the likely values for these critical assumptions and the nature of their effect on the new product introduction. This keenly focuses discussion on assumptions versus solely a numeric forecast. This is a particularly important matter during launch planning. In many cases, managers unfortunately become preoccupied with the forecast value and not how the value was calculated. It is just as important to know how a forecast was derived.

Leading companies also have found that establishing a set of common, consistent assumptions that would apply across all products within a strategic business unit (SBU) helps focus discussion and evaluation of forthcoming introductions. Similar questions can then be asked across projects, allowing for an equivalent evaluation of each project's forecast. In other words, managers can compare forecasts and their assumptions on an apples-to-apples basis. Failure to use common assumptions across projects makes comparisons of forecasts more difficult due to different assumptions. For example, a shipments forecast cannot be equated to a consumer demand forecast; leading companies predetermine which forecast is the focus. Use of common assumptions also can allow for the development of a standardized launch scorecard for evaluating the progress of each new product introduction. Tracking sales along with data for each assumption will provide insight into the nature of market behavior and will answer the question of why the market is behaving the way it is.

Should the new product not reach its target value for sales, then an analysis of the tracking data that corresponds to the forecasting assumptions will likely provide an answer for why the product is not performing well. Sales data alone cannot answer this question.

ASSUMPTIONS MANAGEMENT

New product forecasting is inherently a process of *assumptions management*. When employing an assumptions-based model, the assumptions are clearly defined and made transparent during decision making. New product forecasting occurs throughout the new product development process (Kahn 2006), but it is crucial to establish a final set of assumptions so that a forecast can drive launch decisions such as marketing budgets and supply chain commitments.

Thinking of new product forecasting as assumptions management leads a company to begin tracking assumptions from the point when the first forecast is generated through the point when the new product is introduced. At each review gate between these points, the new product forecast and corresponding assumptions are revisited, verified, and reissued to underlie an updated new product forecast. Risk relative to base, pessimistic, and optimistic cases are noted at each gate, and actions are pursued (such as conducting a market research study) to reduce high-risk assumptions and increase confidence in the proximate accuracy and meaningfulness of the new product forecast. Over time, the tracking of assumptions across multiple projects will create a sizable database of assumptions data on which analysis may be performed to assess the relationship of various assumptions to new product success and failure. This will further validate whether assumptions can be characterized as critical. Internal company benchmarks and guidelines for model assumptions can be established for future projects and new product introductions.

PITFALLS TO AVOID WHEN USING AN ASSUMPTIONS-BASED MODEL

Assumptions-based models are certainly not a panacea for forecasting new-product introduction. It is presumptuous to say that use of an assumptions-based model will immediately result in accurate new product forecast. There are many other forecasting techniques that could be used to forecast a new product introduction, and these might be better suited for the new product forecasting task at hand (see Kahn 2005, 2006). Unlike other techniques, however, an assumptions-based model forces robust discussion over the assumptions and the issues that underlie the new product, resulting in a more thoughtful determination of the new product forecast. Assumptions-based models thus offer transparency to forecast assumptions so that reappraisal and verification of forecasting assumptions can occur across functional departments, where each department can bring incremental data and knowledge about the marketplace and technology capabilities.

Another pitfall is the failure to undertake due diligence to clarify an assumption. If there are data, then some degree of validation and verification of subjective inputs should be undertaken to solidify an input assumption. Such analysis can include past data, surrogate products, and consumer data. Available data and information should be readily analyzed and referenced in the course of building an assumptions-based model. A tendency to rely solely on judgment, anecdotal evidence, and gut feel, in lieu of analysis, can lead management decisions astray. In short, use of an assumptions-based model should not excuse thoughtful data analysis.

Third, employing an assumptions-based model without a mindset of new product forecasting as assumptions management may not result in a meaningful forecast. Such a mindset emphasizes systematic thinking around assumptions, data collection, and forecast calculation. Without an assumptions management mindset, assumptions in the model are not regularly verified, nor are they tracked to gauge consistency. Assumptions not documented or tracked have the greater tendency to meander, be manipulated without notice, and fall suspect to internal company politics.

KEYS TO SUCCESS IN APPLYING ASSUMPTIONS BASED MODELS

It is important to mitigate these pitfalls when developing an assumptions-based model. Thus, several questions are posed to stimulate thinking around the ability to establish a model and manage assumptions. These questions are offered to frame your evaluation of your company's new product forecasting endeavor and readiness to use an assumptions-based model. Although there are no right or wrong answers, the ability to readily generate an answer to each question is an indicator that at least an effort toward developing and establishing an assumption-based model may be possible and worthwhile:

- What factors can be used to forecast a new product introduction?
- What assumptions should be associated with each relevant factor?
- How can/should these assumptions be operationalized?
- What assumptions appear to be common across new product forecasts?
- What prelaunch data sources are available?
- What assumptions/variables can and should be tracked?
- How can these variables be tracked, both prelaunch and postlaunch?
- How can the endeavor for developing a new product forecast be effectively linked to the processes of new product development and sales and operations planning?

SUMMARY

The task of new product forecasting is certainly laudable, but a systematic approach and delineation of assumptions for the purpose of understanding, versus target setting, will result in a meaningful forecast at the point of new product introduction. Through use of assumptions-based models and the practice of assumptions management, the new product forecasting process can be managed and repeated in a valid and reliable fashion. As defined, forecasting is the process of deriving an estimate of attainable demand under a given set of conditions (Kahn 2005). new product forecasting via assumptions-based models clarifies what the conditions should be along with a numeric output. So, rather than a numbers exercise or a computer-generated statistic, new product forecasting is truly analysis aimed at gaining underlying insight.

REFERENCES

Clemen, Robert T. 1996. *Making Hard Decisions: An Introduction to Decision Analysis* 2nd ed. Pacific Grove, CA: Duxbury Press.

Kahn, Kenneth B. 2002. "An Exploratory Investigation of New Product Forecasting Practices." *Journal of Product Innovation Management*, 19 (March): 133–143.

Kahn, Kenneth B. 2005. "Approaches to New Product Forecasting." In *The PDMA Handbook of New Product Development*. Ed. Kenneth B. Kahn. Hoboken, NJ: John Wiley and Sons, 362–277.

Kahn, Kenneth B. 2006. *New Product Forecasting: An Applied Approach*. Ormonk, NY: M.E. Sharpe.

Latta, Michael. 1998. "Using Market Models to Forecast Demand for Ethical Pharmaceuticals." *Journal of Business Forecasting*, Spring: 3–8.

McBride, Sarah. 2006. "Until Recently Full of Promise, Satellite Radio Runs into Static," *Wall Street Journal*, August 15, 248(38): A1–A9.

Moore, Geoffrey A. 1995. *Crossing the Chasm: Marketing and Selling High-Tech Products to Mainstream Customers*, New York: Harper Business.

Rogers, Everett M. 1995. *Diffusions of Innovations*, New York: The Free Press.

Tyebjee, Tyzoon T. 1987. "Behavioral Biases in New Product Forecasting." *International Journal of Forecasting*, 3 (3/4): 393–404.

Wallace, Thomas F. 1999. *Sales and Operations Planning: The How-To Handbook*. Cincinnati, OH: T. F. Wallace and Company.

Weida, Nancy C., Ronny Richardson, and Andrew Vazsonyi. 2001. *Operations Analysis Using Microsoft Excel*. Pacific Grove, CA: Duxbury Press.

Part 3

Strategic Tools For Improving NPD Performance Across the Firm

This section covers two tools that span from applica-tion at the project level to application at the business unit or firm level. However, they likely are more powerful when used strategically at the business unit or firm level. Although they concern issues important to all sizes of firms, small firms who can not yet afford large legal or human resource groups especially will benefit from these chapters.

Chapter 10 presents processes and tools for man-aging and protecting intellectual capital in the NPD realm. In addition to defining the different categories of intellectual property, the chapter shows how to create, maintain, and protect different kinds of intel-lectual property, deploy intellectual property across the firm, and assess and analyze the intellectual prop-erty contained in your technology portfolio. The authors have succeeded in presenting a very com-plicated legal subject in a clear, easily understood fashion.

In Chapter 11, four tools are presented that should promote creativity, motivation, teamwork, and high performance in NPD teams and organizations.

The career ladder *provides a mechanism by which the capabilities and expectations of every individual in the technical organization can be organized. Use of this tool allows the technical expert to advance professionally without having to go into* management. *The* selection model *is a hiring tool that includes a set of robust selection and promotion evaluation criteria. The* performance review *identifies each person's strengths and weaknesses in relation to a previously agreed-on clear set of expectations. Finally, the* action plan *is the set of specific, measurable, and attainable goals that help each individual outline the steps to continue to meet the expectations of his or her current job, and grow into the next level of the organization. Taken together, these tools can help a large company get the most out of its technical employees, and can help a small, more informally managed company move to the next level of organizational sophistication.*

10 Intellectual Property and NPD

Sharad Rastogi
Principal, PRTM

Ari Shinozaki
Director, PRTM

Matthew Kaness
Manager, PRTM

INTRODUCTION

Intellectual property (IP) has emerged from the dusty back rooms of the legal department to become one of the most effective tools for improving the corporate bottom line for innovation-driven companies. These companies have shifted the paradigm for IP from *legal expense* to *strategic asset*. Going forward, leading organizations will be the ones that not only create a strong intellectual property portfolio, but extract real value from it as well. Companies need to approach this in a systematic manner with appropriate management systems, processes, and tools. What separates the winners from the runners-up is the ability to maximize both intellectual property value creation and value extraction or capitalization. The former refers to a company's ability to develop an optimal intellectual property portfolio, while the latter refers to the ability to exploit the portfolio's real and perceived value.

It is important to understand how significant this value can be. For those companies with aggressive IP strategies, the business results can be staggering. IBM, Texas Instruments, and P&G are well-known examples. IBM boosted its patent licensing revenues from $30 million in 1990 to build a billion-dollar business (with very high margins) by aggressively mining and out-licensing its massive patent portfolio (see Rivette and Kline (2000) and Berman (2006)). Texas Instruments followed a similar approach to grow its licensing revenues to $800 million per year by 2000. As another example of IP-based business strategy, Huston and Sakkab (2006) describe how P&G has implemented an "Open Innovation" business model that encourages in-licensing and co-development alliances with third parties, and accounts for 35 percent of new products introduced and 45 percent of its development pipeline. In all of these cases, strategic IP management has been a key enabler for business performance.

In addition to top-line benefits, management of IP also has significant expense implications. Not only can the cost of protecting and maintaining a multicountry IP portfolio approach several hundred thousand dollars *per patent*, but also the liability risk and related expense can be in the multimillion dollar range. With these heavy costs at play, companies face tough decisions regarding the invention disclosures that they should file, the patents they should continue to maintain, the countries in which they should maintain an IP presence, and the IP research (prior art search and clearance studies) they should conduct. Companies that manage their IP portfolios well can reap significant cost savings. Petrash (1998) describes how a disciplined approach at Dow Chemical resulted in savings of $40 million over a 10-year period. Regardless of whether the source of value captured is incremental revenue or cost-savings/avoidance opportunities, companies must take a systematic approach—one that aligns IP, business, and technical strategies, applies effective portfolio management processes, and implements best-in-class IP management tools and techniques.

This chapter is written for two distinct, but related, audiences: (1) senior management and business leaders who are charged with developing and/or approving IP strategies and related investments and (2) middle management, NPD project team members, and IP professionals, who are typically tasked with implementing these strategies. It is important to address both audiences, as the strategy and execution of IP management is interwoven throughout IP creation and value extraction. This chapter has been organized around this premise.

This chapter includes the following topics. The next section sets the stage by providing a brief overview of different types of intellectual property. The following section discusses creation, protection, and maintenance of IP. It highlights activities related to IP in a typical new product development process, provides tips on choosing the appropriate instrument for protecting IP in different situations, discusses how to select countries where IP should be protected, describes the key activities and milestones in the lifecycle of a patent, reviews the criteria for selecting inventions that should be protected, and discusses some strategies and best practices for effective creation of IP. Then different strategic alternatives, cost-benefit characteristics of different strategies, and factors that affect the choice of the most appropriate strategy are discussed. The next section describes an approach for the assessment and analysis of IP portfolios and illustrates it with an example, followed by another example that enumerates some best practices for the overall management of intellectual property. Throughout the chapter, critical tools are described and examples are utilized to help bring the discussion to life. The sections on overview of IP, IP creation, protection, and maintenance and IP best practices will be of interest to readers from both small and large organizations. The sections on IP deployment strategies and portfolio management will be of greater interest to readers from larger organizations that have sizeable IP portfolios.

OVERVIEW OF INTELLECTUAL PROPERTY

An organization's *intellectual capital* can take many forms. It includes ideas, know-how, skills, inventions, technologies, processes, and publications. Intellectual capital items that have been codified and documented are typically called *intellectual assets*. Intellectual assets that have been protected under the laws of a nation are called *intellectual property* (IP). Figure 10-1 shows these relationships in a model adapted from Sullivan (1998).

Most countries have provisions for four types of intellectual property. Yoffie (2005) and Conley and Orozco (2005) also provide good overviews:

- ◆ Patents
- ◆ Trade secrets
- ◆ Trademarks or servicemarks
- ◆ Copyrights

Patents

A patent is a *government* grant extended to the owner of an invention to *exclude* others for a *limited time* from *making, using or selling* the invention and includes the right to *license* others to make, use, or sell the invention. In the United States, the limited time runs for a 20-year period from the date of the filing of the patent application.

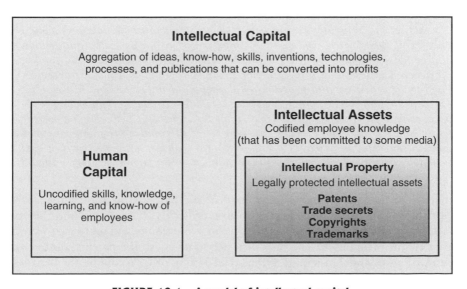

FIGURE 10-1. A model of intellectual capital.

The patent application includes the following:

◆ A written description of the invention and the process of making and using the invention

◆ *Claim(s)* that define the essential elements and scope of the invention

◆ Advantages of the invention that distinguish it from previously known similar techniques or structures, which are collectively called *the prior art*

Once the patent is issued by a government agency, the patentee must enforce the patent without aid of the patent office. Patents are issued for four general types of inventions:

1. Compositions of matter
2. Machines
3. Articles of manufacture (i.e., man-made products)
4. Processing methods (including business processes)

The most important part of a patent application is the *claims* that appear at the end. Each claim is a sentence that defines the essential elements of the inventive concept. The scope of the words used in each claim, as interpreted in view of the written description of the invention contained in the application, defines the metes and bounds of the invention. For example, U.S. Patent No. 6,000,000 (issued Dec 7, 1999, to 3Com) included this first independent claim for an extendable method and apparatus for synchronizing multiple files on two different computer systems. This claimed invention is better known commercially as a PALM ® PDA:

> *"A method of sharing information on a first computer system and a second computer system, said method comprising: connecting said first computer system to said second computer system with a data communications link; providing a library of functions in said second computer system for accessing information on said first computer system; creating a conduit program database, said conduit program database for storing a list of conduit programs that may be executed, registering a first conduit program by placing an identifier for said first conduit program in said conduit program database, said first conduit program comprising a computer program on said second computer system for performing a specific data transfer task; successively executing a set of conduit programs identified within said conduit program database from a manager program, each of said conduit programs accessing said library of functions for communicating with said first computer system."*

Since more than one claim is permitted in the patent, the applicant has the opportunity to describe the invention in words of varying scope and thereby claim the invention as broadly as possible in some claims and more narrowly in other claims. The patent attorney then negotiates with the patent examiner at the U.S. Patent and Trademark Office (USPTO) in order to obtain the broadest possible claims that distinguish the invention from the prior art while still encompassing the crux of the invention.

Trade Secrets

A trade secret is technical or business information that does the following:

- Derives *economic value* from *not being generally known* to other persons who can obtain economic value from its disclosure or use
- Is the subject of reasonable efforts to maintain its secrecy

In other words, a trade secret is "almost any knowledge or information", used in the conduct of business, which is held "in secret":

- The information must be used by the company.
- It must give the company a competitive advantage.
- It must be a secret (i.e., not known generally to the industry).
- There must be a reasonable system of security to protect the secret.

A trade secret can exist until it is generally known in its industry. However, once it is publicly known, it ceases to be a trade secret. Some trade secrets have lasted for decades (e.g., the Coca-Cola formula). Examples of trade secret and sensitive business information include formulas, patterns, compilations, programs, devices, methods, techniques, processes, customer lists, manufacturing instructions, marketing plans, financial performance data, and business strategies.

Trademarks

A trademark is any word, name, symbol, picture, sound, device, or any combination thereof, adopted and used by companies to identify their goods and distinguish them from those manufactured by others:

- The owner of a trademark has the exclusive right to use it on the product it was intended to identify and on related products.
- Servicemarks serve the same purpose for services.
- Trademarks and servicemarks are indefinite in duration as long as they are renewed.

Trademarks are suffixed with a ™ symbol. Trademarks can also be registered with a government agency such as the U.S. Patent and Trademark Office (USPTO). Registered trademarks are suffixed with a ® symbol and provide the owner exclusive right to use the trademark and bring action against any infringers. Unregistered trademarks (designated by the ™ symbol) do not block others from *using* the mark but do exclude others from *registering* the trademark.

Copyrights

A copyright is protection afforded to the author of an original work of art that is *fixed in a tangible medium of expression*. A work is fixed in a tangible medium of expression when its embodiment in a copy, by the author, is sufficiently permanent or stable for it to be perceived, reproduced, or otherwise communicated for a period of more than transitory duration. Works of authorship include writings, graphic arts, motion pictures, sound recordings, architectural designs, and computer software, as examples. A copyright gives the owner exclusive rights to the work including right of display, distribution, licensing, performance, and reproduction. A copyright lasts for the life of the owner plus 70 years. Copyrights are obtained simply by filing an application with the Library of Congress. They do not need to be examined, assessed or granted.

IP CREATION, PROTECTION, AND MAINTENANCE

This section describes the concepts, tools, and examples for the creation, protection, and maintenance of intellectual property.

IP Activities in New Product Development

New product development (NPD) entails several IP-related activities. Figure 10-2 shows the schematic of a typical product development process with the timing, sequence, and duration of major cross-functional activities and milestones, including activities related to the management of IP. NPD practitioners should ensure that these IP activities and decisions are well integrated with the overall product development process.

The first activity, *IP strategy and planning*, is important in early stages of new product development. When evaluating new product concepts and assessing their technical feasibility, the NPD teams should also understand the related IP landscape and develop an IP strategy. This includes understanding IP needs of the NPD effort, existing IP and prior art, competitive IP and its implications for design freedom and design limitations, new IP that might be generated by this development effort, and any IP-related risks. The IP strategy will be shaped by any technical innovations inherent in the concept as well as unmet technology needs. Technology and IP scans should identify what is available externally that can be in-licensed or acquired and what needs to be developed internally. The IP strategy should also define the scope of IP protection (focused on the current development effort or broader) and countries where IP should be protected.

IP strategy implementation includes activities related to the implementation of the IP strategy defined in the earlier stages of product development. This includes the negotiation and execution of agreements with any external

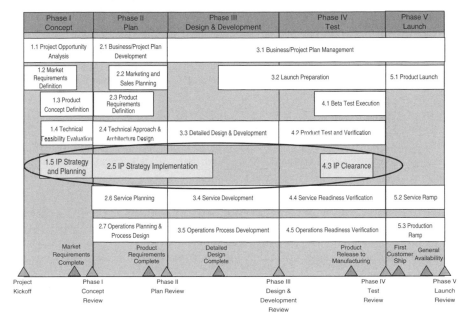

FIGURE 10-2. IP activities in NPD.

parties that might be involved in the development program (suppliers, licensors, co-developers, etc.); the acquisition or in-licensing of external IP that is critical to the development effort; the disclosure of new inventions conceived as part of technology and product development; and the preparation, filing, prosecution, and issuance of IP applications. Inventors should be careful about the timing of recording the disclosures. In countries like the United States with *first-to-invent* rules, if multiple inventors claim the same invention, the timing of invention determines who gets the patent. Furthermore, a patent application must be filed within one of year of public disclosure. By contrast, European countries follow *first-to-file* rules and do not provide the one-year window for filing the application. The later sections in this chapter discuss IP creation and protection in greater detail.

Another key IP-related activity is *IP clearance* prior to product launch. The purpose of this activity is to ensure that any new IP included in the new product is suitably protected. Issuance of patents on new IP is not necessary prior to the launch of the products as long as the patent applications have been filed (i.e., products can be launched with patent-pending status). IP clearance activities should also ensure that the product does not infringe on any other existing IP not owned by the company. Infringement could expose the new product to the risk of an enforcement action by the IP owner. To avoid this risk, NPD Teams should consider alternate designs around the existing IP or consider obtaining usage rights to the IP.

Criteria for IP Protection

Managing the IP protection pendulum can be tricky for practitioners. A company's competitive landscape, IP litigation history, financial performance, and senior management preferences may drive IP protection momentum to one extreme or the other in order to protect any and all intellectual capital that can be construed as IP or to focus very narrowly on a particular product line or technology. While the former drives up expenses and bogs down inventers with IP administration, the latter increases risk and reduces leveragability. In order to balance this pendulum, a well-defined set of criteria is required to decide which IP items should be protected.

In this context, an important question to be addressed is not what *can* be protected but what *should* be protected. Although the difference is subtle, the impact is not. Ideally, a company's critical IP is what should be protected. However, the definition of *critical IP* could be as varied as there are viewpoints in an organization. What is critical to the *company* may not necessarily be critical to the person or team being asked to make the IP protection decision, such as R&D, management, or legal. Therefore, a set of cross-functional objective criteria is required. The best criteria to use will depend on a company's specific situation but, generally speaking, should cover the following categories: criticality of the problem solved by the invention, level of innovation (breakthrough, major improvement, or incremental improvement), level of difficulty to design around the invention, commercial value, breadth of applications, strategic fit, and lifecycle IP costs. The section on portfolio assessment and analysis discusses these criteria in greater detail and provides a tool for structuring them. Companies should use it to define their own system of assessing and prioritizing the invention disclosures.

CHOOSING THE RIGHT IP INSTRUMENT

Different types of instruments are available for protecting intellectual property, such as patents, trade secrets, copyrights, and trademarks. Often, multiple instruments might be applicable for specific IP situations. This section discusses factors involved in three common situations: (1) choosing between patents, trade secrets, and other approaches, such as deciding whether to keep the information company-confidential or to publish it, (2) deciding whether to file provisional patent applications, and (3) using trademarks.

Patents, Trade Secrets, or Other Approaches

In choosing the right instrument for an IP item, the useful life of the IP (period of commercial viability and benefit) and the need to have the IP (the invention and the associated know-how) available in the public domain should be considered. Figure 10-3 shows the best approach for four different scenarios:

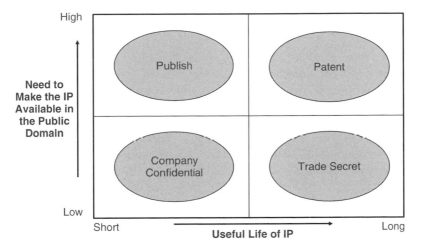

FIGURE 10-3. Choosing the right IP instrument.

1. Patents are best suited for inventions that have a long useful life but need to be made available outside the company, such as in commercial products or in know-how that needs to shared with business partners. Inventions that are embedded in commercially available products are often easy to discern and copy through reverse engineering, so patents provide a protection by preventing others from benefiting from the invention.

2. Trade secret is the best option for IP items that have a long useful life but do not need to be made available in the public domain and can be maintained as internal secrets, such as recipes, formulas, and manufacturing processes. Trade secrets, however, require considerable investment in operational securities and internal procedures to maintain their secrecy. There are different levels for trade secrets based on the level of disclosure required to enable commercialization:

   ◆ *Level 1 trade secrets:* Highly valuable trade secrets, like the Coca-Cola formula. They are maintained as legally defensible with stringent operational controls and protections that can be legally used to show that proper steps were taken to control access to the trade secret and restrict any public disclosure.

   ◆ *Level 2 trade secrets:* Trade secrets for which a controlled disclosure to a third-party (e.g., a non wholly owned subsidiary, a joint-venture partner, or a contractually bound development partner) is necessary to extract commercial value. Though operational protections are extended to cover the third-party to maintain controlled disclosure, this type of trade secret is tougher to defend legally. For example, a manufacturing process developed for a new patent-protected product may not be legally defensible as a trade secret because of prior disclosure to a joint venture partner, but the competitive value may be significant enough to maintain the process as a level

2 trade secret in order to mitigate further disclosure outside of the company's network of partners.

3. For IP items that only have a short useful life, patents or trade secrets may not be cost-effective options (due to the costs for maintaining operational security and internal controls for trade secrets and the costs associated with procuring and maintaining patents). For such items, if there is no need to disclose them externally, it may be sufficient to keep them as company confidential or proprietary information. The operational controls for company confidential information (such as nondisclosure agreements or employee confidentiality agreements) are much less onerous than for trade secret protection. Examples of company confidential information include pricing sheets, employee lists, and customer information.

4. For other IP items with short useful life that need to be in the public domain, the best approach may be to publish the information (e.g., in scientific or trade publications). Although this approach does not provide any protection, it does create prior art that prevents others from claiming rights to the invention. Trademarked or copyrighted recipes are examples of this category.

Figure 10-3 can be demonstrated using two common scenarios.

Scenario 1: A new product technology (e.g., hardware/software) is developed that is easy to reverse engineer but has a long commercial viability. Figure 10-3 would recommend considering the use of a patent to protect the IP (which is common among technology companies). Further analyses show that the technology has low business value because it is in a nonstrategic market. This would indicate that the technology should be protected as a non-core patent, i.e., with controlled investment in IP protection by limiting the geographical scope of protection (fewer countries), number of patents to protect different aspects/uses of the technology, or the duration of protection (stopping renewals when the technology becomes obsolete).

Scenario 2: A new manufacturing technology (e.g., a food processing technique) is developed that is hard to reverse engineer but has a long commercial viability. Figure 10-3 would recommend considering the use of a trade secret (which is common among consumer packaged goods companies). Further analyses show that the technology has high business value but has some disclosure requirements in commercializing the technology due to the need to outsource food processing to contract manufacturers and joint venture partners. Therefore, this technology should be protected as a level 2 trade secret.

Provisional Patent Applications

Another consideration about the choice of appropriate IP instrument is the use of provisional patent applications. In the US, inventors are allowed to file a simple, low-cost provisional application without detailed claims in order to establish a date for the invention. The applicant is then given up to one year from the provisional application filing date to submit a formal and detailed application (including claims) for the invention.

It is important to note that the provisional application by itself cannot be the basis for patent issuance—a complete application must still be filed before the patent can be issued. The provisional application is merely a bridge that buys the applicant time while establishing the earliest possible invention date, referred to as the priority date for the invention. Often, the value and strategic fit of the invention is not known at the time when the invention is made. The provisional application, therefore, provides a cost effective way of gaining a priority date on an invention disclosure without having to define the invention with claims and incurring the costs of procuring a patent.

Although the use of provisional applications has grown since its introduction, this instrument is still underutilized. Since the United States follows first-to-invent rules (i.e., the patent is awarded to the inventor who can provide evidence for the earliest date of invention), as opposed to first-to-file rules, some companies view the provisional application as an unnecessary step between invention disclosure and filing a complete patent application. However, in the absence of a provisional application, the company has to rely on less reliable means such as internal records (e.g., lab notebooks) to establish the date of invention. Therefore, in situations where patentability and/or commercial issues are still being assessed, a provisional application should be considered.

Trademarks

In addition to patents and trade secrets, trademarks are a useful instrument for protecting IP. Trademarks (and servicemarks) can help a company in building a marketing advantage by articulating the value of the underlying patents, trade secrets, and other know-how for its market offerings. If a company consistently delivers market offerings with superior competitive differentiation, the associated trademarks become increasingly more valuable. Over time, market offerings become obsolete, patents expire and trade secrets may be independently discovered. But, trademarks can be held forever and sustain value well after other forms of IP protection lose their strength. Also, properly maintained, a trademark is essentially unassailable—a patent, by contrast, can often be challenged, and either found invalid through legal means, or can diminish in value through the patenting action of a competitor. To this end, it is critical that trademarks be protected to ensure that all brand and logo iterations are covered. Trademarks and brands/logos related to critical patents and trade secrets should be properly registered and consistently commercialized with the symbol ® so as to protect against brand dilution or the mark becoming ubiquitous.

Patent Plus Trademark

The coupling of trademarks with patents forms a more formidable set of IP than either by itself. For example, one company with a strong offering in materials had patents on the composition of materials as well as on the method of making the materials. But, more importantly, these materials also had a strong trademark. Over time, this trademark became synonymous with the best available material in the minds of many customers. Customers would often ask for the trademarked materials regardless of their performance attributes. Other competitors could create similar or even technically better materials, but only this company had the trademark that was recognized throughout industry. As another example, the ubiquitous trademark of Xerox® was named upon the original patented xerography process. Although the original patent has long since expired, the Xerox® brand still remains.

SELECTING THE COUNTRIES TO PROTECT

NPD professionals frequently face the question of deciding which countries they should protect their IP in. Most countries have their own laws and regulations for IP protection and an inventor is required to obtain separate patents in each country where he or she wants to protect inventions. In order to simplify the process of applying for patent protection in multiple countries on a single invention, the World Intellectual Property Organization (WIPO) created an application instrument via the Patent Cooperation Treaty, the PCT application. This instrument allows an inventor to file the application with his or her national office, but have the invention be examined for patentability in multiple offices based on the same application.

A similar mechanism is provided by the European Patent Office for European countries. However, there are fees and costs involved for the procurement and maintenance of patents in each country. Procurement costs include the costs of drafting and filing patent applications, responding to office actions by the Patent Agency, and the official fees for patent issuances. Maintenance costs include costs for renewing the patents (annually or less frequently depending on the country) for the life of the patent (typically 20 years). Table 10-1 shows the illustrative costs involved for selected countries. Actual costs depend on several factors, such as the complexity of patent application (number of claims, drawings, and pages), translation costs, use of outside counsel and legal support, number of office actions by the Patent Agency.

It is clear from Table 10-1 that the costs of protecting and maintaining the IP grow significantly in proportion to the number of countries selected. Most companies are faced with a trade-off between the costs and breadth of foreign protection. Figure 10-4 depicts a framework with two critical dimensions for making this choice. One dimension is the business value of the country for the IP that is being protected. Business value, typically, depends on factors such

TABLE 10-1.
Illustrative Costs ($) of Patent Protection/Maintenance in Different Countries

Country	Official Procurement Fees	Typical Additional Procurement Costs	Total Patent Procurement Cost	Maintenance Fees	Total Patent Lifetime Cost
Australia	762	4,163	4,925	12,449	17,374
Brazil	476	6,845	7,321	17,360	24,681
Canada	1,444	3,273	4,717	11,506	16,223
China	883	4,501	5,384	16,080	21,464
France	1,131	6,391	7,572	13,327	20,899
Germany	526	6,315	6,841	24,440	31,281
India	816	2,620	3,436	10,718	14,154
Italy	127	5,463	5,590	0	5,590
Japan	1,758	7,807	9,565	15,681	25,246
Mexico	1,330	5,360	6,690	10,115	16,805
Spain	1,440	6,445	7,885	13,492	21,377
United Kingdom	380	3,455	3,833	11,730	15,563
United States	2,600	6,400	9,000	7,000	16,000

Note: This table shows approximate costs for selected countries. Actual costs depend on the complexity of the patent application (number of claims, drawing, and pages), translation costs, use of outside counsel and legal support, number of office actions by the patents agency, etc.

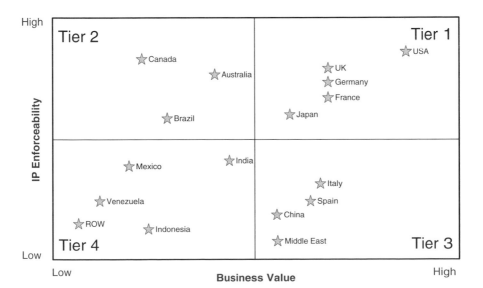

FIGURE 10-4. Prioritizing countries for IP protection.

as market size, growth rate, market share, and competitive advantage for the technology and products associated with the IP in question.

The other critical dimension is the enforceability of IP. This depends on factors like the supportiveness of a country's legal system for asserting IP rights, cost and timeframe of enforcement litigation, probability of success, and the cycle time for obtaining a patent. Managers in one company used this framework to segment the countries into four tiers. They decided to focus their IP investments mostly in Tier 1 countries (high business value and high enforceability) and selectively in Tier 2 countries (low business value, high enforceability). They decided to not invest in IP protection for Tier 3 and 4 countries due to low enforceability of IP in those countries. Note that the placement of countries in Figure 10-4 is only an illustrative example. The users of this tool should populate it for the countries that are relevant to them with the business value and enforceability for their own IP.

PATENT LIFECYCLE PROCESS

A patent, like other assets, goes through its own lifecycle, starting with the initial invention conception and ending with patent expiration. Figure 10-5 provides a quick-reference overview chart of the patent lifecycle process with typical durations for each step. This process can be broken down into five major stages and can be compared conceptually to a manufacturing process for a physical good.

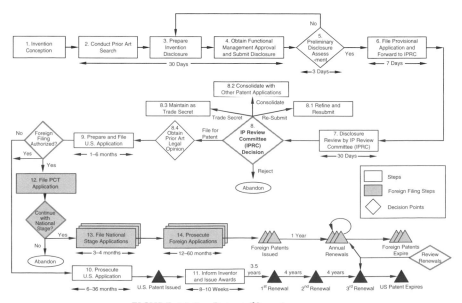

FIGURE 10-5. Patent lifecycle process.

The patent lifecycle process has raw material inputs (e.g., invention disclosures) and value-added processes (e.g., evaluation decision steps, patent application creation, filing and prosecution activities) that may result in a finished good (e.g., a patent) with an expected useful life that must be managed (e.g., patent renewals). The different stages in the patent lifecycle process are:

1. Disclose
2. Evaluate
3. Decide
4. File and prosecute
5. Renew/maintain

Each stage has one or more steps, as discussed next.

Stage 1: Disclose (Steps 1—4)

The initial steps in the lifecycle of a patent relate to the conception of an invention, research to determine its patentability, and formal disclosure of the invention to someone other than the inventor(s). Driven by the inventor(s) and/or R&D function, these initial activities are like creating the raw materials for the patent development process.

- *Step 1: Invention conception.* Whether generated via R&D discovery, marketing ideation, or some other internal capability, the patent lifecycle begins on the day when the invention is first conceived (this date is particularly important in countries with first-to-invent rules).

- *Step 2: Conduct prior art search.* In order to confirm that the invention is patentable, a prior art search is conducted to understand the related invention landscape and ensure that the invention concept is truly novel—prior art is anything that was publicly disclosed or would be known to someone of general skill in the particular category of invention (or art) at the time of invention conception.

- *Step 3: Prepare invention disclosure.* This is the initial documentation (in a format that is easily transferable to a patent application) of the invention conception that fully describes the invention and how it differs from the prior art known to the inventor(s).

- *Step 4: Obtain approval to submit disclosure.* Since the patenting process can be long and costly, many companies include an initial filter by the inventor's functional management to ensure that the invention disclosure is complete before submitting it for broader internal evaluation.

Stage 2: Evaluate (Steps 5—7)

The next stage in the patent lifecycle involves the evaluation of the invention disclosure to determine the best approach for protecting the IP.

- ◆ *Step 5: Preliminary disclosure assessment.* This initial decision milestone uses high-level criteria to make a go/no-go decision on whether to file a provisional patent application (PPA). A PPA can be compared to work in process (WIP) in manufacturing. As discussed already, provisional applications have been widely adopted by firms that submit a large volume of applications due to their cost-effectiveness and the value of establishing a priority date with a one-year grace period to continue commercial and technical evaluation before submitting a formal and detailed patent application. Note that provisional applications do not eliminate the ability to utilize trade secret protection for the invention, since these disclosures are held in confidence with the patent office and are not published until sometime after the filing of the formal application.

- ◆ *Step 6: File provisional application.* Once the invention disclosure is approved to file a provisional application, a patent counsel prepares and submits the provisional application with the local patent office.

- ◆ *Step 7: IPRC review.* In the next step, the invention disclosure is reviewed by an IP Review Committee composed of senior, cross-functional decision makers (typically from R&D, marketing, legal, and other relevant functions). The IPRC evaluates the invention disclosures within the context of broader IP, technical, and business strategies. It is critical to consider both the value and costs related to the IP for the entire lifecycle. The IPRC assesses the strategic value of the invention, evaluates how the invention fits within the current IP portfolio, determines the appropriate instrument for IP protection, approves the funds required for patent or trade secret protection, and deploys the resources necessary to continue or expand any ongoing commercial or technical development of the invention. The IPRC utilizes various strategies, tools and techniques discussed in this chapter in order to make the appropriate assessment.

Stage 3: Decide (Step 8)

Following the IPRC review, this is the critical decision node in the process where go/no-go decisions are made for the invention disclosures. If IP protection criteria and the invention evaluation indicate value in protecting the IP, the appropriate IP instrument is selected. Besides a no-go (or reject) decision, the IPRC may make the following go decisions:

- ◆ *8.1 Refine and resubmit.* If the disclosure is incomplete or the timing for formal protection is premature, the IPRC should give specific guidance on the deficiency and timetable for re-submitting the invention disclosure for consideration at a later time.

◆ *8.2 Consolidate with other patent applications.* For firms that submit large volumes of patent applications, this is an opportunity to consider consolidating the disclosure with other pending applications.

◆ *8.3 Maintain as trade secret.* As discussed earlier, the IPRC may determine that trade secret protection is the optimal instrument for IP protection considering the long useful life of the IP as well as the minimal need to make the IP publicly available to extract value. Choosing this protection strategy requires a different set of protection and maintenance activities (including constraints on use of external partners during development and commercialization) that the firm should fully understand before making this decision.

◆ *8.4 File for patent/obtain prior art legal opinion.* Last, but probably most common for the IPRC, is the decision to file the formal patent application, including getting an external legal opinion on prior art. This initial step for formal patent filing is a common practice used to both expedite the patent prosecution process but also to meet certain legal standards set by the patent office.

Stage 4: File and Prosecute (Steps 9—14)

This stage includes activities for the preparation and filing of patent applications and negotiation with the patent agencies to obtain the best possible patent (called patent prosecution). Depending on the geographical scope of filings, these activities may be undertaken in parallel in different countries or regions. The outputs of this stage are patent issuances which are akin to the finished goods in a manufacturing process.

◆ *Steps 9, 10: Prepare, file, and prosecute domestic application.* These steps involve the preparation of formal patent application, filing the application with the domestic (e.g. US) patent agency, and negotiating with the patent agency to obtain the best possible patent. These 'legal activities' are typically managed by patent attorneys and agents.

◆ *Step 11: Issue awards.* Following the successful issuance of patents, some firms issue awards to the inventor(s) as a form of recognition and financial incentive.

◆ *Steps 12, 13, 14: File and prosecute foreign applications.* A legal process, similar to steps 9 to 10, is followed for extending the application for patent rights to other countries. A common practice is to determine, at the time of filing the domestic application, whether foreign protection is required. If so, as discussed earlier, the Patent Cooperation Treaty (PCT) application is a convenient and cost-effective way to do so.

Stage 5: Renew/Maintain

Just as a mature product in the marketplace must be managed until its end-of-life, so too must a patent. Most countries require periodic renewal of patents to maintain patent rights. The last stage of the patent lifecycle process involves regular decisions to continue maintenance and the payment of associated fees.

IP CREATION STRATEGIES AND BEST PRACTICES

Many organizations, whose IP management practices are unstructured, do not have a clear strategy for protecting their IP or for developing a coherent portfolio of IP. In such organizations, the decisions to protect IP are made on an ad hoc and one-off basis. Leading organizations, on the other hand, have a well-defined IP creation strategy with clearly articulated strategic objectives that help to guide day-to-day decision making. Often, this IP creation strategy is linked with other R&D and business strategies. In these organizations, the view of intellectual property is not a loose, disjointed collection of patents but a strategic portfolio of business assets and the creation of IP has evolved from random, ad hoc, quantity-driven patenting to strategic, quality-based patenting. This section discusses examples of strategies and practices for building a strong IP estate:

♦ *Ensure protection of inventions related to core/defining technologies that enable sustained vectors of differentiation for products.* Although this concept may be obvious, many organizations fail to clearly define product attributes where they will seek to drive differentiation on a sustained basis and underlying technologies that will enable the differentiation.

♦ *Protect alternate design options in core areas to make it difficult for competitors to design around protected IP.* In new product development, designers often consider multiple design options to implement different product features and select the option that is best suited for the product requirements. Many organizations choose to protect only the selected design option. However, alternative design options may be equally good and, if left unprotected, may provide competitors an opportunity to achieve comparable performance in their products. In such situations, it may be desirable to also protect alternative design options. For example, Rivette and Kline (2000) describe how during the development of Gillette's Sensor shaver, the design engineers came up with seven different design versions for mounting the twin blades in the shaver. Gillette selected the design option that the potential competitors would have the most difficulty getting around but patented all seven options.

♦ *Protect next generation improvements.* A related practice is to protect next-generation improvements on a fundamental patent. A patent provides the holder the ability to exclude others only from the patented invention, but may not provide adequate protection for the right to use subsequent improvements. This may become a problem if a competitor patents a desirable improvement and blocks access to it even to the inventor of the original technology. As a hypothetical example, imagine if a more stable form of an electronic device is developed by an inventor that improves on a device that was originally patented by a different inventor that has stability issues when used in commercial applications. In this example, the inventor of the stability improvement cannot practice the improvement without the ability to utilize the original patent and the original patent holder cannot utilize the commercially more viable improvement.

♦ *Build a patent wall (often called a picket fence) around products by identifying and patenting distinguishing design features and different product elements that are part of a whole product solution.* As an example of this strategy, Gillette filed for 22 patents for its Sensor shaver that protected different aspects of the shaver including the cartridge, handles, packaging, and even the manufacturing process. This strategy is called *clustering.*

♦ *Expand the application footprint of the inventions by seeking broad claims and filing supporting patents for other applications of the technology that may even be unrelated to the core business.* Such patents may, in future, support business expansion or licensing revenue.

♦ *Bracket competitors by patenting the design landscape around their patents to restrict their design options and improvements.* This blocking strategy enables competitive advantage by increasing competitors' cost of workaround and time to market.

♦ *Explore opportunities to extend duration of protection.* For example, establish later conception dates with new but related patents that provide some of the protection of the original patent. In some cases, it is possible to come up with new patentable ideas around some critical aspect of existing patents, for example a manufacturing process. In this way, although a competitor may gain access to the fundamental technology after a patent expires, it may be excluded from making it in the most economical fashion.

IP DEPLOYMENT STRATEGIES

Organizations that are prolific creators of intellectual property are frequently faced with the problem of determining how they can make the best use of their IP. There is a spectrum of IP deployment strategies, ranging from being defensive to being highly aggressive. In a defensive strategy, companies use their patents primarily as legal shields to protect their innovations, to block others from acquiring patents on their innovations, and to gain design freedom. Under aggressive strategies, companies also view IP as a source of competitive advantage and incremental revenue by actively enforcing and licensing their patent rights.

Leading companies aggressively pursue myriad ways of extracting value in deploying their IP. These include out-licensing to generate incremental revenue streams, negotiation of favorable terms and access to leading technology in cross-licensing, locking out competitors to gain competitive advantage, and using IP to gain equity interest in new businesses, to influence industry standards and to boost corporate valuations. Elton et al. (2002) suggest a rule of thumb that in companies with more than 450 patents and $50 million in annual R&D spending, about 10 percent of the patent portfolio has the potential for generating incremental value. They also cite a study by McKinsey & Company that estimates that such assets could generate 5 to 10 percent of these companies' operating income.

Types of IP Deployment Strategies

Figure 10-6 shows a number of strategic options that are available for deploying IP. Reitzig (2004) describes examples of these strategic options in pharmaceutical, semiconductor, telecommunications, chemicals, and consumer packaged goods industries. In order to determine the appropriate IP deployment strategy, it is helpful to first segregate the IP into different categories based on the relevance of the IP to an organization's current business focus.

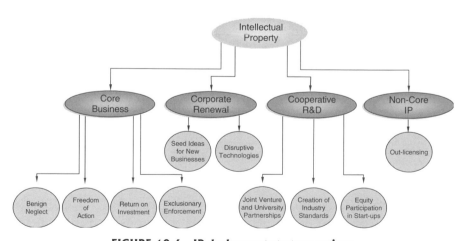

FIGURE 10-6. IP deployment strategy options.

CORE BUSINESS   This category includes IP that is related to the core businesses of the organization. Typically, most of an organization's IP would belong to this category because of the focus of product and technology development efforts in these areas. Four sub-options exist for deploying this kind of IP. Table 10-2 compares these options in terms of their benefits, drawbacks, and infrastructure needed to implement the options, and the circumstances under which each of the options are applicable. Briefly, these options are:

◆ *Benign neglect.* This is the simplistic do-nothing option in which an organization adopts a passive approach to IP management. Protection, deployment, and enforcement of internally developed IP are not pursued and infringement analysis of external IP is minimal. This is a low-cost but high-risk strategy, as it fails to exploit the benefits of internal IP and exposes the organization to enforcement actions from the owners of external IP that might inadvertently be infringed. This strategy is typically followed by emerging companies or new industries. However, companies that follow this strategy are often jolted out of their complacency by the enforcement action of a competitor or a third party. This spurs patenting activity in an effort to build defenses against further attacks and recognition for the need to have a more sophisticated IP strategy. The dispute between Research In Motion (RIM) and NTP over Blackberry patents provides a recent example of the significant risks of simplistic IP strategies. RIM's inability to take potential infringement actions seriously at an early stage led to an extremely expensive settlement eventually. It has been suggested in the business press that RIM's failure to respond to a warning letter from NTP and its failure to consider an early settlement resulted in a roughly $600 million settlement, instead of a much smaller amount if an earlier settlement had been negotiated.

◆ *Freedom to operate.* In this option, the key objective is to have design freedom without being constrained by concerns of infringing on other IP. When the ownership of IP in an industry is diffuse and shared among many companies, it becomes complicated to design products without infringing on others' IP. In such situations, companies often get into cross-licensing agreements with each other to gain design freedom. This is a popular approach in high-technology industries where IP ownership is very diffuse.

◆ *Return on investment.* In this strategy, an organization seeks to realize a return on its investment in developing its IP. This strategic option is viable when an organization owns IP that is needed by others. The return may be in the form of access to external IP through cross-licensing, balancing payments to compensate for relative strength of the IP portfolio in cross-licensing agreements, or royalties. High-tech companies such as IBM and TI have used this approach to assert their large IP portfolios and generate incremental revenue streams.

TABLE 10-2.
Comparison of Core IP Deployment Strategies

Strategy	Benign Neglect	Freedom of Action	Return on Investment	Exclusionary Enforcement
Description	No cross-licensing, in-licensing, or active enforcement	Broad cross-licensing	Cross-licensing with balancing payments/other business benefits or out-licensing for fees/royalties	Competitive advantage (suppression) by denial of access to technology/patents
Benefits	◆ Ease of implementation ◆ Faster Time-To-Market (TTM)	◆ Greatest design flexibility ◆ Access to future technology ◆ Faster TTM, lower development cost ◆ Ease of implementation ◆ Cordial industry environment (competitors, customers, and vendors) ◆ Lowest risk of infringement and litigation ◆ Low risk of scrutiny by regulators and ease of regulatory approval of acquisitions	◆ Freedom of Action strategy benefits ◆ Additional revenues ◆ Medium implementation costs	◆ Weaker competition ◆ Greater market share
Drawbacks	◆ Extremely high risk ◆ Highest vulnerability to lawsuits, especially from stronger firms and firms that go bankrupt ◆ High litigation costs	◆ Inability to exploit technological superiority ◆ Low return on R&D investment (low morale) ◆ Difficult to control access to proprietary, differentiating technology by "unlicensed" competitors/vendors	◆ Less cordial and trusting relationships ◆ Higher negotiation costs ◆ Need for portfolio valuation and determination of licensing terms	◆ Hostile industry environment ◆ Greater need for product clearance studies ◆ High cost of infringement opinions ◆ Higher possibility of the need to "design around"

	• No benefits from or enforcement of own patents	• Need for product clearance studies until cross-licenses are signed (leading to increased product development cycle times and cost) • Need for monitoring the patent activity of others • Renewal uncertainty		• High infrastructure and implementation costs • High litigation costs • Diversion of R&D resources to litigation support resulting in TTM slips and lower employee morale • High risk • Need to be aware of anti-trust implications
Infrastructure Needed	• Minimal	• Portfolio valuation of licensor and licensee (PRBs, Engineering time) • Clearance studies by engineers • Licensing department to identify and qualify licensees and to negotiate and manage agreements • Portfolio managers to analyze and evaluate the portfolio	• Minimal	• Portfolio valuation of all parties involved • Clearance studies by engineers • Active infringement detection • Enforcement department to conduct infringement analyses and to notify infringers • Litigation department • Process and mechanisms for gathering and retaining evidence
Appropriate When	• Almost never (usually followed by emerging companies due to ignorance of IP issues)	• IP portfolio (quantity and quality) is generally stronger than that of others but it does not contain any critical keystone patents • There is a need to maintain friendly relations with cross-license partners (e.g., vendors, customers)	• Patent ownership in the industry is diffuse • IP portfolio is weaker than that of cross-license partner (however, the portfolio should be strong enough to attract others) • Access to future technology developments is critical • Others have deeper pockets and greater resources	• IP portfolio is strong with clear edge over the competition on key technology(ies); quality of patents is more important than quantity • Access to future technology developments by rivals is unimportant • Need to maintain friendly relations is not critical (e.g., pirates, competitors)

◆ *Exclusionary enforcement.* In this aggressive strategy, an organization seeks to gain competitive advantage by denying access to its proprietary technology and IP to others in the industry and is prepared to pursue enforcement actions against others who infringe on its IP rights. This strategy relies on the strength of the IP portfolio and may frequently lead to litigation. For example, in the pharmaceutical industry, multibillion-dollar blockbuster drug businesses are based on patents on the composition and manufacturing process for underlying compounds. Companies in this industry zealously pursue exclusionary strategies to gain exclusive access to markets.

CORPORATE RENEWAL   This includes IP that is not related to the core business but could form the basis for new business ventures for the organization. Frequently such IP is for emerging technologies that could disrupt the basis of competition in the existing business.

COOPERATIVE R&D   This includes IP that a company may not want to use exclusively but that can be used to jointly develop new technologies with third parties. Firms often donate or share such IP with universities, in return for access to other technologies being developed in academia (e.g., Dow Chemical), share it with other companies in industry or trade associations in order to create industry standards in an attempt to standardize their technologies (e.g., Motorola and Rambus), or share their IP with start-ups in return for equity in emerging companies.

NON-CORE IP   Frequently the research and development organizations of companies develop IP that is unrelated to their core business and does not have a good strategic fit. Often, such IP remains unutilized. However, some companies institute active programs to search prospects that may use this IP and attempt to generate a revenue stream by out-licensing their non-core IP.

COST-BENEFIT CHARACTERISTICS OF STRATEGIC ALTERNATIVES FOR CORE IP

As discussed earlier, a large proportion of a company's IP is typically related to its core business, and there are four strategies for deploying such IP. The costs and benefits of these alternatives differ significantly. Companies need to understand these carefully before selecting an option (see Figure 10-7).

On the cost side, two dimensions drive the cost of a strategic alternative—IP portfolio management costs and transaction and infrastructure costs. *Portfolio management costs* are the costs of creating and maintaining the IP—not including the research and development expenses but including the costs of prior art searches, filings, legal fees, and maintenance fees. For large IP portfolios and broad international coverage, these costs can be significant. The transaction and infrastructure costs are primarily the legal and preparatory

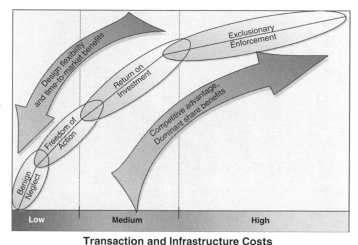

Portfolio Management Costs (Creation and Maintenance)

Transaction and Infrastructure Costs

- Portfolio valuation
- Negotiation
- Contract management
- Competitive and infringement analysis
- Clearance studies
- Litigation

FIGURE 10-7. Costs and benefits of IP strategy options.

costs associated with more aggressive IP strategies. These include costs of portfolio valuation, negotiation and contract management, infringement analyses and clearance studies, and litigation fees and expenses. Although portfolio management costs increase linearly with the size of the portfolio, the transaction and infrastructure costs increase exponentially for more aggressive strategies. IP litigation can be extremely time consuming and expensive. As an example, patent litigation statistics cited by Conley and Orozco (2005) show that the average time to resolve patent suits is 1.1 years and just the legal fees and expenses for cases with $1 million to $10 million at stake average $1.3 million for each side (plaintiff and defendant).

On the benefit side also there are two dimensions. Defensive strategies lead to greater design flexibility and faster time to market for products because the product developers are less concerned about the complexities of the IP landscape and have more freedom to design the best products. More aggressive strategies, if successful, may result in order of magnitude greater benefits in the form of large licensing revenue streams, compensatory payments for infringements by others, or greater market share and competitive advantage by excluding others from the market.

The next section discusses other factors that help in choosing an appropriate strategy.

Choosing the Appropriate Strategy for Core IP

This section earlier discussed four alternatives for deploying core business IP. So how does a company decide which is the optimal option for its situation?

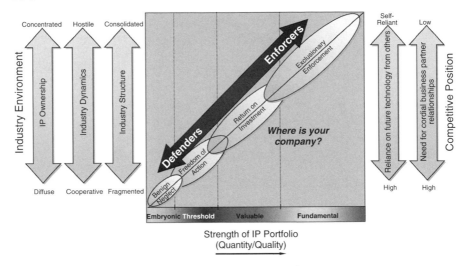

FIGURE 10-8. Factors affecting the choice of IP strategy.

Figure 10-8 shows the four options as a continuum from Benign Neglect to Exclusionary Enforcement and the three factors that drive the appropriateness of different options.

Strength of IP Portfolio

The strength of the IP portfolio depends on the quantity (i.e., number of patents) as well as quality (i.e., strength and breadth of claims, economic value of technology applications) of IP in the portfolio. The strength of portfolio can be characterized as *embryonic* (small portfolio of limited value), *threshold* (minimum level of strength at which it becomes viable to enforce the IP), *valuable* (when others would be willing to offer some benefit in return for access to the underlying technology), or *fundamental* (when the IP is related to a core technology and is strong enough to be desirable to others). Figure 10-8 shows that the stronger the IP portfolio, the more aggressive the IP strategy can be.

Industry Environment

The prevalent industry environment also determines the suitability of the different strategic options. The industry environment can be characterized by the concentration of the existing IP ownership, the level of fragmentation in the industry, and the prevalent level of cooperation and willingness for co-existence among the industry players. If the ownership of IP is distributed diffusely among the industry participants (e.g., semi-conductors, computers, telecommunications), the environment is more conducive to cross-licensing and goals of freedom-to-operate. However, if the IP ownership is concentrated with one or two dominant companies, they will tend to be more aggressive in enforcing their IP and demanding return on their R&D investments from other industry players who may need access to their IP.

Similarly, in fragmented industries, companies tend to adopt less aggressive strategies in enforcing their IP. The prevalent industry dynamics also drives the strategy. In some industries, competitors adopt a strategy of IP détente when it comes to flexing their IP muscles and prefer to compete on their products and services. In others, aggressive enforcement of IP leads to frequent legal battles and competition in the courts. For example, Lanjouw and Schankerman (2001) found that average patent suit rates vary from a low of 11.8 per thousand chemical patents to 25 to 35 per thousand computer, biotechnology, and non-drug health patents.

Competitive Position

The third dimension that drives the choice of strategy by a company is its competitive position in the industry. If a company expects to gain access to future technologies of other industry players (competitors, suppliers) for its products and needs to maintain cordial relations to do so, strategies on the defensive end of the spectrum are more appropriate. However, if a company expects to be self-reliant for future technologies, it can adopt a more aggressive stance against other industry players in protecting and enforcing its IP.

IP PORTFOLIO ASSESSMENT AND ANALYSIS

In the course of their research and development activities, companies and their inventors routinely create IP items. Over a period of time, their portfolio of IP grows. However, there is usually a wide variation in the quality, strength, and value of different IP items. Most IP items are of limited value; only the occasional ones are real jewels. As companies accumulate more IP and their portfolio grows, many lose track of what is in their portfolio and how valuable it is. For such companies, it is useful to periodically analyze their portfolio, prune (i.e., phase out by discontinuing maintenance fees) the low-value IP, and deploy high-value IP to generate incremental value for the company. The challenge in this exercise is to define a consistent and effective methodology for IP assessment and portfolio analysis.

Portfolio Assessment Approach

Traditional measures of portfolio strength are largely based on patents as a numbers game. Measures typically used include, numbers of patents issued:

R&D expenditures per patent, licensing revenue supported, cross-licensing success, and density of patent landscape. These measures may indicate some level of patent portfolio strength, but high numbers themselves do not ensure the portfolio is meeting the required strategic objectives. More systematic portfolio analyses consider two dimensions of value: *intrinsic value* and *strategic value*. Intrinsic value refers to the fundamental strength or value of the IP inherent in the associated invention. Strategic value captures the value of the IP to the company. A regular and repeatable portfolio assessment approach requires technical and business judgment, applied in a disciplined manner to assess the portfolio along these two dimensions.

INTRINSIC VALUE   Conceptually, intrinsic value depends on the following factors:

- Fundamental technology impact

  - *High:* If a fundamental technical problem is solved in a unique way with many potential applications with no known technically practical alternatives, there is long productive life left in patent, and/or detection of infringements is easy. Note that productive life is shorter of years left until patent expiration or technological irrelevance due to other factors.
  - *Low:* If many technical alternatives are available, IP is only incremental to existing prior art, there is limited life left in patent, and/or detection of infringements is difficult.

- Market application breadth and size

  - *High:* If IP supports broad applications in large markets that could create new product lines or niche applications that are unique, underserved, and tap into cost-insensitive markets
  - *Low:* If IP supports narrow applications and/or its application is limited to a niche or limited set of needs

- Business enablement

  - *High:* If IP supports breakthrough in cost, responsiveness, or quality of manufacturing; enables a new business approach or radically improved value-chain positioning; or enables higher margins, growth rates, and/or market share for holder.
  - *Low:* If IP has no or limited effect on business approach, improved value-chain positioning, or business metrics such as cost, responsiveness, or quality.

STRATEGIC VALUE Strategic value of IP for a given company depends on the following factors:

◆ Strategic fit

  ◆ *High:* If IP directly contributes to core business, market, technology strategy, sales to core category, and/or active exclusion of competition.

  ◆ *Low:* If IP does not align to core business, unlikely to contribute to significant sales, and/or provides no protection (passive or active) against competition.

◆ Market/Product expansion leverage

  ◆ *High:* If IP opens up prioritized new market opportunities or greatly expands existing product applications in areas of interest.

  ◆ *Low:* if IP has limited use in any new markets or application areas of interest.

◆ Competitive positioning, suppression, and freedom-to-practice

  ◆ *High:* If IP is highly protective or exclusionary, materially impacts competitive position, and/or would be highly coveted by competitors.

  ◆ *Low:* If IP is of no interest to competitors, has no ability to exclude, nor has any value in cross-licensing in existing or target markets.

Table 10-3 provides an example of a practical tool based on these concepts. It includes different rating criteria for intrinsic and strategic value, suggested weightings to capture the importance of the criteria, and definitions of rating levels. Each company may have to customize this tool based on its specific situation. The tool may be applied to existing patents and pending patent applications.

PORTFOLIO ASSESSMENT AND ANALYSIS EXAMPLE

As an example of how such a portfolio analysis may be used, consider the situation of a company that had a large and robust portfolio of patents with about 1,200 active patents, 1,000 pending patent applications, and 500 unfiled disclosures. The portfolio included more than 10 different technology categories and several subcategories. Company management was concerned about the increasing costs of maintaining this portfolio. It was unclear about what was actually in the portfolio and the business value of owning the portfolio. It was also interested in monetizing the IP but was unsure about what to keep and what to out-license to generate incremental revenues. The company applied this portfolio analysis approach as follows to gain a better understanding of its patent portfolio and to determine specific strategic actions to achieve its objectives:

TABLE 10-3.
Rating Criteria for Patents

Intrinsic Value	Weight	Rating					Score
		1	2	3	4	5	
Is it a key patent, i.e., Is it a pioneering/defining technology or incremental improvement (# of forward citations)	20%	Not likely (0)	Maybe (1)	Possibly (2–4)	Probably (5–10)	Definitely ()10)	
How difficult is to engineer around the patent or are alternative solutions available?	20%	Very easy	Somewhat easy	Possible	Somewhat difficult	Very difficult	
How easy is it to detect infringement?	20%	Very difficult	Somewhat difficult	Possible	Somewhat easy	Very easy	
What is the useful life of technology?	10%	Obsolete	Short (1–2 years)	Medium (3–4 years)	Quite long (5–6 years)	Very Long (>6 years)	
Number of patent references (How well was the prior art researched?)	10%	0–1 (Not done) or >25	2–4 (Not well) or 21–25	5–8 (Moderately)	9–12 (Very well)	13–20 (Extensively)	
Age of patent (since date of filing)	10%	>15 years	>10 and ≤15 years	>5 and ≤10 years	>2 and ≤5 years	>2 years	
Does the patent have significant and unique applications in other industries?	10%	Not likely	Maybe	Possibly	Probably	Definitely	
Total	100%						

Strategic Value	Weight	1	2	3	4	5	Score
Has it been used in a Company product?	10%	No				Yes	
Does the patent enable freedom to operate for key applications of interest to the Company?	20%	Not at All	Weakly	Somewhat	Strongly	Critical	
Is the patent useful to others in this industry and potentially restrict their freedom to operate?	10%	Not likely	Maybe	Possibly	Probably	Definitely	
Is there potential for use in a future Company product or opening new markets?	20%	Not likely	Maybe	Possibly	Probably	Definitely	
Is further research continuing in this area at the Company?	10%	No		Low priority		Yes	
Does this patent support a key technology strategy or area of product differentiation	20%	Not at All	Weakly	Somewhat	Strongly	Critical	
Is the patent part of a cross–license?	10%	No				Yes	
Total	100%						

Step 1: Develop a technology classification scheme for the different technology categories and subcategories.

Step 2: Group all existing patents and pending applications according to this technology classification scheme into patent families.

Step 3: Charter a small cross-functional team, including R&D, business, legal representatives and external experts (if applicable), to rate all the patents and pending applications in the portfolio, using the rating scheme for intrinsic and strategic value described in Table 10-3.

Step 4: Aggregate individual patent scores for all the patents in different patent families

Step 5: Plot the patent families on a 2×2 framework, as shown in Figure 10-9. This framework has two dimensions—intrinsic value of IP and strategic value of IP to the company—and four quadrants, depending on whether the intrinsic and strategic values of IP are low or high. Patent families in quadrant 1 (low intrinsic value and low strategic value) were candidates for divestment. The company decided to stop creating new IP in these areas and discontinued maintenance of existing IP, thereby saving on maintenance costs. Patent families in quadrant 2 (high intrinsic value but low strategic value) were candidates for harvesting. New investment in these areas was stopped and efforts were initiated to find out-licensing opportunities for these patents. Patent families in quadrant 3 (high intrinsic value and high strategic value) were well suited for leverage, either by commercializing unused IP or by taking an aggressive approach for enforcing IP rights against

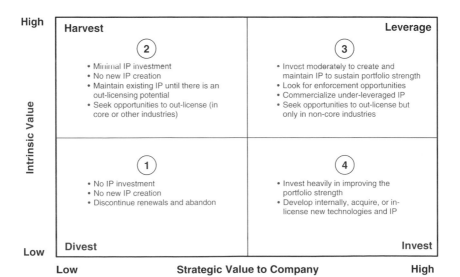

FIGURE 10-9. Portfolio analysis framework.

infringers. Patent families in quadrant 4 (low intrinsic value but high strategic value) were ideal for augmenting the strength of the portfolio in those areas, by creating new IP, finding in-licensing opportunities or acquiring IP. The tools in Table 10-3 for portfolio assessment and in Figure 10-9 for portfolio analysis allowed company management to identify patent families where they could save maintenance costs, find out-licensing opportunities, be more aggressive in enforcing their IP, and identify areas where they needed to invest to strengthen the portfolio.

PORTFOLIO ANALYSIS PROCESS

Beyond portfolio analysis, each organization should develop a business process to systematically evaluate the portfolio. Depending on the scale of the organization and the IP strategy, the process for managing the portfolio may be more or less formal. Regardless of the formality, Table 10-4 represents the principles that should be evaluated and implemented in any portfolio analysis business process.

TABLE 10-4.
Portfolio Analysis Business Process Principles

Principle	Typical Considerations for Portfolio Analysis Process
Accountable portfolio decision makers	Often a multi-functional committee structure (sometimes multi-level for large companies) that integrates research and development, legal, strategy, business development, marketing, and product management perspectives.
Decision cycle linked to business planning and individual projects	IP Portfolio reviews that precede key technology planning milestones to provide input on R&D focus and investment. Product development processes that require patent reviews for potential disclosures and freedom-to-operate issues early.
Robust strategic inputs	Market and product strategy inputs along with clear decisions on basis of differentiation for the business to provide inputs for strategic fit analysis.
Relevant portfolio boundaries and "strategic buckets"	Clear technology/IP categorization that is linked to product platform strategies and areas of strategic innovation.
Well-defined ranking criteria	Commonly agreed upon method and definitions for evaluating the portfolio (See Table 10-3).
Clear prioritization approach	The stepwise process to bring decision makers, information, and criteria together to generate portfolio analysis and follow-up decisions. When the portfolio is large and complex, the process is necessary to enable repeatable and consistent execution, and to establish expectations on the time and effort needed by all participants.

IP MANAGEMENT BEST PRACTICES

Best-in-class companies have well-defined management processes for IP creation, protection, maintenance, and deployment. This section describes some of the best practices for the key aspects of IP management processes. As always, competent legal counsel must conduct a legal review of IP-related processes to ensure that IP rights are not jeopardized unintentionally.

Best Practice: Align IP and Business Strategies

In leading organizations, IP strategy receives as careful attention as do corporate and R&D strategies and IP becomes a pervasive part of the company culture. A company that believes in innovation as its main competitive weapon will invest more in R&D and other innovative business capabilities than its competitors. However, within a level of gross innovation investment, there are other factors that affect the choice to invest in inventive capabilities:

- ◆ *R&D strategy.* How R&D is focused and structured should be in alignment with business strategy. Further, a clear separation of investment within the R&D strategy between enabling research, product development, and new concept development should be clear. In particular, investment in new concept development should not be confused with general investment in product development projects. The focus on new concept development should be toward exploration and novel problem solving—separated from the immediate pressure to commercialize. If the selection of new concept work is aligned to business strategy, certainly the patentable ideas that result will have higher potential value.

- ◆ *Linkage to targeted ideation and capability improvement programs.* Closely related to R&D strategy, a company should have investment in programs to drive targeted ideation and capability improvement. Such programs should be generated in alignment with business priorities for growth into new markets or applications. Facilitators and managers of such programs should be aware and trained in the basics of IP management, and enable protection of ideas generated.

- ◆ *Invention time for non-R&D functions.* Providing for time or the ability to request time for investigating new ideas for people outside of the R&D function can spark new ideas, particularly from people whose area and contact with customers corresponds with a strategic area of interest to the business. Combining such people with resources from R&D to listen to customers, and invent new approaches can create a source of strategically aligned ideas. Everyone should have access to invention submission programs, including those from non-R&D functions.

Best Practice: Reward and Reinforce Behavior that is Consistent with the IP Strategy

Communications and vision setting with all employees is the key to implementing a business strategy successfully. IP strategies are no different. In particular, a company should make clear its interest in IP, the opportunities for employees to participate, the rewards and other recognition that results, and a vision of what types of business the IP should enable.

Leading companies actively encourage disclosures through effective inventor recognition and incentives, inventor training on IP concepts, person/group who beats the bushes for new inventions, and in some cases, performance objectives based on number of disclosures (especially for researchers and senior engineers). IP experts work closely with R&D to train and encourage inventors to disclose their inventions, to help in the documentation of disclosures, to evaluate the disclosures, and to improve the technical and legal quality of patent applications.

Screening criteria used for assessing disclosures for content and filing investments should be aligned to the business strategy. Similarly, the reward and recognition to individuals for submitting the disclosures should be linked to the successful screening and acceptance of those disclosures. Depending on the decision to file patent applications, the reward may be increased to recognize the merits of the idea. Additionally, recognition and reward should accrue to the inventors of patents that are used in product development and that are eventually commercialized by the company.

Best Practice: Implement Structured Processes and Practices for IP Management

◆ Leading companies have well-documented process guidelines for intellectual property management (creation, protection, and maintenance) in order to promote their consistent interpretation and execution across the company. The guidelines describe the patent lifecycle process including key steps, their timing, deliverables, and roles and responsibilities for inventors, patent review committees, IP legal staff, and others involved in the IP process. It is critical that the IP process be nonbureaucratic; be easy to understand, communicate, and use; and be communicated to key employees in all relevant functions (technical, marketing, manufacturing, etc.). The process should also include tools and templates for inventors to make it easier for them to document their disclosures, such as Web-based process guidelines with online forms where inventors fill in the blanks.

◆ Lab notebooks should be maintained religiously. A lab or engineering notebook is a critical legal document that provides evidence of the scope and dates of new inventions. A well-kept engineering notebook can serve as a source of information to assist inventors in developing a solution to a technical problem from the conception of an idea to the construction of a working model or simulation which demonstrates the operation of the concept. The basic requirements of laboratory notebooks are to provide evidence of what was done, what was understood, and when things were done or understood. It maintains a record that is admissible evidence in court and in the U.S. Patent Office. Engineering notebooks also provide an evidentiary record in support of patent applications, engineering awards, and defensive publications. Questions of who invented what, and when, can be resolved with the aid of engineering notebooks.

◆ Careful prior art searches should be conducted. They help increase the patent issuance rate and quality of claims writing, avoid filing/prosecution expense of frivolous disclosures, and raise awareness of state-of-the-art and competitor filings. Many companies use a Technical Information Systems Group or patent liaisons (patent agents, technical librarians, retired engineers) to assist inventors in prior art searches.

◆ Product clearance studies should be an integral part of the product development process to ensure that all inventions are appropriately protected and that the company stays clear of infringing on others' patents.

◆ The marketplace and competitors should be monitored actively to identify infringements of intellectual property by third parties.

◆ Proactive steps should be taken to avoid becoming a victim of *patent trolls*. In this situation, a group, commonly referred to as patent trolls, accumulates IP solely for the purpose of enforcing IP rights against accused infringers, but does not manufacture products or supply services based on the IP in question; see Varchaver (2001). Often, such groups buy patent estates at fire-sale prices from bankrupt companies. They go out and file infringement notices with companies to get them to settle out of court. Many companies conclude that it is cheaper to settle with some amount of payout for a license than to litigate and win (or possibly lose). The trolls don't want to cross-license and they are not concerned if they cannot practice their patent—they never want to use the patent. To avoid future trolls, sometimes it is better for companies to monitor IP that may end up in the possession of trolls and to preempt them by buying or licensing such IP proactively.

Best Practice: Implement Clear Decision-Making Mechanisms

◆ Disclosures should be reviewed by senior, multi-functional patent review committees of subject matter experts that evaluate the technical, business, and legal merits of the disclosures. Disclosure reviews should include presentations by inventors (often as rigorous as peer review of journal articles and defense of doctoral dissertations) to ensure the quality of disclosures.

◆ Patent review committees should decide if a disclosure should be protected. They should also select the appropriate legal instrument for protection (patent, copyright, trade secret, consolidation with other applications, or publication). Leading companies have well-defined practices in place for choosing the most appropriate tool for protecting their nonpatentable IP with clear guidelines for trade secrets, copyrights, publications, and trademarks. They also implement effective security practices for protecting and defending trade secrets.

◆ Patent review committees should also proactively manage patent portfolios by making value-based patent renewal and foreign filing decisions that are consistent with business strategy.

Best Practice: Conduct Regular Portfolio Reviews

Leading firms conduct regular (every 6 to 12 months), cross-functional reviews of their IP portfolio to do the following:

◆ Assess strengths and weaknesses of the portfolio.
◆ Compare portfolios with competitor strategies and attempt to block key patents of competitors.
◆ Identify gaps and develop a patenting strategy to close those gaps by building patent fences around own key patents.
◆ Identify out-licensing opportunities.
◆ Review budgets and patent maintenance practices.

It is important to see portfolio management as a dynamic and living process that enables an organization to adapt to changes to the business, changes to the environment, and the inevitable evolution of the portfolio itself. At the most basic level, the evaluation of the IP portfolio should be used to adjust future investments to improve the alignment of the portfolio with future business needs. Merely flagging gaps in the portfolio will not generally change behavior, unless it is accompanied with a clear signal from the management by tying resources and funding to their decisions. IP portfolio reviews should identify approaches for closing portfolio gaps, internally or externally. Internal approaches could include adjusting future R&D funding or launching specific innovation and ideation initiatives targeted at closing the gaps. External approaches could be based on getting access to external IP through co-development alliances, joint ventures, licensing programs, mergers and acquisitions, and partnerships.

CONCLUSION

Intellectual property portfolios, in technology-based companies, represent unexplored treasures that have grown out of cumulative R&D investment over the years. These assets could be mined for incremental value streams flowing straight to the corporate bottom line. In the last decade, senior managers across industries have begun to recognize this opportunity and have embraced the need for a systematic approach to capturing this value just as they have traditionally done for hard assets.

This chapter has described the key strategies, tools, and approaches for a systematic and comprehensive management of IP including (1) creating, protecting, and maintaining IP assets, (2) deploying these assets, (3) assessing and analyzing portfolios of IP assets, and (4) business processes for managing IP. Each company will have to calibrate its approach to all these elements based on its industry and competitive environment, business needs, and legal ramifications. Without a systematic approach, a company runs the risk of not only forgoing IP-based opportunities but also providing the competition a window to exploit IP-related weaknesses. IP-savvy companies fully leverage their innovation capabilities and minimize these risks by investing in their internal IP management capabilities and formalizing the related processes and systems.

REFERENCES

Berman, Bruce. 2006. *Making Innovation Pay: People Who Turn IP into Shareholder Value.* New York: John Wiley & Sons.

Conley, James G., and David M. Orozco. 2005. "Technical Note: Intellectual Property—The Ground Rules." Kellogg School of Management, KEL140, August 30.

Davis, Julie L., and Suzanne S. Harrison. 2001. *Edison in the Boardroom: How Leading Companies Realize Value from Their Intellectual Assets.* New York: John Wiley & Sons.

Elton, Jeffery J., Baiju R. Shah, and John N. Voyzey. 2002. "Intellectual Property: Partnering for Profit." *The McKinsey Quarterly Special Edition: Technology,* 59–67.

Huston, Larry, and Nabil Sakkab. 2006. "Connect and Develop: Inside Procter & Gamble's New Model for Innovation." *Harvard Business Review* (March), Reprint R0603C.

Lanjouw, Jean O., and Mark Schankerman. 2001. "Enforcing Intellectual Property Rights." Conference on New Research on the Operation of the Patent System, December.

Maxwell, Roger. 2002., "Smart Patents." *Harvard Business Review* (April), Reprint F0204A.

Parr, Russell L., and Patrick H. Sullivan. 1996. *Technology Licensing: Corporate Strategies for Maximizing Value.* New York: John Wiley & Sons.

Petrash, Gordon. 1998. "Intellectual Asset Management at Dow Chemical." In *Profiting from Intellectual Capital: Extracting Value from Innovation.* New York: John Wiley & Sons.

Reitzig, Markus. 2004. "Strategic Management of Intellectual Property." *MIT Sloan Management Review*, 45(3): 34–40.

Rivette, Kevin G., and David Kline. 2000. *Rembrandts in the Attic: Unlocking the Hidden Value of Patents*. Boston: Harvard Business School Press.

Rivette, Kevin G., and David Kline. 2000. "Discovering New Value in Intellectual Property." *Harvard Business Review* (January–February), Reprint R00109.

Sullivan, Patrick H. 1998. *Profiting from Intellectual Capital: Extracting Value from Innovation*. New York: John Wiley & Sons.

Varchaver, Nicholas. 2001. "The Patent King." *Fortune* (May 14).

White, Edward, P. 1997. "Licensing: A Strategy for Profits." Licensing Executives Society, Inc.

Yoffie, David B., 2005, "Intellectual Property Strategy", Harvard Business School, April 7–9, 704–493.

11

Mad Scientists or Brilliant Inventors? How to keep your staff running like a well-oiled Invention machine

Douglas Neff
President, Toucan Learning Systems

Kimberly Houchens Ph.D.
VP of Product Development, Amcor

"Many of the familiar principles of Quality management amount to an elaboration of this simple truth: an innovative, healthy organization requires that we work with people rather than do things to them."

> —*Alfie Kohn, American lecturer and author in the fields of education and psychology*

HANG GLIDING TO MARKET

Bringing new products to market can feel a lot like hang gliding, especially for a product development manager! Consider the major difference, though. In the case of a product development manager, the glider, instead of being purchased from a reputable hang glider manufacturer, was designed and constructed from nothing. The team invented it, developed it, and built the prototype in the lab. They ran simulations and tests, and though all of the data say this thing will soar like an eagle, the job of the product development manager is to launch it—and find out.

One day soon, this manager will strap herself in, start running as fast as she can toward the edge of a cliff, and see what really happens. Once her feet leave the ground, she will discover the true meaning of the term *innovation*,

315

as she gets to truly experience what kind of product she and her team have developed. Is their new product soaring like a bird or dropping like a stone?

If that scenario sounds familiar to you, you're not alone. The looming reality of that launch date (just like the edge of a cliff) can have disastrous effects on your product development team. Innovators, in order to *fly*, must not be afraid of failure. In order to launch successful products in the marketplace, they must be willing to strap in and run full-tilt toward that cliff edge, regardless of the possibility of failure.

Sadly, some members of the team will often stand off to the side and just point at the abyss, complain about the sloppy stitching on the wing fabric, or chisel your name into a tombstone, then stand by with a shovel! Left unchecked, these attitudes will infect the rest of the product development team, eventually spreading to the entire organization and eroding your ability to invent.

If innovation requires an attitude of fearlessness, then what is the secret to creating and maintaining it on your work team? Surely the success of your company depends on the performance of the product development team, and yet this strange mix of motivation and imagination so often seems out of your control. Despite your best efforts, the social structures and expectations inherent to the business world—along with the very real threat of failure—far too often squash the same creative energies needed for any team to produce outstanding results. (Some studies suggest that six out of seven product launches fail!)

And yet, innovation is possible. Perhaps you know this because you see glimpses of it in certain people on your team. Or you remember the enthusiasm and motivation that surrounded your last product launch. But what about now? With cutbacks on resources and higher pressure to meet sales forecasts, "innovation on demand" is more important than ever. And yet, the more you *need* innovation, the more elusive it becomes. How do you bring it back? And once you find it, how do you sustain it?

This chapter will outline four tools that, when integrated into the working life of the team, will provide a solid foundation that promotes creativity, motivation, teamwork, and high performance. Consider these tools cornerstones for your *invention machine:*

1. *Career ladder*—A transparent, equitable, concise model that shows every level in your technical organization and the rest of the company as rungs on a ladder, including how they relate to each other and the necessary skills required, as well as the different career paths available to each level.

2. *Selection model*—A hiring tool that includes a robust set of criteria and evaluation tools for sorting through the pile of resumes on your desk and potential promotions in your existing team.

3. *Performance review*—An easy-to-understand review process that allows each person to see their strengths and weaknesses in relation to a clear set of expectations. This model uses the same grid for every position in the organization, so each employee can see their performance in relation to positions higher and lower on the career ladder.

4. *Action plan*—A set of specific, measurable, and attainable goals that help each individual outline the steps they will be taking to continue meeting expectations. Perhaps the most valuable tool a manager can use to help employees grow in new directions.

WHO CAN BENEFIT FROM THIS CHAPTER

These four tools constitute a comprehensive employee selection and evaluation process that will be invaluable to R&D managers, especially in regards to the overall structure of the organization. But project leaders within the R&D department can certainly benefit from examining these tools, particularly the performance review, as a guide to effective behaviors that should be developed within any innovation team. (From here on, the word *team* will be used to refer to your work group, whether it is an R&D department of 150 people or a 5-person project team.)

This chapter could be used very effectively by a manager tasked with building a new product development (NPD) team with a solid foundation, and while this system is a good blueprint for building a new team, it also provides many advantages to the renovation of existing ones. Not only does it help you map out your strengths and weaknesses as a team, but it also allows you to plan effectively for the future. Most importantly, though, it creates a working environment that values and nurtures the specific qualities you need in order to keep inventing.

This chapter was written based on the assumption that development tools designed to be employee-centric lead to employees who feel more empowered and are more effective in their work. Experience has also shown that keeping these tools transparent throughout the organization encourages individual growth and career planning, as well as self-motivation (which will keep your staff running full-speed with you all the way through the launch date).

Finally, all of the ideas in this chapter are rooted in the belief that empowerment leads to product innovation, and a set of development tools designed around building up the people around you will result in a much higher level of innovation on your team and greater success for your organization.

A BRIEF TOUR

Before getting into the specifics of each part of the system, it might be helpful to look at the process as it would be used in an actual organization. So, imagine yourself the R&D manager for an ultralight aircraft manufacturer, and you are responsible for building a highly inventive team that will create the organization's newest model of hang glider. Where will you start? How will you choose the best inventors? How are you going to build a winning team from the ground up?

Before hiring a single employee, you will need to develop a *career ladder*. This means first evaluating the needs of your team, which are dictated in part by the product you will be inventing. For instance, since you're inventing hang gliders, you'll have no need for medical doctors on your team—(hopefully!)—but you will definitely need people skilled in aeronautics, physics, and engineering. Your needs will also be determined by your strengths and weaknesses as a coach and manager. Maybe you work best with independent-minded people because you have a hands-off approach. And maybe you have a teaching style of management, so you prefer team leaders with less experience who you can help grow into strong managers.

Your career ladder will be a visual representation of what you want your team to look like, and a constant guide to you as you reevaluate your needs throughout the years. As you build your new team, it will serve as a blueprint, always bringing your attention back to the skills and traits you need most. For those of you managing existing teams, the career ladder will be an invaluable resource for showing you the gaps between what you have now and what you will need for tomorrow.

Your next step will be to acquire some new employees. The *selection model* is most useful here. Using your career ladder, you will make a list of the positions you need to hire, including all of the skills and experience you will require for each position and how many of each position you need. Armed with your selection model, you'll start recruiting new employees. The model will keep you focused on the things you really need so you can fully engage in the interview process. Furthermore, when you find yourself with too many applications and not enough time to read them all, your selection model can be used (by you or others) to screen prospective employees before spending time on a full interview.

Now you have your winning team—or, at least, your winning-*quality* team. (The proof will be at the edge of that cliff on launch day.) Now, though, you will need to set expectations by conducting a *performance evaluation process*. Certainly you don't expect the same level of performance from a technician as from a machinist or an inventor? You might expect the same attitude and drive, but a technician doesn't need to create ideas from thin air, right?

Chances are, you'll evaluate the technicians on their detailed work and follow-through, the machinists on their welding and ability to accurately follow drawings, and the inventors on their ability to create new products. Since our career ladder is transparent and *integrated* with the evaluation process, each person can see where she or he stands in relation to every other position in the company, as well as where they would need to grow in order to earn a different position. Who knows, maybe a technician has been secretly working on a new invention!

Refining and grooming the team is where the *action plan* comes in. Once you've evaluated the individuals on your team, you'll notice that some of them are achieving their goals and some are falling short. All of them need a new set of goals (an action plan) to get them where they want to go (or where you need them to go) next. For underperformers, the goals will be dictated by the gaps they need to close in their own work, and employees who are on target will have more freedom to set their own goals and focus on career advancement. Both, however, will be evaluated in the future, in part by the goals they have set for themselves and their ability to follow their action plan.

A team development system is not new, and parts of this process can already be found, in some form, in every team or company. But when used in an integrated way, as suggested here, these four tools enable empowerment, innovation, and teamwork to exist on your team in a lasting way.

THE CAREER LADDER: BLUEPRINT FOR YOUR DREAM TEAM

Consider this scenario: John is a chief engineer at Acme Ultralight Aircraft who has been with the company since before you were hired. Your predecessor had a great deal of respect for him, and John's seniority and dependability earned him promotion after promotion. Today, he is your chief engineer, the highest-ranking engineer on your team. You find him to be a hard worker and a dedicated employee, but you have also started to notice some gaps in his abilities. As you hire additional engineers, some of them have more experience than John, others have better problem-solving skills, and many of them demonstrate a greater capacity for leadership.

Your own opinion of John begins to decline when he becomes a common face in your office, frequently pointing out what's wrong with the team, how every product about to launch is going to be a disaster, and what he thinks you should do about it. You begin to feel frustrated, especially when you notice that his abilities seem to end at problem-identifying. You are forced to rely on lower-ranking engineers to create solutions while John spends his time pointing fingers.

Furthermore, he has difficulty negotiating among the other departments in the company (marketing, sales, accounting, etc.) and usually ends up creating trouble whenever he tries to communicate outside his team. This effect is annoying to you, but it is devastating to the rest of your team, as John has become a negative role model, causing division and confusion throughout the group.

Does this sound like anyone you have managed? The frustration you might be feeling with John is due to the fact that his position is not commensurate with his skills, and he is not performing at the level at which you need him to. (It's a pretty safe bet that John is also frustrated by this situation.)While you and the rest of the team are diligently preparing for launch (fastening your helmets and running full-speed toward the cliff), John is standing off to one side, megaphone in hand, describing to you (and anyone who will listen) the rocks at the bottom of the ravine.

Since there are no full-time openings for sports announcer on your team, how did this happen? How did your predecessor promote John to such an important position when he obviously didn't have the skills or aptitude for the job?

This is where the career ladder comes in. If one was in place and functioning properly, John would not have been able to advance to his current role merely because he had been there the longest or was next in line for the job. First, he would have needed to show aptitude for some very important skills and characteristics.

A career ladder exists in every organization, whether explicitly defined or not, and it determines the required skills, competencies, and authority of each position. It also provides a guideline for salary, bonus, and other compensations. One simple example of a career ladder would be the organizational chart you use for your team. Typical organizational charts are designed to show lines of responsibility, but are not usually able to communicate much to an employee who wants to advance in the company. You might think of them as *manager-centric*.

But with any luck, your organizational chart will answer a few simple questions for you:

- ◆ What positions do I have in my organization?
- ◆ How do those positions fit together?
- ◆ Who reports to whom?

If your chart is like most, though, it will probably not answer the next layer of questions:

- ◆ How is salary/bonus/etc. determined between different levels of responsibility?
- ◆ What are the educational/technical/experience requirements for each position?
- ◆ How are positions related horizontally, especially when in separate fields?
- ◆ What is required to move up or change fields in the company?

And a common question among technical folks:

- ◆ Does that sales guy with a B.A. in English really make three times the salary of the Ph.D. in engineering who launched three products this year?

These questions *can* be addressed by building a more employee-centric model, as follows: Imagine a ladder with three rungs. (The rungs themselves will be important later, but for now, think about the four spaces created in between.) Label them one through four. Unlike the organizational chart, you aren't going to populate your ladder with employees, but rather, with specific positions. Figure 11-1 shows how you divide up the job titles.

Begin to think about which of your positions belong in which level. Some will be obvious, but others may not seem clear yet. It will probably be helpful at this point to make a list of all the positions on your team. Add the ones you can to your career ladder, then continue reading to learn how to add the rest.

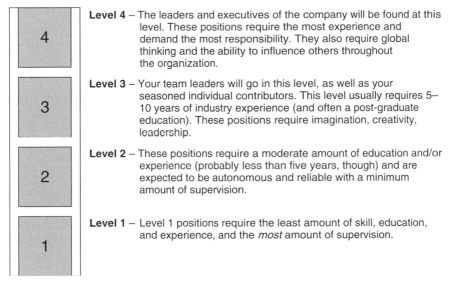

Level 4 – The leaders and executives of the company will be found at this level. These positions require the most experience and demand the most responsibility. They also require global thinking and the ability to influence others throughout the organization.

Level 3 – Your team leaders will go in this level, as well as your seasoned individual contributors. This level usually requires 5–10 years of industry experience (and often a post-graduate education). These positions require imagination, creativity, leadership.

Level 2 – These positions require a moderate amount of education and/or experience (probably less than five years, though) and are expected to be autonomous and reliable with a minimum amount of supervision.

Level 1 – Level 1 positions require the least amount of skill, education, and experience, and the *most* amount of supervision.

FIGURE 11-1. Employee-centric ladder model.

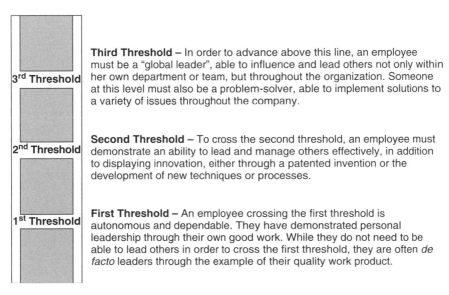

Third Threshold – In order to advance above this line, an employee must be a "global leader", able to influence and lead others not only within her own department or team, but throughout the organization. Someone at this level must also be a problem-solver, able to implement solutions to a variety of issues throughout the company.

Second Threshold – To cross the second threshold, an employee must demonstrate an ability to lead and manage others effectively, in addition to displaying innovation, either through a patented invention or the development of new techniques or processes.

First Threshold – An employee crossing the first threshold is autonomous and dependable. They have demonstrated personal leadership through their own good work. While they do not need to be able to lead others in order to cross the first threshold, they are often *de facto* leaders through the example of their quality work product.

FIGURE 11-2. Thresholds of performance.

Notice that the ladder has three rungs. You might think of these rungs as thresholds for performance. That is, an employee must demonstrate certain skills or aptitudes in order to move from one level to the next. This is true of any promotion, of course, but these thresholds are a little more significant to you as a manager. Think of them as shown in Figure 11-2.

Refer back to the example at the beginning of this section, and you can start to see how an effective career ladder might have kept John from advancing into a job for which he was not prepared; in this case, one requiring global leadership. If his manager had a well-planned model of a high-performance team, they would have realized that John had significant skills in certain areas, but was not yet ready for a Level 4 position. (He also might have helped John create an effective *action plan* to increase those skills.) From this perspective, you can begin to see John's frustrating behavior as a result of poor management. John has found himself out of his league and is doing his best to keep up. It's not very surprising, in fact, that his communication is often negative, critical, and blaming of others. You might recognize, from your own experience, that people who find themselves in the wrong job often resort to criticism, complaining, and stirring up trouble.

Now look at the example from a different perspective. What if John came into *your* office and asked you for the promotion to Chief Engineer? Consider that John has shown nothing but high quality and dependability in his many years of service. You know him well and want to reward him for his hard work, and John makes a compelling argument. What do you do? Do you have a clear and objective picture of what is actually *required* for a promotion to Chief Engineer? Unless it happens to be a position you yourself have held, most people probably do not have this information at their fingertips. And yet, most managers can remember the mistake of promoting someone who wasn't yet ready for the job, (hoping they would grow into it), or deciding to nominate the best employee in the group to be the chief, only to realize later that the new role model for their team doesn't model the best behaviors.

The career ladder creates an opportunity for a reasonable conversation—for both parties. As his manager, you are relieved of the burden of *guessing* whether John will be able to do the job, as he is being asked to demonstrate certain abilities and traits prior to promotion. And from John's perspective, he now knows *exactly* what he needs to do if he still wants that promotion. And now you can focus your energies on helping him develop the necessary skills to really earn his promotion.

One more step to go, and you should be able to populate your career ladder completely. You're now familiar with the three thresholds and four different levels of the career ladder, but remember that there is probably more than one department or functional group in your organization. Thus, there is more than one way up the career ladder. Take a look at the Acme Ultralight Aircraft Company career ladder shown in Figure 11-3

The Acme career ladder has been split into three different tracks: scientist, engineering, and management. Obviously, this provides a much clearer picture to the career-conscious employee, especially one who might be looking to switch tracks. After implementing a career ladder like this, you will find yourself having meaningful career conversations with members of your team. For instance, Associate Scientists will give serious thought as to whether they want to work in Project Management or continue along the individual contributor scientist path.

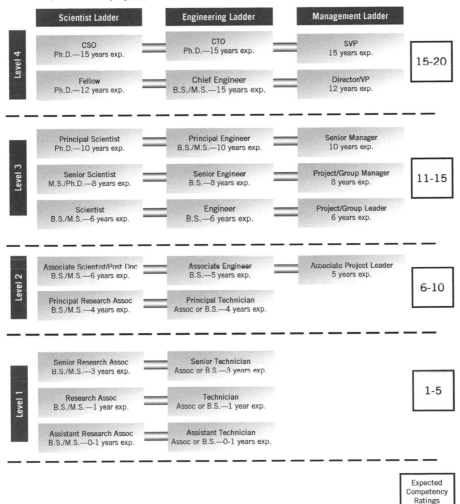

FIGURE 11-3. Sample Technical Career Ladder.

Far too often, the decision of whether to pursue a management role comes down to a question of salary. In many companies, employees on the management or business track are compensated better than those on other tracks, so the top individual contributors (all of those brilliant scientists and engineers building your hang glider in the lab) find themselves pursuing a management position just because it pays better. Unfortunately, that strategy produces a lot of leaders who aren't really suited to (or interested in) leading others. The best alternative is to use the career ladder to set salary ranges that are equal throughout the different functional areas. For example, a senior scientist, senior engineer, and project manager should be in the same salary range. If you find yourself with a lot of managers who don't really want to be managers, then it's worth asking this question: Why are you driving them into management by dangling their financial prosperity (and that of their families!) in front of them as an incentive? Shouldn't they have incentives to invent great products? An organization with pay ranges based on career ladder levels is definitely a more equitable environment in which to pursue one's career, and will lead to individual contributors and leaders who *want* to (and are able to) be in those positions.

Now you should be able to figure out where each position in your department or organization should be placed on the career ladder. Having completed this blueprint for your department or organization, it's time to make up your shopping list. If you were building a hang glider product development team, you would write down, next to each position on your ladder, exactly how many team members you need. (12 technicians, 4 project leaders, 2 associate engineers, etc.)

Once you complete your list, you're almost ready to start hiring!

THE SELECTION MODEL—HIRING THE BEST

Armed with a clear and concise career ladder, you're ready to start hiring the best performers you can. Before long, your inbox will be filled with promising resumes, and your phone won't stop ringing. The best employees in the world will be lining up outside your door, hoping for the chance to interview with you. There's just one problem. For every star performer hiding in your inbox, there are another dozen mediocre performers hiding in there, too (not to mention a few disasters waiting to happen).

With all of these possibilities, how are you supposed to choose the group of people that will produce the best invention team? Don't believe the managers who tell you they look for a sparkle in the eye, or simply trust their gut instinct. Those are both important tools to have with you in the interview, but there are plenty of people out there who have learned how to fake a good sparkle, and the truth is that the mediocre folks usually look surprisingly similar to the star performers. Furthermore, your gut instinct is probably attuned to recognize talent. And talent is great, but you just spent all of this time identifying exactly what positions, skills, and abilities you need on your team. Hiring twelve talented people at the wrong level would be just as disastrous as not having them at all. After all, what would an orchestra sound like with 85 talented violinists? You might benefit here from some extra help in maintaining focus throughout the hiring process.

You have already designed your career ladder, so you know exactly what positions you need, and you have a very solid sense of what you expect from employees at each level, regardless of their specific job duties. Now you're going to build a selection model to help make sure your new employees come equipped with all of the technical skills and experience these positions will require on a daily basis and also fit well into the team. Begin by selecting a position (project manager, for example) from the career ladder.

The first question to ask, and perhaps the most important, is *What deliverables will this position be accountable for?* In other words, what am I going to be paying them for? There should only be one or two of these, and they should be large in scope. For example, in the case of our project manager, there are two deliverables:

1. Manage the development of new and unique commercializable products while meeting R&D deadlines from project definition to manufacturing readiness.
2. Lead a team in the creation of unique intellectual property and commercial products.

Think of these as a kind of *mission statement* for the position. Don't just write down the job description, but spend some time thinking about what you actually want this person to produce for you. The more thought you put into these deliverables, the easier time you'll have in the rest of the process.

For the next step, take a look at the first deliverable. For a project manager, the first deliverable is all about project development, and it will have a different set of expectations or requirements than the second deliverable. Now consider what experiences, credentials, qualifications, or skills your future project manager must possess in order to be successful at their job. While you're listing them under Deliverable #1, give them a letter grade according to their importance. Use the following scale:

Q—*Qualifier*—These will be very helpful in the screening process. The applicant won't even be considered unless they possess these qualifications. Use the letter Q on the selection model to remind yourself that this is not an ordinary qualification, the absence of which will immediately disqualify a potential candidate. Normally, qualifiers are things like advanced degrees, years of experience, and so on.

A—*Must-have*—These are things that may not be perfectly clear on a resume or application. They are qualities that are essential to the operation of your team and show that the applicant will fit in with the team's unique style.

B—*Should-have*—Not as essential as the first two qualifications, but still valuable to your team.

C—*Nice-to-have*—Use this label for any characteristics, skills, or abilities that would help to round out your team.

The scale is simple enough to understand, but grading the qualifications is really key here, as they will help us stay focused during the screening and interview process. Some examples for our project manager position:

Deliverable #1—Lead development of new and unique commercializable products while meeting R&D deadlines from project definition to manufacturing readiness

Q—Advanced degree or substantial experience in aeronautics and small plane development

Q—Demonstrated ability to meet development deadlines

Q—Experience managing a team of developers

A—Proof of commercialized products

B—Experience in materials—sufficient to become in-house expert

C—Experience in specific fabrics for wing components

C—Expertise with safety systems (just as a backup!)

Deliverable #2 —Demonstrate ability to lead a team in the creation of unique intellectual property and commercial products

Q—Demonstrated ability to provide a culture of controlled chaos in order to support the creative process

Q—Demonstrated success in managing the performance of scientists, researchers and other lab employees

A—Good verbal communication skills

A—Excellent written communication skills

B—Proof of walking around management style

C—Ability to speak multiple languages

Once you have a complete list of qualifications, you can begin to construct the interview process. For each qualification, write one or two questions that would tell you whether someone has that qualification. When it comes to determining whether someone meets the requirement related to product development industry experience, the question could be as simple as, "Tell me about your industry experience." When it comes to finding out how the applicant manages others, the question will need to be more complex. For example, "Tell me how you have managed conflict on your team." Or, "Tell me about a stressful product launch and what role you played during that process."

Repeat this step for each of your deliverables. (The qualities you define later in the creation of your performance evaluation tool will be a great resource in reminding you what key attributes will be required for success. It will be helpful to refer to it often while developing the selection model.) If the career ladder is a blueprint for your dream team, the selection model is a blueprint for your star performer.

This can be a great tool for phone screening, by the way, especially if you don't have the time to do it all yourself. However, be sure to write in some key words and phrases that you're looking for so your interviewer can be more helpful to you. They can help you find the right person for the job, but only if you are explicit in your requirements. For instance, underneath the stressful product launch question in the example, add the following:

Listen for a commitment to the success of the whole team, not just in terms of product, but also with regard to employee development, morale, and effective leadership.

Your screening person will be able to take notes on that initial meeting specific to what you are looking for, enabling you to make an informed decision about whether to invite the candidate in for an interview. This tool has been used in the past with an external hiring firm, and even though they didn't have technical expertise, the selection model enabled them to screen out candidates who did not meet technical and leadership requirements of the team.

It's worth noting here that personality plays a large part in the development of any team. Countless books and seminars have been devoted to the subject of different personality models, so they won't be discussed here, but it is very helpful to the hiring process to get familiar with one. Any insight into the personalities on your team gives you an edge in learning to anticipate how different people will react to hurdles and conflict. Two popular examples are the Meyers-Briggs Type Indicator (MBTI) and the Personality Enneagram, both of which are easily found with a quick Internet search.

Remember that the successful hiring process ultimately comes down to making good choices. Develop a thorough selection model for each position, and you'll be best-prepared to make those choices.

THE COMPREHENSIVE PERFORMANCE REVIEW

When it comes to evaluating the performance of your team, there are many different models from which to choose, all of which exist on a wide spectrum of thoroughness. Some are simple and direct, providing a clear ranking system that is common to every employee, a small set of criteria, and a section for additional comments. Others are extremely thorough, involving ratings on dozens of criteria, eliciting comments after every rating, and resulting in numerous graphs and figures related to the employee's overall score.

In developing a performance review that is both comprehensive and effective, it is necessary to create a balance between thoroughness and simplicity. The most useful performance review should meet the following criteria:

◆ Provide useful information and helpful feedback, to both employee and manager.

◆ Be simple enough to complete in 20 minutes.

◆ Be integrated with the career ladder and selection model, and reflect the management style of the team.

◆ Force tough conversations about topics that would otherwise lead to disciplinary issues.

◆ Give each party a starting point for talking about career changes.

Traditional performance reviews are often weighted heavily toward either technical abilities or leadership skills. As we've seen in our career ladder, though, technical ability alone isn't enough to cross the different thresholds, and a performance review that focuses solely on relationship skills and leadership provides little value to a product development manager. Surely it's more useful to you as a manager to evaluate the qualities you actually need, both in terms of technical and interpersonal skills.

If you were to divide your needs into five broad categories, what would they be? For this exercise, think of the big picture. What big qualities do you look for in a new hire? For the Acme Ultralight Aircraft Company, the categories are as follows. Your culture and needs will be different, so you will need to develop your own set of desired qualities. However, this example is a good guideline:

- ◆ *I know how we win.* I set and achieve goals effectively within my team's unique style.
- ◆ *Together we win.* I work within a team environment, supporting my co-workers in achieving their goals. I set a good example with my attitude, my punctuality, and my work product.
- ◆ *I am flexible.* I cope well with plans and changes, and I display adaptability and resourcefulness when it comes to the changing needs of my work team.
- ◆ *You can count on me.* I am a trustworthy and credible resource to my team. I make reasonable commitments, then manage my work so that I produce consistent results.
- ◆ *I bring my best game.* I have technical abilities related to my position. Also, I have the ambition to bring the best of those skills to my work on a regular basis.

Some review models might stop right here and ask you to evaluate the employee on those five criteria, perhaps on a scale of 1 to 20. Obviously, that would yield a minimum amount of information regarding a specific employee, but it *could* be useful when comparing the strengths and weaknesses of a group of employees. For instance, in Figure 11-4 which employee would you promote to project manager?

Generally, you will probably want your project manager to have higher ratings in the people-oriented areas, like "Together we win," "I am flexible," and "You can count on me" (the right side of the radar graph). Well-rounded would be nice, too, but if you had to choose between technical ability and leadership ability, you will probably choose the latter for this particular role. Therefore, Sue is probably the best choice in this case. While she may not be the best scientific contributor in your department (Rob and Marty are probably better choices in that regard), she is definitely competent in her technical ability, and she is already displaying the leadership skills of a project manager, which is an absolute necessity for this role. So, a simple five-question performance review already has the power to give you some important information.

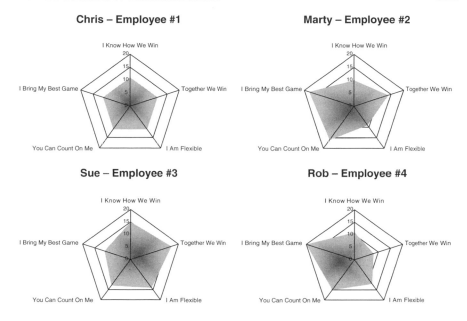

FIGURE 11-4. Employee radar graphs.

But the review process becomes even more valuable when integrated with the career ladder. Did you notice the boxes along the right side of the Acme career ladder? On a scale of 1 to 20, those boxes indicate the expected competency ratings for each level of the career ladder. For instance, a project manager would be expected to receive a score between 11 and 15. For that particular position, a score of 15 would be excellent, a score of 19 would be way above expectation, and a score of 5 would indicate a serious problem, as shown in Figure 11-5.

As you can see, this performance evaluation will accommodate every employee in your organization, while maintaining appropriate levels of expectation for each position. How is this valuable to the employee? Consider, for a moment, the chief engineer, John, from earlier in the chapter. If John received a traditional performance evaluation, he probably would have scored very high, since he is a very dedicated worker. If, for example, he received a 20 on every rating in every category (which is certainly possible, since he is only being rated in relation to himself) he would have nothing with which to compare

		Project Manager		
Level 1	Level 2	Level 3	Level 4	
1–5	6–10	11–15	16–20	??
You can count on me				

FIGURE 11-5. Project manager level 3 evaluation.

his ratings. He wouldn't know how he was performing in relationship to, say, a chief engineer! However, if each employee is rated in relation to the *career ladder*, then they have a much clearer idea, not only of what is expected of them, but what would be expected of someone at the highest levels of the company (and what is expected for that promotion they want). In John's case, he probably wouldn't have received 20 on every rating, but he probably would have received a lot of ratings in the 11 to 15 range, which would have given him much more useful information.

Even this version of the performance review is relatively simple, though. It still evaluates only five broad categories, which don't yield much information on their own. Those categories would be more useful if they were described more thoroughly, and if ratings were somehow based on a variety of criteria within each category. For example, if you subdivide "I am flexible" into some more specific phrases, you might come up with:

- I cope well with change.
- I have critical and strategic thinking skills.
- I can identify and solve complex problems.
- I am resourceful and creative.

Just like the previous example, you could rate your employees on each of these subdivisions. You could also use the 20-point scale from the career ladder, merely considering which level of critical and strategic thinking (for instance) the employee has demonstrated. Now, if your project manager receives a 12 in this category, you both know that she's performing within expectations.

The tool can be made more helpful, though, by going one step further. For each subdivision, it's clear that expectations will be entirely different for each level of employee, which would render a phrase like "I am resourceful and creative" rather useless. Do you really expect an entry-level technician to be creative? And does *resourceful and creative* really describe the kind of ingenuity required from an executive in your organization? In order to give these subdivisions more meaning, it's necessary to rewrite them for each level of the career ladder. Figure 11-6 shows an example of "I am resourceful and creative" as it appears on the performance evaluation.

Level 1	Level 2	Level 3	Level 4	
1–5	6–10	11–15	16–20	
I am resourceful	I am creative	I am resourceful and creative, and I am inspiring to my team. I create new products and intellectual property.	I develop creative strategies that shape the business and support customer needs.	14

FIGURE 11-6. ``Resourceful and creative´´ performance evaluation.

Think of each level as cumulative, in that a rating of 14 (as in the example) would assume that the person has also demonstrated all of the qualities in levels 1 and 2. Here's where the evaluation gets interesting. What if this was an evaluation for a first-year assistant technician? And what if it was the evaluation of your chief scientific officer? Either way, you as a manager have an interesting conversation ahead!

Take a look at the sample performance review from Acme, shown in Figure 11-7

This evaluation tool consists of five pages, followed by one summary page. Each rating page contains between five and eight ratings on it. The entire system was created in Microsoft Excel, and is used for every employee, from chief

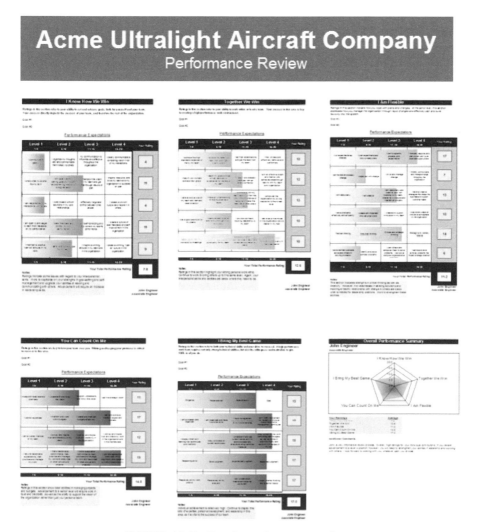

FIGURE 11-7. Sample performance review.

engineer to lab assistant. The review has also been used in a 360-degree setting. This means that for every person being reviewed, five evaluation surveys are sent out; to their manager, themselves, and three co-workers. Yes, this adds up to a lot of evaluations, but since they only take 20 minutes to complete, it ends up being well worth the effort. And the combined feedback is often much more informative and useful than just the perspective of one manager. The employee's self-evaluation, for instance, is often a great place to begin conversations about career changes. The summary page of the review contains average scores as well as a radar graph. The graph is useful in taking a macro-level snapshot of the individual's performance. If your team is large enough for anonymity to be possible, the summary page may also include the radar graphs of other employees at the same level (without identification) in order to help the employee see what their competition for a promotion looks like.

Remember that the evaluation is really only a starting point for a conversation. An effective review will help force a conversation about any specific issues that arise, so that they might be talked about in a supportive and objective framework, rather than being attributed to hearsay or rumor. As morbid as it may sound, you'll know your performance evaluation is working if the next person to be fired from your team had been fully aware of their situation, and the qualities and characteristics that led to their termination were called out in their last review.

In most reviews, especially those that may lead to disciplinary action, some sort of further action will be required by the employee if they wish to make any lasting changes as a result of the feedback they have just received. This is where the final part of this team development system comes in.

THE ACTION PLAN

No formal review process would be complete without an action plan. Indeed, the tools that you've learned here and the effort you have put into developing a high-performance team will all have been for nothing if your employees have no way to translate all of that information and feedback into concrete results for themselves. Most likely, if you've worked your way up to a management position, you are no stranger to setting and achieving your own goals and those of your organization. How is an action plan different?

For the employee going through this process, an action plan is only slightly different than the goal-setting practices with which they might already be familiar. Most will contain a list of things the employee intends to accomplish within a specific period of time, and to be effective, those steps will be specific, measurable, and attainable. An action plan, just like any goal, will help produce something in the end. In the context of these development tools, though, the action plan has a much more specific purpose: to produce a desired result in the next performance review.

Traditional performance reviews, especially those geared toward technical ability, will often generate very simple action plans. For instance, if an employee is rated below expectation for not having welding skills, then a plan of action is relatively straightforward (welding classes). But when we begin talking about personality traits and leadership qualities, classes might not be enough. Those kinds of changes will usually require a more thoughtful approach.

Don't underestimate the impact of this step. The greatest managers and leaders are those who have learned to evoke the best in others, and the performance review process creates an incredibly effective environment for practicing that skill.

Before we begin, remember that an action plan will not be needed for every employee, although everyone in the company is accountable to some set of predefined deliverables, and those are certainly worthy of an action plan. But some employees will be right on track with their own (and your) goals, and will not need the extra assistance. And while an action plan may not be *required* for every employee, most of them will be able to identify some area in which they want to grow or improve. Some of them will even be *hungry* for the chance to grow in new ways. The action plan is a great tool for helping each of those employees create something new for themselves.

Construction of an action plan should always begin with identifying the desired outcome. Consider the following questions:

◆ What is it that your employee would like to change before their next performance review? New skill sets? A promotion?

◆ What is it that you (as their manager) need from them in order to continue running a high-performance team?

◆ What general themes were raised during the evaluation process? What comments were repeated more than once?

Let's use the example of John, your engineer. At the end of his performance review, you told him that he wasn't ready yet for the chief engineer position (thank goodness you had that career ladder!), and John made it clear that he wanted to do whatever it took to earn that position. Great! Now you're in a position to help John acquire the skills and qualities that will turn him into a *successful* chief engineer. Here's how to approach his action plan.

It was clear from John's performance review scores that communication is an issue for him. Specifically, look at the section that says "My communication is influential and effective throughout the organization." He's simply not communicating well outside his own work team. In order to create the possibility of a promotion for himself, John will need to demonstrate to you that he has learned how to do this. John also received low marks on his ability to negotiate conflict. Clearly, this will need to be improved before he's ready to function at an executive level. So you can begin by assuming that these two areas are the main barriers between John and that promotion.

Now that you've identified the two barriers, it's easy to rewrite them in terms of desired outcomes:

◆ Influential and effective communication throughout the organization.

◆ Effective negotiation of conflict in his own work relationships and throughout the rest of the team.

For each of these desired outcomes, you will need to collaborate with John in designing a series of actions or steps that will help him create a new result for himself. As his manager, your role is to help him create those steps, but you'll find the process ineffective if you attempt to do it for him. For the action plan to succeed, John must be fully committed to the two desired outcomes and any actions he'll be using to accomplish them. For the sake of this example, imagine that you worked with John to create the following action plan:

Result #1: Influential and effective communication throughout the organization

1. Read a book on the Myers-Briggs Type Indicator, then start a journal to begin noticing different ways that colleagues prefer to communicate.

2. Work with a mentor, someone at the VP level, who is good at this particular skill. Ask for tips, advice, and so on, and meet with the mentor every two weeks.

3. Register for a class in effective group communication. Meet with your supervisor at least three times during the course to discuss your progress.

Result #2: Effective negotiation of conflict in my own relationships and throughout the team

1. Attend a course or seminar on conflict resolution, then write a review that integrates the material learned with examples from your own work relationships.

2. Find a partner (someone you respect) and have the partner role-play conversations with you, especially ones that have turned out badly in the past. Ask for feedback and assistance in approaching conversations differently.

Obviously, none of these steps will *guarantee* that John will grow into a chief engineer, but if he is truly committed to something new, these steps will help him begin his journey. (And the level of commitment to these steps, by the way, is good information for you when someone is asking for a promotion to a leadership position.) Remember also that it will be up to you, as his manager, to continue giving him feedback throughout the process.

Notice that each step on the action plan is specific, measurable, and attainable. These are essential when writing a plan. Don't settle for phrases like *communicate better* or *improve leadership skills*. Nothing specific will ever come out of a vague action plan!

CONCLUSION

Directors, managers, and team leaders of departments and work groups of all sizes can gain something by using the tools described in this chapter. Anyone who wants to effectively manage a group of inventors has something to gain by embracing the concepts that gave rise to these tools in the first place—namely, the idea that employees work harder, better, and more creatively when you approach them with openness and fairness, not only on a personal basis, but also with the policies and systems with which you evaluate their performance. The use of this system will create and sustain an environment where innovation can thrive and inventors and leaders can be evaluated according to the attributes necessary for innovation.

Before concluding, take a look at the ways each of these four tools can make your current job easier, even if you choose not to implement the entire system:

1. *Career ladder*. Your organization already has a career ladder, even if no one went to the trouble of putting it to paper. Before you do anything else, find out what it is! This is an absolutely essential tool. As a manager, you should know where your direct reports fall on this ladder, and you should also know where *you* can be found on it. Even on its own, a career ladder is a great tool for career planning or salary decisions.

2. *Selection model*. Obviously, this tool is only useful if you are involved in a hiring process. But if you are a manager, chances are good that you will need to hire somebody in the next year. You will find it very useful, helping you to stay focused and making sure you get all the qualities you need.

3. *Performance review*. Of the four tools, this one clearly takes the longest to develop, but it has also evoked a tremendous response from employees who have experienced it. Many remarked that it was the best evaluation they had ever experienced, and that they received really valuable information in it. If you're dedicated to a useful and effective review process, then consider using this tool. It is well worth the time it takes to develop.

4. *Action plan*. The action plan is the most common of these tools, and the one you are probably already most familiar with. Writing and implementing action plans is an art in itself, and worthy of its own study. If you are having difficulty in helping an employee grow from one position into another, this is a good place to start. Remember to keep every item on the action plan specific, measurable, and attainable.

Leadership is a lifelong journey, and these tools will help support you along the way. Used properly, they will provide a smooth foundation upon which to build, or rebuild, your invention machine, and ensure that your organization doesn't kill the innovative spirit for which they hired you and your team in the first place.

May you approach every cliff edge with the confidence and determination that comes from having an empowered, innovative team of professionals working with you and committed to flying. Happy landings!

The authors would like to acknowledge the contributions of Julie Ganim at VisionaryHR and Marilou Myrick at ATALX Corporation for their contribution of content and support during the development of the innovation team at Nano-Tex, Inc.

Part 4

Strategic Tools For Improving NPD Project Performance

This last section of the ToolBook *emphasizes tools that ensure that the NPD project is being managed strategically. The tools of Part IV start with ones that must be implemented before the project starts, such as those for setting strategy, assembling the team and planning the project. It then provides a tool for increasing learning during (and after) the project, and concludes with a tool for measuring the outcomes.*

As more and more NPD projects involve a defined strategic partner, understanding how to formulate a collaborative development, or co–development, strategy such that both participants "win" is imperative. Chapter 12 presents a codevelopment strategy planning process for first, analyzing the situations in which a codevelopment strategy may provide benefit to a firm, and then formulating the particulars of the strategy. These particulars include defining the objective behind partnering, determining how many partners are needed and how deep our relationships with each partner should be, and constructing a secondary strategy, should an appropriate partner for the first strategy not be identified. Again, this is a powerful chapter for any size organization.

Chapter 13 presents a tool to successfully launch and develop a high-performance team. This tool leads the new team through the steps of orienting, organizing, taking action, and achieving results. The emphasis of the tool is on getting the up—front processes *associated with the initial launch and team introduction, analysis of the business, organizational and environmental situations, and development of mission statements, goals, and operational strategies in line before worrying about defining the behavioral norms for the team. The time spent in developing these up-front processes reduces execution time later in the performance part of the process.*

Rolling wave planning, *or "plan a little, do a little," is the tool presented in Chapter 14. Rolling wave planning is one example of an "agile tool," one that requires an open-minded adaptive work environment. Although it may not be appropriate for incremental projects, where the path to completion is obvious and there are no real unknowns involved with getting there, it is highly useful for more uncertain projects where multiple unknowns still exist and the full path to completion is unclear. Thus, in rolling wave planning, detailed planning is completed for a shorter time horizon, but not for the full length of the project. As each unknown is made more certain by the tasks undertaken in the current work module, the detailed planning is rolled forward to the next short time period, until at last project completion is in sight.*

Chapter 15 details the after action review process, *which helps a team determine what went right with the project and what went wrong. In addition, it uncovers why something went wrong, what we could have done to prevent it, what we need to do to prevent it from happening in the remainder of the project, how this information could help other teams right now, and how to get this information to those other teams. It is essentially a knowledge management process for NPD. However, unlike most knowledge management solutions, implementing it does not depend on creating and maintaining large databases of information and investing in a large information technology infrastructure. After action reviews differ from the typical NPD project postmortem in that they are short and fast, and should be used at various points throughout the project to maximize ongoing team learning, rather than just once at the end. Perhaps one of the more significant values of this tool is its ability to change the organization's culture to a more sharing, learning mindset.*

"That which does not get measured does not get done." Fittingly, ToolBook 3 ends in Chapter 16 with a metrics determining process for measuring the outcomes of NPD. The measurement process first starts with defining the improvement goal and identifying a metric that is properly aligned with measuring that goal. Next, the actions that lead to achieving each goal are determined and the metric owner for those actions is assigned. The final step in the process is reviewing the set of metrics for the project and ensuring that the team is focused just on the critical few that will help them obtain the desired outcome. An important message of this chapter's tools is that the team (and individual) metrics must be linked (contribute) to the larger organization's metrics.

12 Formulating a Strategy for Codevelopment

Kevin Schwartz
Director, PRTM

Jennifer Abell
Manager, PRTM

INTRODUCTION: WHAT IS CODEVELOPMENT?

During most of the twentieth century, companies funded and conducted their new product development (NPD) internally and regarded this area as off-limits to collaboration, for competitive reasons. Attitudes toward this closed model of innovation began to change in the 1990s, however, and in today's global economy outsourcing and collaboration are commonplace in most business areas—from the supply chain to core research and product development.

Recent surveys and research confirm that this is a growing trend. According to a 2004 cross-industry survey by The Performance Measurement Group, LLC (PMG), over 75 percent (a 30 percent increase from 2004) of the survey respondents reported that at least 20 percent of their NPD efforts involve collaboration with a major strategic partner. Moreover, almost all respondents expected this ratio to increase significantly by 2007 (Figure 12-1). This move toward wider collaboration in R&D is discussed in-depth in Henry Chesbrough's book, *Open Innovation: The New Imperative for Creating and Profiting from Technology* (Chresbrough 2003). The book explains the historical context and the reasons for the emergence of more open innovation, and includes a number of excellent examples from industry leaders, such as Intel, IBM, Xerox, and Lucent.

What is codevelopment? We define it as a strategic partnership between two or more external parties working together to develop a new product, service, or technology for a mutual benefit. In terms of commitment and joint decision making, codevelopment partnerships are in the center of the relationship spectrum between the traditional arm's-length transactional relationship (e.g., supplier) and a fully merged organization (Figure 12-2). Thus, a simple agreement to purchase an off-the-shelf component from a supplier as part of a new product design does not constitute R&D collaboration. If, however, the supplier redesigns the component to meet specific design requirements, then a

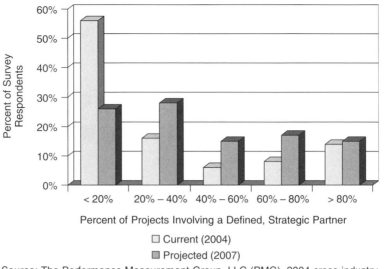

Percent of Projects Involving a Defined, Strategic Partner

☐ Current (2004)
■ Projected (2007)

Source: The Performance Measurement Group, LLC (PMG), 2004 cross-industry survey.

FIGURE 12-1. R&D collaboration on the rise.

FIGURE 12-2. A spectrum of business relationships.

codevelopment relationship has occurred. Should the two parties establish a formal agreement to codevelop a full product line, they would have entered a collaborative relationship that requires significant time commitment, joint decision making, and—typically—intellectual property sharing.

As shown in Figure 12-2, there can be different types of alliances on the relationship spectrum: codevelopment, coproduction, comarketing, co-licensing, and so on. In practice, an alliance between two companies often involves more than one of these elements. Partners who codevelop a product may choose to enter into a co-production or co-marketing agreement as well. For example, a company may work with a partner on the development of a product and choose also to co-brand with that partner. Though we are focusing on codevelopment in this chapter, the tools and techniques discussed here can be applied more broadly to other facets of collaboration strategy. In formulating a codevelopment strategy, or codev strategy for short, it is often helpful to keep in mind the larger context for collaboration.

Even more significantly than the increasing trend toward collaborative development is the relatively recent emergence of a set of best practices for effectively using external partners to achieve R&D objectives. In Chapter 6 of *PDMA ToolBook 2* (Deck 2004), the author observed that "Optimal codevelopment performance results from process excellence in three dimensions: strategy formulation (where to partner and why), project execution (how to execute projects with partners), and partner selection and management (how to initiate and cultivate productive partnerships)." That chapter focused specifically on the third dimension—selection and management of external partners for new product development. In this chapter, we will discuss a set of practices for use in the first area—formulation of codevelopment strategy. We will examine the definition of codevelopment strategy, how it fits in with an overall corporate strategic vision, and a series of tactical tools for formulating and communicating the strategy in order to guide day to day partnering decisions and choices.

WHAT IS A CODEVELOPMENT STRATEGY?

In simple terms, *a codevelopment strategy* is a clear statement that describes why, where, and how an organization intends to use partners in meeting its R&D and business objectives. This may seem like a straightforward concept, but the reality is that most organizations have not defined an explicit strategy to guide their partnering decisions, particularly around new product development. Instead, these decisions generally are made opportunistically by individual development teams, depending on their specific preferences and circumstances. Even companies that have a defined set of criteria and a standardized approach for selecting individual partners (partner selection and management practices) often have not addressed the more fundamental issues of strategy to guide when and where to consider partnerships.

Yet without such an explicit roadmap, how can you ensure that the internal and external resources will be deployed most rationally and effectively to support your business strategy? Without a clear understanding of objectives, collaboration decisions happen opportunistically and may be inconsistent with overall business strategy. In addition, you may not have the infrastructure in place to manage the relationship effectively.

An effective codevelopment strategy stems from the organization's core strategic vision—the fundamental statement of intent that defines a sustainable level of differentiation for the company and guides all functional and cross-functional decisions. A well-defined core strategic vision helps to identify growth opportunities in the company's target markets based on a solid understanding of its competencies and intellectual property. It also allows management to align the company's strategies with its business goals for growth and performance. An effective core strategic vision addresses three fundamental issues, which should guide the strategy formulation process for codevelopment:

1. *Why will we be successful?* Possible answers include competitive positioning, strategic business objectives, basis for differentiation, and relevant cultural attributes.
2. *Where are we going?* This is a determination of intended direction, product and market focus, and financial objectives; what the company will and will not be.
3. *How will we get there?* Strategic themes, an investment gauge, and a results scorecard could be used.

A codevelopment strategy based on this type of strategic vision statement will rationalize the use of external partners in meeting the company's business objectives. It will guide when and where to use partners: for which product/technology components; in what areas of R&D; and to what extent and purpose.

FORMULATING A CODEVELOPMENT STRATEGY

In this section we discuss the major steps for formulating a codevelopment strategy that is aligned with a business strategy or core strategic vision. We focus on the issues and tools that are specific to NPD collaboration, rather than on the general topic of strategy development (which is well covered in other literature). The four-stage approach presented here represents a toolkit that has helped a number of companies across industries to guide their R&D partnering decisions.

Why Will We Be Successful? Establishing Business Objectives for Partnering

An explicit set of business objectives is the critical first step in creating a codevelopment strategy, because they determines its purpose and, ultimately, guides the selection of individual partnerships. To determine these objectives, you must consider three key questions:

1. What are your high-level business goals (cost cutting, technology innovation, flexible R&D capacity, etc)? Do you expect R&D partnering to help realize some of these goals?

2. What are the key strategic differentiators for your business versus competition? Do you expect R&D partnering to help deliver on certain differentiators in particular?

3. How critical is partnering going to be to your overall NPD/innovation strategy?

Without a clear understanding of objectives, opportunistic collaboration decisions may be counterproductive to your business strategy. Consider, for example, the case of a consumer goods company with a premium product line selling at gross margins between 60 percent and 70 percent. This product line involved a cleaning product that was perceived as being extremely high-end and had a robust and solid-feeling external package (bottle) that reinforced this brand image. The general manager's strategy was to drive additional top-line growth from this blockbuster product line. However, the division's R&D director had recently been promoted from responsibility for a lower-end product line. where cost reduction was considered a critical objective. Thinking that cost reduction could only be a good thing, even on this higher-end product line, he formed a partnership with a supplier to develop a new, much cheaper, packaging for the product. The new supplier was based in Asia and had expertise in low-cost materials and manufacturing technologies. Hence, it was able to design a new bottle to the R&D director's specifications that met the same size and shape requirements as the original packaging at a much lower cost.

Although the new packaging reduced costs by millions of dollars, it led to a subtle but critical change in the consumer's experience—the new bottle felt flimsier in the consumer's hand and didn't convey the same sense of high-end quality. The unintended result of this change was that it caused revenues to drop by tens of millions almost overnight, as consumers reacted to the perceived change in the product represented by the flimsier packaging and opted for lower-priced options on the retail shelves.

The critical issue here was not that the R&D director had chosen to engage a partner to redesign the bottle, it was that he had engaged in the partner selection for the wrong reasons. If he had instead approached the partnering selection with the goal of increasing the user's perception of product value, he might have selected a different partner who had expertise in consumer marketing and user experience, as well as low-cost manufacturing. The resulting design would likely have cost more, but could have driven additional sales through a positive change in the consumer's perception of the product. Alternatively, the R&D director might have simply decided to focus partnering efforts on other aspects of the product, such as the fragrance or the chemical formulation, that could have an impact on the product's perceived value and could hence drive increased sales. As this company learned the hard way, different interpretations of business objectives can lead to different day-to-day decisions about codevelopment—causing costly mistakes.

Once business objectives are clearly defined, it is important to explicitly articulate their implications for codevelopment. For each business objective,

TABLE 12-1.
Translating Business Objectives into Codevelopment Goals

Business Objective	Codevelopment Goal
Profitability improvement	Partner/outsource for less critical components where others have an advantage in low-cost design capabilities.
Increased innovation	Develop strategic research partners and focus on components with fast-paced technology change.
Flexible R&D capacity	Establish strategic R&D partners to absorb spikes in bottleneck areas.
Market access	Partner for complementary R&D capabilities to tailor existing offerings for new markets.
Cycle-time improvement	Partner/outsource for less-critical components where others have existing designs or a time-to-market advantage.

there should be a clear statement of the resulting direction for partnering (Table 12-1).

As the examples in Table 12-1 indicate, the codevelopment goals do not relate to specific product components or individual partnerships. The purpose of these goals is to provide the high-level direction for making more specific codevelopment strategy decisions. Although there may be multiple objectives for forming a collaborative relationship, we recommend selecting the most important ones to keep the organization focused.

In the telecom electronics industry, for example, key business objectives for partnering include cost reduction and flexible development capacity. This has made the role of partners very clear. According to Lucent Technologies' Dave Ayers, VP for Platforms and Engineering, outsourcing some development makes sense because it allows engineers to concentrate on next-generation technologies. Ayers said, "This frees up talent to work on new product lines." "[For us,] outsourcing isn't about moving jobs. It's about the flexibility to put resources in the right places at the right time" (Ayers 2005).

Or, consider an Internet-based company trying to expand its market into China. Normally, the company considers Web-page development as its internal R&D competency. But, to reach its objective, it decided to partner with a Web-content development firm with Chinese language skills. This partner may be either a local Chinese company or a foreign entity—whoever offers the best prospects for opening the door to China. By being clear about its business objectives in the Chinese market, the company was able to make the right decision to reverse its normal approach and engage a partner for this central development work.

Where Are We Going? Identifying the Targets for Collaboration

How do you decide which elements of the development process should be maintained internally, which can be outsourced completely, and which are candidates for some level of collaborative development? Although codevelopment goals provide some guidance, typically you will need to conduct a more in-depth analysis of your products and markets to decide the following issues:

♦ What are the elements or components that make up the whole product or solution that you deliver to your customers? Which of these components are you best suited to develop internally? Which could be better addressed by partners?

♦ What technical skill sets are required to effectively maintain or grow your pipeline of new products? Are some of these skill sets candidates for outsourcing to R&D partners? If so, in whole or in part?

The starting point for this analysis is to break down the company's complete set of products or services (*the whole product*) into its component building blocks. These building blocks can consist of physical components, elements of a service offering, or types of research and design work required in the development process. Regardless of the vector used for this breakdown, it is important to disintegrate the company's products far enough so that concrete decisions can be made about where to assign development responsibilities—internally or externally. However, if you get too granular, the problem can become overwhelming. As a result, it is usually necessary to experiment with various options to find the appropriate breakdown structure. A simplified example of this breakdown for an automotive company might look something like this:

♦ Audio components
♦ Drive train
♦ Chassis
♦ Safety systems (belts, air bags, etc.)
♦ Engine
♦ Exhaust system

Also, in defining your components for this analysis, you should strive to include a complete view of the overall value proposition that your product or service offers to customers—the customer value chain—even if some pieces of that value proposition are not directly provided by your company. In other words, try to think of your product or service in its largest sense from the customer's perspective before starting to break it down into component parts.

Once an approach has been established for breaking down your company's value offering (*whole product*) into discrete components, you need to make a strategic decision about whether each component should be developed internally, outsourced, or placed somewhere in-between (e.g., considered for potential collaboration under specific circumstances). In order to capture these strategic direction decisions, we will use a simple classification system of *core*, *critical*, or *contextual* to indicate the range of development options that should be considered for a given component.

CORE AREAS   *Core areas* are those truly differentiating, strategically and economically important components, skill sets, or activities that provide the foundation of the company's competitive advantage and shareholder value. For this reason, companies invest in these activities and, generally, keep them in-house. There are, however, exceptions.

For example, a company may choose to partner in a core area to access new technology. Cisco Systems has made a regular practice of this type of partnership; it partners with (and often eventually acquires) small companies with new, breakthrough technologies in its core networking product areas. This strategy allows Cisco to stay on the cutting edge of technology without developing all capabilities internally. Despite the exceptions, however, designating a product component or development capability as core indicates that, as a strategy, you are going to focus on maintaining research and development expertise for this area in-house and partnering selectively to enhance your internal capabilities.

CRITICAL AREAS   *Critical components* or capabilities are elements that are critical to delivering value to customers but that are not considered core to the company's differentiation in the marketplace. This is where most true codevelopment happens because the critical nature of these product/service elements forces a company to keep a close involvement in their development. Actually performing that development, however, is not the best use of the organization's resources.

An example of this might be the video screen on a cell phone for a handset producer that has defined its core value proposition as providing the highest-quality call reception and phone functionality. Hence, the objective of video screen technology development is simply to maintain parity with the competition. So, if good partners are available that specialize in screen technology and manufacturing, engaging in a codevelopment relationship may be much more cost-effective and beneficial to the company overall than attempting internal development.

Thus, designating an element of your overall customer offering as *critical* indicates that your strategy will allow for significant use of external partners in this area, but with an expectation of fairly close relationships to maintain some connection to the relevant technologies.

CONTEXTUAL AREAS   The last classification, *contextual,* is used for components or capabilities at the other end of the strategic spectrum from "core" and represents the activities that support the company's product or services but are not critical to their value perception by customers. These are good candidates for outsourcing to well-chosen, well-managed suppliers for whom they do represent core capabilities. Instead of investing capital, resources, and management time in activities that someone else can do better for less, the company can focus on its core (and critical) activities that drive its competitive advantage.

Though we are indicating that partnering is desirable for both critical and contextual areas, the distinction is important because it helps the organization understand how closely it needs to manage the development partner (e.g., much less closely for non-critical than for critical context components.) An example in this area might be the development of downloadable ringtones for the cell phone handset provider. In this case, the additional ringtones are not critical to the consumer's ability to use his or her phone, but they do represent some added value to the consumer related to the handset company's product. So, a light relationship could be created with one or more external companies to drive creation of ringtones, ensuring ringtone availability but without the intricacies of a full codevelopment relationship. (These distinctions of relationship depth will be discussed more fully in the following section.) So, identifying a component or capability as *contextual* indicates a strategic decision to actively avoid internal research and development in this area and instead leverage extended partners to drive development.

Next, we need an approach to making the core-critical-contextual assignment for each component. Since such an analysis is particularly important in industries with complex products that involve multiple components and require a range of R&D competencies, companies at the leading edge of practices in R&D collaboration have begun to create analytical models for making these strategic decisions. Figure 12-3 illustrates one such model that facilitates assignment of the core-critical-context classification (as a guide for collaboration/outsourcing decisions) based on a set of qualitative criteria.

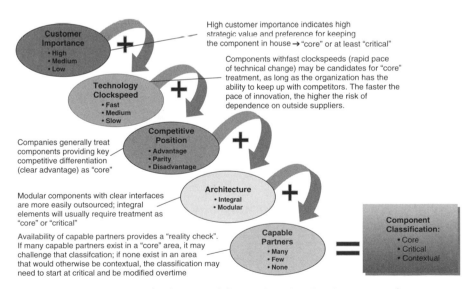

FIGURE 12-3. Criteria for deciding preferred codevelopment options.

In this model, each component of an overall product offering is evaluated against five criteria:

1. *Customer importance.* How will the sourcing of the component affect customer perceptions of quality, safety, or efficacy? High customer importance indicates high strategic value and preference for keeping the R&D work in house—i.e., develop internally or at most, develop with a closely held partner.

2. *Technology clock speed.* How fast will the underlying technology change through time? The faster the pace, the shorter the competitive advantage from a given technology and the higher the dependency on outside partners.

3. *Competitive position.* How does the company stack up in its ability to design and develop the component (cost, quality, technology leadership, etc.)? A strong position argues for keeping development in-house.

4. *Product architecture.* How does the component fit into the overall product architecture? Modular components are more easily outsourced. Integral components are more risky and difficult to outsource, and should be developed internally, or potentially with a partner if close communications and collaboration can be ensured.

5. *Capable partners.* How many potential partners have the right mix of R&D skills, capacity, financial health, and location? The greater the availability of viable partners, the safer outsourcing is. This also provides a helpful reality check: Too much availability in an area where the other indicators are leaning toward internal development, or too little in an area leaning to outsourcing, may mean you've misclassified some of the other criteria.

Although it is possible to create algorithms to calculate the appropriate classification based on numeric scorings of the five criteria (or other more quantitative criteria), this is not absolutely necessary. At the simplest level, these criteria can be considered by a strategy team and used as guides to make a rationale decision about which of the three categories (core, critical, or contextual) should apply for each product or service component. In the next step, we will look at how these classifications are captured and translated into an actionable strategy for codevelopment.

How Will We Get There? Working out the Details of Your Codevelopment Strategy

The ultimate goal of a codev strategy is to provide a clear and detailed picture of corporate intent regarding the development of each component across the customer value chain. So, this step represents the main tool and resulting output of the codev strategy exercise. With the core-contextual analysis complete, the question now becomes: How do we make it operational? Below we describe

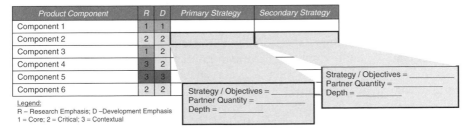

Product Component	R	D	Primary Strategy	Secondary Strategy
Component 1	1	1		
Component 2	2	2		
Component 3	1	2		
Component 4	3	2		
Component 5	3	3		
Component 6	2	2		

Legend:
R – Research Emphasis; D –Development Emphasis
1 = Core; 2 = Critical; 3 = Contextual

Strategy / Objectives = _____
Partner Quantity = _____
Depth = _____

Strategy / Objectives = _____
Partner Quantity = _____
Depth = _____

FIGURE 12-4. Codev strategy matrix.

the development of an operational *codev strategy matrix*—a summary table that captures the organization's plan for building R&D partnerships across the components of its product or solution. This is a concrete document that captures the strategic intent for development (internal versus external) on a component-by-component basis, as well as a rough plan for how to realize that strategic intent.

Figure 12-4 shows the basic structure of a typical codev strategy matrix. The first three columns capture information generated from the discussion in the prior step on identifying targets for collaboration: (1) the names of the product or service offering components; and (2) the core, critical, or contextual classification for each component. In this version, we have included two columns for the core-critical-contextual classification to distinguish between the strategic direction for research and development emphases, respectively, in that component. The reason for this distinction is that some companies may choose to treat development capabilities for a given product component as more or less *core* to their business than early-stage research in that same area (or vice versa). If this distinction is not important to your business, it is possible to combine these two columns into one.

In order to understand how to fill in the codev strategy matrix, we need to discuss the thinking behind the last two columns of information—the primary and secondary strategies. This additional information provides the critical operational details that your organization needs to implement a codevelopment strategy. In this section, we will discuss an approach to filling in this additional information, organized into four distinct steps that must be applied to each distinct line item (product/service component) in your matrix:

1. Determine the primary intent (strategy/objectives).
2. Estimate the number of desired partners.
3. Determine the appropriate depth of partnering relationships.
4. Consider a secondary codev strategy.

PRIMARY INTENT FOR CODEVELOPMENT This first piece of information is really just an elaboration of the thinking behind the core, critical, or contextual decision that you made for each component using the guidance from the previous step. This information should be captured as the *strategy/objectives* in the primary strategy column for each component in your codevelopment matrix.

The reason for this elaboration is that simply labeling a component as core, critical, or contextual is not enough guidance for day-to-day partnering decisions, since the implications of a classification may vary within a given component category. To provide clarity to the larger organization and to make your classification actionable, you must define (and clearly articulate) your primary purpose for collaborating (or not) in a given component area based on the desired level of control.

This definition should also specify the benefits of collaboration for that specific product or service offering component. For example, is there a cost advantage? Is there intellectual property to be gained? Will the collaboration help your market presence? Are you collaborating because you just don't have the skills available internally? What do you really expect to gain from collaborating (or not)?

When analyzing the potential benefits, consider several aspects of product and market maturity:

◆ *Customer access to technology.* Will collaboration eliminate a technology bottleneck/gap in your market? Will collaboration accelerate a product through its technology adoption lifecycle? (See discussion box below.)

◆ *Standardization.* Will collaboration help you standardize the technology components in your customer value chain?

◆ *Competitive landscape.* Does the number and size of the competitors preclude you and a collaboration partner from doing business in that space?

◆ *Partner landscape.* Can you find a partner who would collaborate with you on elements in your customer value chain?

One other point to note is that this elaboration of your strategic intent for each component will provide a "reality check" for your original core-critical-contextual assignment. For example, if you classify a particular component as "contextual" and there are no viable partners in the marketplace, you may as well rethink your strategy sooner rather than later.

NUMBER OF DESIRED PARTNERS To assess how many partners you'll need, an analysis of the relative risk profile and importance of each component area is required. For contextual components, it is often desirable to have multiple partners to ensure that those parts of the whole product will be available, despite issues with any given partner. Multiple partners are also preferred when the risk of sharing intellectual property (IP) is low and the integration with other technology components is well-defined.

For critical and core components, most companies choose to limit the number of their relationships to allow for the deeper level of investment (in people, communication frequency, infrastructure, and often funding). It's better to limit the number of partners when the risk of sharing IP is high, when integration points with other components are ill-defined, or if there is a complete gap in technology for a particular customer value chain element (usually a high IP risk as well).

For example, one medical device company has been considering a development partnership(s) to incorporate its technology into another company's offering for hospitals in the United States and Europe. To allow for a high degree of technology IP exchange, the company would prefer a single partner. But to protect its ability to get to the market if this relationship fails, the company would rather have two or three competing partners, possibly on both continents. Thinking through such trade-offs is one of the most important strategic partnering decisions.

APPROPRIATE DEPTH OF PARTNERING RELATIONSHIPS   Like any relationship, codev partnerships vary in depth. A *light* relationship requires only infrequent and generally formalized interactions and little, if any, IP sharing. A *deep* relationship is more strategic and therefore more "high-maintenance" by nature. Typically, it requires significant interactions between enterprises—such as when the outsourced design components are critical to the company's overall product solution, or when complex technical interfaces are involved.

For each component in the codev strategy matrix, you should identify the level of depth that is expected for the partnerships in that area (see Figure 12-6, later in the chapter). Of course, first you will need to define what light, medium, and deep relationships mean to your company, so that these categories become tangible and meaningful across the organization.

Core and critical components tend to require deeper relationships than contextual areas, but other factors should also be considered in determining the depth of a desired relationship. Figure 12-5 illustrates criteria that can help you determine how deep a given set of relationships should be.

The criteria for determining the depth of your relationship are often not equal in importance: some are more critical than others and should be weighted accordingly. For example, one medical diagnostics company uses five criteria to define relationship depth, but weights the first two more heavily:

◆ Criticality of solution component (core vs. contextual)
◆ Annual revenue impact
◆ Number of elements of the customer value chain affected

Criteria		Light	Medium	Deep
			Relationship Depth	
Criteria	1. Importance to product offering	1.Low —————————→		1. Critical component
	2. Strategic Value(potential $, competitive value, market expansion, IP access)	2.Low —————————→		2. High strategic value
	3. # of CoDev projects & relationshipduration	3.Few and/or short term —→		3. Many and/or long term
	4. Solution modularity	4.Highly modular —————→		4. Highly integrated
	5. Functional touchpoints	5.Low —————————→		5. Many functions

FIGURE 12-5. Partnering relationships: from casual to committed.

- ◆ Level of commercial cooperation
- ◆ Equity investment in partner

This analysis needs to be conducted properly because, ultimately, it will guide partner-selection teams in identifying the right relationships and in communicating to those partners the expectations for the relationship. It will also help to set up appropriate management structures to execute the relationship and manage it on an ongoing basis—deeper relationships require more infrastructure, touch points, resources, and investments than light relationships.

Microsoft's ecosystem of relationships illustrates the full range. At the light level, there is an extended network of independent software developers who are constantly creating new software applications to expand the company's product offerings. In the middle, Microsoft has relationships with hardware companies, such as Dell and IBM, which require more direct interactions. On the deep level, there are few partners (e.g., Intel) with whom collaboration may involve an extensive exchange of information in development of a new platform.

SECONDARY CODEV STRATEGY In many situations, there is no one-size-fits-all answer. This is especially true for product components or value-chain elements classified as *critical*, which implies that they could be developed either internally or with partners. So, a complete codev strategy matrix identifies not only a primary plan for codevelopment (intent, target number of partners, and target relationship depths) but also allows for a secondary strategy that applies under certain clearly defined circumstances. For example, a company may decide that its primary strategy is to minimize internal investment in a certain area by developing a network of partners. But it also wants to maintain some internal development capability to drive the next-generation versions of the technology (secondary strategy).

Western networking or telecom equipment companies often adopt such a dual approach in order to develop cutting-edge products, while also taking advantage of lower-cost overseas R&D resources. The last column in the codev strategy matrix (Figure 12-4) provides a place to capture management's thinking with respect to these fall-back strategies for individual product or solution components. By providing this information, you effectively anticipate the "what now?" question that will arise if the organization can't find the right partners to support its primary strategy.

HOW WILL WE COMMUNICATE OUR STRATEGY? DOCUMENTING AND SOCIAL-IZING THE CODEV STRATEGY MATRIX Once you have worked through each of the steps above, you should have a blueprint for putting together a detailed codev strategy matrix. To help demonstrate what this will look like, Figure 12-6 shows an illustrative example of a resulting matrix using the sample automotive components discussed earlier, with representative strategy details for one of the "critical" components.

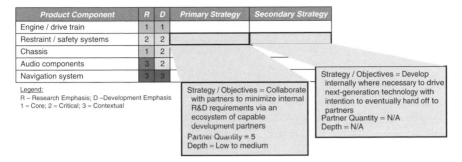

FIGURE 12-6. Illustrative codev strategy matrix example.

Of course, actually creating a codev strategy matrix for your organization is easier said than done. A useful approach is to develop drafts based on your current collaborative relationships and experience, existing product roadmaps, and the strategic vision. You can use such drafts to facilitate discussions and build the internal consensus. Areas of discrepancy can be resolved in workshop sessions with executive management and R&D leaders as part of the portfolio update process. Table 12-2 provides some guiding principles to facilitate these discussions.

A Strategy Matrix or Strategy Matrices?

Companies may consider developing multiple strategies if the customer value chains (or *whole product* offerings) that they are targeting vary significantly from market to market. Be careful, though! Remember that the strategies and the infrastructure that supports them need to be scalable. Where possible, strive to align customer value chains and address them with as few as possible strategy matrices, so that maintenance, communications, and the pursuit of codevelopment relationships are as straightforward as possible for the company.

Obviously, no matter how good your codev strategy, it must be well communicated to the organization to make a difference. You need to present your codev strategy guidelines to others as clearly as possible, since managing relationships is a shared responsibility across multiple functions:

- ◆ Business development—fosters relationships that align with codev strategy
- ◆ Product development—works directly with partners on development efforts and may recommend or select new partners
- ◆ Alliance management—supports and monitors relationships with codev partners
- ◆ Supply chain/procurement—makes sourcing decisions that will align (or conflict) with the intent of development partnerships

So, the final step in building your codev strategy is establishing vehicles to communicate to key stakeholders in the organization, and perhaps even more

TABLE 12-2.
A Short Guide to Formulating Codev Strategy

	Core Components	Critical Components	Contextual Components
Recommended Partnering Emphasis	Focus on in-house development; consider cultivating a few highly strategic partnerships.	Develop strategic partners to provide key capabilities that are not core to our vision and business strategy.	Develop a network of company-qualified partners for whom this capability is core. Ensure redundant capacity.
Target # of Partners	Very few	A manageable number of key partners to provide back-ups without spreading IP sharing or partner management staff too thinly.	Many
Relationship Depth	Deep	Mid-to-deep	Low-to-mid
Secondary Strategy	Might select strategic partners to drive R&D for specific purpose (e.g., new market/application).	Develop internally as needed (e.g., to ensure in-house expertise, or when lacking capable partners).	Develop in-house only when absolutely necessary.

importantly, to monitor whether it is being followed. Since this document and the strategic decisions it contains may represent some fairly confidential information, the codev strategy matrix is generally not something that companies post on a general Web site, but you do need to make sure it is readily accessible to decision-makers across the organization, down to individual contributors who may be making partner arrangements for R&D support on a day-to-day basis.

Having a central owner for the codev strategy matrix is another good way to ensure clear two-way communications regarding your collaboration strategy. Logical choices for this owner include senior individuals within Corporate Strategy, a central Alliance Management function, corporate planning, Research and Development, or Business Development. The specific functional location is not as important as just having someone clearly identified to drive periodic updates to the strategy document, to ensure that it is clearly, consistently, and repeatedly communicated to the organization, and to watch for issues where someone may be pursuing a relationship that conflicts with the existing strategy.

CODEV STRATEGY IN PRACTICE—CASE STUDY

To help understand how a codev strategy is formulated and used in practice, let's discuss a case study of a company that has used this methodology. This company is a mid-sized corporation that is a market leader in the development and commercialization of systems that help scientists perform genetic research. The company's customers include pharmaceutical, biotechnology, agrochemical, diagnostics, and consumer products companies, as well as, academic, government, and other non-profit research institutes. The company had experience with several codevelopment relationships, but lacked the infrastructure to support it and companywide understanding of how codevelopment should be executed.

The company needed to bring consistency to how it selected, managed, and executed on outsourced or codeveloped components of its products. Customers' increasing emphasis on whole product (the entire customer value chain), made it critical for the company to tightly manage all the product elements, whether developed internally or externally. The company had successful experience with codev relationships in the past, but recently had experienced some setbacks on critical projects that involved complex R&D collaboration. The company's difficulty stemmed from the fact that its partnering strategy was not clearly articulated and communicated. The company lacked clear "owners" for partner relationship management and had no consistent criteria and processes for selecting partners. In addition, there were no specific performance measures to assess the success of codev capabilities, and no indication that the product development projects had a codev component.

The company's management decided to develop a codev strategy matrix. The project team wanted to conduct core-contextual analysis at the technology component level, so the first order of business was to define the customer value chain and the applicable technology component(s) to clarify the company's make-buy strategy at a more granular product component level. Since the company had codev partners, it also had some preexisting strategies, but these were undocumented and inconsistent. As a result, the organization was not aligned around a single codev intent.

Through a series of one-on-one sessions with R&D leads, the project team drafted a codev strategy matrix, which, after several revisions, was subsequently agreed upon. Critical to the success was the alignment of key executives around the strategy. With executive support, the R&D organization at all levels was able to plan projects as prescribed in the codev strategy matrix. This served as the basis for, and a critical input to, the product portfolio planning process. The various product portfolio managers were able to develop roadmaps with an appropriate balance of both internal and collaborative development projects. Most significantly, the strategy effort resulted in a concrete consensus to shift internal R&D investment away from one part of the company's product offering suite (instrumentation) to focus more heavily on other elements that provided more strategic differentiation (applications). This shift was made possible because the company established a strategy of developing a small number of deep codev partners to take on the majority of development work for the instrumentation components.

The company is currently implementing its codev strategy as defined in the matrix by doing the following:

◆ Putting in place processes for partner selection, initiation, management, and termination.

◆ Securing resources to adequately manage partners of any depth.

◆ Executing codev specific activities for successful product development.

The strategy of establishing partners for instrumentation development has led to a rapid increase in the internal R&D investment (and engineers) dedicated to the development of breakthrough new applications. This shift has helped the management team to achieve its overall goals of driving growth through new products while maintaining a lean cost structure.

CONCLUSION: REALIZING THE FULL BENEFITS OF THE CODEV STRATEGY

The codev strategy matrix provides direction to the organization's R&D efforts in a variety of ways. In product portfolio planning, for example, the matrix can facilitate discussion among the R&D leaders, and ensure that all the product roadmaps are aligned with the strategy. With codev objectives clearly stated across the customer value chain, product development leaders can adjust their resource mix and requirements, as well as project timelines, objectives, and approaches accordingly. In fact, the roadmaps should all explicitly identify projects that involve collaborative development partners—visible reminders that product plans are taking advantage of collaborations as intended by the codev strategy.

The strategy matrix framework can also be leveraged in other infrastructure-related aspects of collaborative development. For example, it can be used to inventory the portfolio of existing and potential partners—as a one-page directive for business development to secure or terminate collaborative development relationships. Similarly, the framework can be used to evaluate partner portfolio performance, including the strategic fit, fulfillment of the target number of partners, and relationship depth. In addition, it can be used to determine the resource requirements for supporting the collaborative relationships.

Clearly, taking the time and effort to create a thoughtful codev strategy—and refreshing it periodically as part of the product portfolio planning process—is essential to success in R&D outsourcing or collaboration. And, since this innovative new product development paradigm is clearly here to stay, companies that excel at integrating codevelopment into their business and product portfolio strategy will be well ahead of the curve. For companies that are just starting down the codevelopment path, rapid implementation of the fundamentals described here will facilitate the launch of their first initiatives. For companies that have many partnerships already in place, a rigorous approach to codev strategy formulation will ensure that their partnerships are providing maximum business value—and minimize the chance of experiencing an outsourcing disaster.

REFERENCES

Ayers, Dave. 2005. "Outsourcing Innovation", *BusinessWeek* (March 21).

Chesbrough, Hentry William. 2003. *Open Innovation: The New Imperative for creating and Profiting from Technology*, Boston, MA: Harvard Business School Press.

Deck, Mark. 2004. *PDMA Toolbook 2*. Hoboken, NJ: John Wiley & Sons, Chapter 6.

The Performance Measurement Group, 2004. cross-industry survey by , LLC (PMG).

13 Team Launch System (TLS): How to Consistently Build High-Performance Product Development Teams

Douglas A. Peters
President, DS Performance Group

The Team Launch System (TLS) is a comprehensive system that consistently develops high-performance teams within six weeks of team launch. TLS accomplishes this by organizing team development into four phases with specific milestones, tasks, and deliverables for each phase. This creates a defined process—with high levels of accountability for performance—that can be defined, measured, analyzed, improved, and controlled.

STARTING NPD TEAMS IS CHALLENGING

Teams are complex. They typically operate within highly complex organizational and political dynamics and start out in a somewhat chaotic situation due to lack of consensus on team rules, roles, and structure. Teams are intense. Motivation and commitment among team members can vary widely, with some not even attending the initial team meetings, and others remaining passive until they are able to get an accurate assessment of the situation.

Our research on teams has identified 10 team dynamics that are key success factors in predicting team performance. Any one team can have all, none, or some of these variables. This creates over 1,200 different combinations of variables that any one team may start with. And these variables are not binary—they are not necessarily just "present" or "absent." This means that within these 1,200 combinations, each team can have a different range of performance on each variable. In short—team dynamics are extremely complex.

In addition to this internal complexity, each team also faces a different set of external variables that will affect its performance. These variables will be different for every team, even teams within the same organization. One product team might find, for example, that a given function is uncollaborative because it pays a high price for supporting the team. Another team may find that same function highly collaborative because it benefits from that team. Therefore, what the team is doing, and the costs and benefits it creates for others in the organization, will combine to create a different set of circumstances for every team.

Given this extreme level of complexity, the answer to every question about team development is the same—*It depends!* In this environment there is usually great variation in team performance within and across organizations, and it is not unusual to take months to develop a high-performance new product team. By organizing these dependencies into *phases, milestones*, and *tasks*, TLS creates high-performance teams faster, better, and cheaper.

DESCRIPTION OF THE TOOL

TLS was developed through over 20 years of applied research into high-performance teams in 75 different divisions and staff groups at 3M. The best practices have been applied to a wide range of teams, companies, and industries to validate their general applicability. Based on the learning from over 400 team training and team-building sessions, TLS captures best practices of high-performance executive, business, new product, and major project teams.

TLS is not rigid set of rules, but a set of guidelines that provide a team with a clear roadmap for achieving high performance. Each team must decide how much time and effort it should invest in each task, and teams should add or delete tasks based on their unique situations.

Regardless of the size, composition, or purpose, all teams go through four phases of development to achieve high performance. Figure 13-1 shows the four phases of the Team Launch System and the major milestones in each phase.

The *orienting phase* is focused on creating a situational analysis to determine the importance of the team to the organization, the urgency with which the team must act, and the level of power, influence, and support the team will have within the organization—the team figures out what it is "really getting into" and what challenges lay ahead. The learning from this phase will determine how much time and effort the team should put into the next phases of team development.

In the *organizing phase*, the team gets organized by defining its mission, goals, strategies, and structure. These activities turn a collection of individuals into a *team*. This phase builds ownership and commitment to the team through active participation in the team-building process.

In the *action phase*, the team starts implementing its goals and strategies. The team becomes the integration point for all new product activities and

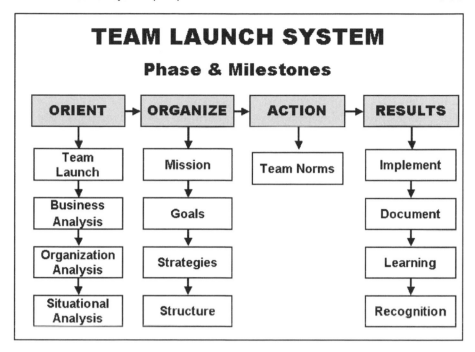

FIGURE 13-1. Team launch system.

a driving force for overcoming the many obstacles and challenges that will arise throughout the new product development (NPD) process. It establishes a collaborative relationship with management to resolve conflicts and manage the inevitable trade-offs between costs, quality, and cycle time.

In the *results phase*, the team maximizes its results by assuring its plans, decisions, and solutions are fully implemented. The team maximizes the return on the organization's investment in the team by documenting results, sharing learning, and providing recognition for contributions to the team success.

The orienting and organizing phases require a significant investment of time, energy, and resources to complete, but they do not yield any tangible results in terms of developing the product. Therefore, there is often a great deal of pressure to skip or move quickly through these phases to get to the action phase. However, skipping or short-changing these first two phases typically results in false starts, inconsistent performance, and a great deal of scrap and rework in the form of bad decisions, damaged relationships, and wasted time and energy.

The Team Launch System expedites teams through the orienting and organizing phases and maximizes team performance in the action and results phases. This chapter provides an overview of the tasks associated with each milestone and some examples of recommended processes for achieving these tasks.

PHASE 1: ORIENTING A TEAM

Phase 1 gets "everyone on the same page" by creating a common orientation to the business, the organization, and the situation. The first thing people want to know when they are assigned to a team is, why are we doing this, and what's in it for me? To answer these questions, team members must understand where the team fits into the business, the organizational dynamics that will support and inhibit team efforts, and the importance and urgency of the situation.

In the beginning, the team is engaged in putting together something that resembles a jigsaw puzzle; each team member has a different piece of the puzzle. To complicate things, some pieces of the puzzle are missing, and no one has a picture of what the final puzzle should look like. Therefore, the first thing the team must do is to get all the team members to put their pieces of the puzzle on the table. The team might also have to reach outside the team to gather other pieces of the puzzle. Team members do not need to reach agreement or consensus in this phase, they just have to share their perspectives to create the most accurate and complete picture of what they are getting into. Once the picture is completed, it is likely that there will still be different interpretations of it, but at least the team is looking at the same picture.

From a process standpoint, most of the tasks in phase 1 are best accomplished in work sessions that are focused on in-depth discussions. Some teams may be able to accomplish all of the tasks in phase 1 in a single session, but most will require several work sessions.

During this first phase, there is a strong dependency on the leader to get things going. Therefore, the team leader must establish a clear process for moving the team through the milestones and tasks in the phase. The team leader must also actively facilitate the group discussion that must take place to achieve the tasks in this phase. It is the team leader's role to keep the team focused on the task and to create balanced participation in order to maximize ownership and commitment through active participation in these team-building activities.

Milestone 1: Team Launch

A mandatory and formal team launch process demonstrates management's commitment to the team and ensures that the team gets off to a good and quick start. This first milestone has tasks for both management and the team leader. Together, the team leader and management can give the team a running start and establish an action orientation for the development of a high-performance team.

TASK 1: TEAM ASSIGNMENTS Management controls the resources. Therefore, management must assign members to the team who are willing, able, and have the opportunity to perform. Since it is easier to remove people from the team than to bring on new team members, it is strongly recommended that management err on the side of assigning too many people to the team at the beginning of the team-development process. The team will adjust its membership at the end of the organizing phase, so team members who are not needed on the team can exit at that time.

TASK 2: SCHEDULE WORK SESSION   To accomplish this task, the team leader must schedule the initial team meetings. The leader must organize the first team meeting by creating an agenda and scheduling work sessions to achieve milestones 2, 3, and 4 in phase 1. The agenda for the work sessions is made up of personal introductions and the phase 1 milestones. It may take several work sessions to complete the agenda.

Work Session Agenda

1. Introductions
2. Business analysis
3. Organizational analysis
4. Situational analysis

Scheduling work sessions can be extremely difficult, especially when team members have major responsibilities in addition to the new product development team. A typical new product development team will require two to four work sessions of four hours each. Scheduling half-day work sessions every other day allows team members time to digest information and think about issues in between work sessions. This format should allow most teams to complete phase 1 in one week.

TASK 3: TEAM MEMBER PREPARATION   To accomplish this task the team leader must prepare team members to participate in the first meeting and work sessions. These sessions will be dominated by in-depth discussions about the business, the organization, and the situation. Typically, team members have a range of knowledge and understanding about these topics, depending on their position and experience. Therefore, sharing information such as business plans, technology plans, marketing presentations, and organizational charts prior to work sessions will establish a more uniform understanding and eliminate the need to bring some people up to speed while others sit through information they already know.

Many of the activities in this phase start with brainstorming, list building, and sharing initial thoughts and reactions. Soliciting and sharing this type of information prior to a work session dramatically increases productivity. Instead of spending time sharing initial thoughts or building lists, the team can focus on higher-level discussions to create greater understanding. This shortens the time needed to complete phase 1.

Web- and software-based document-sharing tools can significantly increase the productivity of work sessions. They allow team members to interact in a virtual environment, sharing thoughts and influencing each other's perspectives prior to the first work session and between work sessions. They make the best use of peoples' time by allowing them to participate when it is most productive for them.

TASK 4: FIRST MEETING INTRODUCTIONS   To accomplish this task, the team leader must prepare team members to introduce themselves during the first team meeting. Starting the first work session with personal introductions initiates a relationship-building process among team members. Providing a format for personal introductions in advance reduces stress by allowing team members to prepare in advance.

The following questions share information team members will need in order to understand each other and make accurate judgments about the team's competence and commitment.

Past Experience

   ◆ What has been your past experience with this business, product, and/or technology?
   ◆ What are your major areas of expertise and experience?

Knowledge Base

   ◆ What have you heard, been told about, or know about this new product?
   ◆ What thoughts, conclusions, or opinions do you have about this new product?

Motivation

   ◆ What about this team excites you? Why?
   ◆ What about this team concerns, worries, or scares you? Why?
   ◆ What do you want out of this team for yourself? Your Function?

Actions

   ◆ What experience have you had with other teams like this?
   ◆ What actions are you taking now that will affect this team?
   ◆ What plans or commitments do you have in the near future that will affect this team?

Some participants will find the level of self-disclosure in these personal introductions to be risky and uncomfortable because of their personal situation. Inexperienced team members, for example, often find this uncomfortable because they cannot answer many of the questions. This may create the impression they are a liability because they will have to go back and check with someone before they can actively participate in team discussions or decisions, causing delay when the team is under pressure to make a decision or take action. But issues like this do affect team performance; therefore it is better to deal with it upfront than to have it explode when the team is under pressure to take action. This will require a great deal of skill and tact on the part of the team leader.

The team must take risks in order to build trust. Personal introductions create the level of risk taking necessary for the team to start building trust and strong work relationships. It begins to establish a norm that the team will talk to each other about issues and not about each other. This activity will be the first real test of the level of work relationships within the team—do people trust each other enough to risk sharing and discussing issues that put team members at risk?

Milestone 2: Business Analysis

The second milestone of phase 1—orientation is conducting a business analysis that results in a "compelling business reason" for the team's existence. An accurate business analysis empowers a team to do the "right thing for the business" and to become self-directed in its actions. To achieve this milestone, the team must achieve the following tasks.

TASK 1: MARKET ANALYSIS To accomplish this task, and assure the team does the "right thing for the business," the team must analyze current market dynamics, including customers, competition, opportunities, and the metrics used by the business to measure performance. This understanding turns each team member into a *businessperson*.

- **Understanding customers**—Who are our customers, and how will this new product affect them?
- **Understanding competition**—Who are our competitors, and how will this new product affect our competitive position?
- **Understanding opportunities**—What current or future opportunities will this new product address or create in the market?
- **Understanding the key metrics**—What are the key metrics we use to measure this business, and how will this team affect those metrics?

Routing these questions to team members for response prior to the first work session will allow the leader to assess the level of business understanding among team members and determine how much time will be required to get everyone at the level of understanding necessary to achieve this milestone. Initial responses should be summarized and routed back to team members for thoughts and reactions. Several rounds of summarizing and sharing can move the team a long ways toward creating a common orientation prior to the first meeting.

TASK 2: STRATEGIC ANALYSIS To accomplish this task, the team must analyze its alignment with larger business plans and strategies. Organizations that chase every good opportunity often spread themselves so thin that they fail to take full advantage of any one opportunity. To avoid this problem, organizations develop strategic plans, business plans, and functional plans and budgets to set organizational priorities and maintain focus. Teams must be aware of where they fall into these larger plans and strategies to assess their importance to the organization.

To prepare for this discussion team members should be provided with copies of relevant business plans and strategies and be asked to answer the following questions. Responses to these questions should be gathered and shared with team members prior to the first work session.

The First Why—Project Level

♦ Why is this project important to each of the organizational units involved?

♦ How does this effort align with the plans and tactics of each organizational unit?

The Second Why—Business Level

♦ Why is this new product important to the business?

♦ How does this effort align with larger business plans?

The Third Why—Strategic Level

♦ Why is this new product important to the organization's long-term strategies?

♦ How does this effort align with the organization's long-term strategic plan?

By understanding their alignment (or lack thereof) with larger strategies, business plans, and tactics, the team can estimate the level of priority, power, influence, and support it will have in the organization. Strong alignment indicates that the team will be working with organizational units that are already in alignment with the team, and can justify providing the team with significant resources and support.

TASK 3: COMPELLING BUSINESS REASON To accomplish this task, the team must draw on the learning from tasks 1 and 2 to articulate a compelling business reason that justifies the investment of time, energy, and resources into this project. A strong, compelling business reason creates high levels of motivation to make the team successful. For example, if it is clear that this is not just some management program designed to find someone to blame for failure, but is really addressing a critical need for the organization, then clearly this team can make a difference.

A good way to start this process is to ask team members to answer the following question:

◆ If challenged by your boss, how would you explain the business reasons that justify your investment of time and energy into this new product team?

Answering this question requires team members to synthesize all of the information and the understanding they have gained through the business analysis. Putting the question in the context of responding to a challenge from their boss forces people to put the compelling business reason into their own words. This assures this is not something that is just written down and soon forgotten.

Milestone 3: Organizational Analysis

The third milestone of phase 1 is to accurately assess the organizational dynamics that the team will experience as it develops the new product. This analysis allows the team to assess what it is really getting into and how much time and effort will be required to successfully launch the new product.

TASK 1: KEY STAKEHOLDER IDENTIFICATION To accomplish this task, the team must identify the key stakeholders it must work with to achieve success. In practical terms, key stakeholders are *the larger team*, and the people assigned to the product team are the *core team*. The role of the core team is to involve the right key stakeholders, at the right time, and in the most efficient manner in order to maximize collaboration and performance throughout the new product development effort.

Key stakeholders are those who are put at risk, who have to contribute resources, or absorb the costs and/or benefits of team actions. To prepare for this discussion the team leader can create a table to route to team members prior to the first work session (Table 13-1). Several rounds of sharing should quickly identify major key stakeholders and would allow the team to focus on fine-tuning the list during the work sessions.

TASK 2: COST AND BENEFIT ANALYSIS To accomplish this task, the team must analyze the costs and benefits that will flow to key stakeholders. This analysis identifies who benefits from team actions (winners) and who pays the costs (losers) of team actions. The team can call on those who benefit from team actions for support. The team can provide those who pay a price for team actions *with* support and protection from the negative consequences of doing

TABLE 13-1.
Key Stakeholder Identification

	Function	Person	Why?
1			
2			

the right thing for the team. But the team cannot do either unless it knows who the key stakeholders are and how the costs and benefits will flow to each key stakeholder.

The prepare for this discussion the team leader should select out those key stakeholders that appear on most participant's lists and create another table that asks team members to identify the costs and benefits that will flow to each function and/or person (see Table 13-2). The team leader should make this a confidential activity—the leader will share the results, but not who said what. Confidentiality often creates more open responses, especially negative responses about specific functions or individuals.

TASK 3: FORCE FIELD ANALYSIS   To accomplish this task the team must analyze the organizational dynamics (culture and infrastructure) that will support and inhibit team performance. These dynamics can be seen as a set of driving and restraining forces. Driving forces such as management support, rewards, recognition, adequate resources, etc. help move the team forward by supporting and encouraging its efforts. Restraining forces such as lack of prioritization, punishing reward systems, inadequate resources, and so on, create obstacles to team performance and success.

Understanding the balance of driving and restraining forces provides team members with a clear idea of what they are getting into, how much effort will be required to achieve success, and the probability of achieving success. It helps answer the question, "What's in this for me?"

The following directions and model can be routed to team members prior to the first meeting to begin building the list of drivers and restrainers. Since this process may raise politically sensitive issues, it is best if confidentiality of responses in maintained. Several rounds of sharing will allow the team to spend time in its work session assessing the consequences of these dynamics.

Force Field Analysis Instructions   Looking at our organization's culture (the way we do things around here) and infrastructure (e.g., people, processes, rewards, measurements, structure, politics), what forces do you see that will support this team and what forces do you see that will create roadblocks to our performance and success? Create a table like the one shown in Table 13-3:

TABLE 13-2.
Cost/Benefit Analysis

	Key Stakeholder	Costs	Benefits
1			
2			

TABLE 13-3.
Drivers and Restrainers.

Support (Drivers)	Inhibit (Restrainers)
•	•
•	•
•	•

Milestone 4: Situational Analysis

The fourth milestone in phase 1—orientation requires team members to draw on their learning from the previous tasks to create a situational analysis. Developing a high-performance team requires a significant investment of time and energy by team members and by the organization. Team members must have an accurate assessment of the importance of the team, the urgency with which the team must act, the strength of team dynamics, and the probability of achieving success to determine, and to justify, the level of time and energy they will commit to the team.

TASK 1: PROJECT IMPORTANCE To accomplish this task the team must determine the importance of the team to the organization. This analysis will help determine the level of effort and resources that the project will likely justify. The more important the effort is to the organization—the organization has much to gain by success and/or lose by failure—the higher the priority the project will be given by team members and by the organization.

The leader can initiate this discussion by routing the following questions to team members prior to the work session in which this topic will be discussed.

- ◆ What would be the impact on the organization if this new product effort fails?
- ◆ What would be the impact on the organization if this new product effort succeeds?

TASK 2: PROJECT URGENCY To accomplish this task the team must determine the urgency with which it must act. The more urgent the situation, the faster the team must act. In situations where there is very little urgency the team can work more slowly to achieve results. Adjusting the level of effort to match the urgency of the situation contributes directly to the organization's productivity by freeing up resources that can be used on other critical and urgent projects.

The leader can prepare team members for this discussion by routing the following questions to team members prior to the work session in which this topic will be discussed.

Fixed Dates

 ◆ Are there fixed dates that will significantly affect this project, such
   as trade shows, successor project launch dates, customer deadlines,
   marketplace events, or a specific sales cycle?

Window of Opportunity

 ◆ How large is the window of opportunity for this project to succeed?
   New product efforts often address specific situations that have a window
   of opportunity that is limited in time.

Predecessors/Successors

 ◆ Are there other organizational efforts that are predecessors or successors
   to this effort, and, if so, how does this affect the timing of this project?

Consider what projects or efforts are predecessors or successors to this
project. Predecessor projects may be counting on the team to take advantage
of a limited window of opportunity they have created. Or other projects may
be waiting for this project to be finished to build on its success or utilize its
resources when the project is finished.

TASK 3: TEAM DYNAMICS   To accomplish this task the team must analyze its
personal, interpersonal and group dynamics. Teams that have strong dynamics
can aggressively move through each phase. Teams with weak dynamics will
need to invest more time to strengthen team dynamics before they will be able
to become a high-performance team.

The following team dynamics are critical to team success:

 ◆ *Personal Acceptance*—Teams with high levels of personal acceptance
   will understand and value individual differences and eliminate unproductive conflict over personality and style.
 ◆ *Professional Respect*—Teams with high levels of professional respect
   will understand and value functional differences and minimize turfiness
   over conflicting functional goals, objectives, and priorities.
 ◆ *Strong Work Relationships*—Teams with strong work relationships
   establish high levels of trust and personal risk taking. This creates an
   environment where the team can create a norm of open, honest, and
   direct communications.

Some teams will discover they are set up for success, and little time and
energy will be required to build strong team dynamics. Some teams will find
they are set up for failure, and will require a significant degree of repair before
they can even begin to operate as a team. Most teams will be somewhere
between these extremes.

TASK 4: PROBABILITY OF SUCCESS To accomplish this task the team must assess the probability of achieving success given the unique business, organizational, and team dynamics the team finds itself in. This activity requires each team member to think through and synthesize all of the information and analysis generated in phase 1.

A process that works well for achieving this task is to have each team member summarize his or her view of the probability of success by picking a percentage between 0 percent and 100 percent—0 represents no chance of success and 100 represents certain success. The team leader should place individual responses on a flip chart and lead a discussion to understand differences in ratings.

Teams that see a high probability of success can feel comfortable moving forward to the next phase of team development. Teams facing a low probability of success will want to make management aware of the level of risk the team will be taking to ensure there are no surprises if the team runs into trouble. Teams that have a wide range of responses on this activity are clearly seeing things differently and have not yet reached a common orientation to the situation. They will need to have more discussions before moving forward.

PHASE 2: ORGANIZING A TEAM

Phase 2 organizes the team for success. In this phase the team must reach *consensus* on its mission, goals, and strategies, and then organize the team in a manner that provides it with the best chance of accomplishing them. Achieving consensus through highly participative processes that balance participation and create open and honest communications will maximize ownership and commitment to the team and ensure the team does the right things.

Milestone 1: Mission

The first milestone in phase 2—organization is to establish a team mission that limits the scope of team action and defines ultimate success for the team. The team mission statement creates focus and determines what the team will do and what the team will not do. Without it, the team becomes a wandering generality that chases every hot issue. It will appear to be arbitrary and capricious in its actions and it will be subject to the forces of personality and politics. Eventually, it will become caught in an "activity trap" where it does more and more to achieve less and less!

The team mission statement should contain the following elements:

Action Verb:	To expedite
Subject:	Development of product X
Measurement:	To thwart a competitive attack on our core business.

ACTION VERB New products will be developed with or without teams. Therefore the action verb defines what contribution the team will bring to the product development process. Will the team expedite, integrate, coordinate, define, monitor, or drive the new product development effort? In this example the action is to "expedite" the development of product X. The scope of the team, therefore, is to consider any actions that will expedite the development of this new product.

SUBJECT The subject determines the focus and scope of the team. The subject of the team in the previous example is "Development of new product X." Contrast this with a team whose subject might be to "Commercialize a portfolio of X solutions." The scope of this second team is much broader and might include the development of several new products.

MEASUREMENT The measurement in the example is to "Thwart a competitive threat." This measurement indicates that team success will be measured not by a fixed date, but by hitting a window of opportunity. If the team succeeds in launching this product, but it does so too late to thwart the competitive threat, the team will have failed to achieve its mission. In contrast, the measurement for "Developing a portfolio of X solutions" might be the number of new products that the team can generate from the technology.

A final check on the mission statement is to make sure the measurement and action verb match. For example, if the action verb is to expedite, but the measurement is sales volume ($100 million in sales), the mission statement sends a mixed message as to what the team is trying to accomplish—is it to expedite development or to maximize revenue?

PROCESS The following process is recommend to balance participation and maximize ownership and commitment when developing a team mission statement:

1. Break the team into groups of two to four, depending on the number of team members.
2. Provide each group with a flip chart and ask that they identify each element of the mission.
3. After each group has its initial thoughts down, put all of the flip charts next to each other and conduct a trends analysis.

   ◆ Ask each group to explain what is on its flip chart.
   ◆ As each group presents, the leader should underline areas of agreement on each flip chart.
   ◆ Encourage questions, comments, and reactions to move thinking forward.

4. Create a new flip chart that captures the areas of agreement from round one.

5. Put team members back into their groups and, using a new flip chart, try again to reach agreement on the mission.

6. Repeat process until consensus has been achieved on major items.

This same process can be used for the development of team goals.

Milestone 2: Goals

The second milestone in phase 2—organization is to establish specific goals that must be achieved in order to fulfill the team's mission. Successful teams, like successful individuals, are goal driven. Goal setting, therefore, is a critical step in organizing the team and getting the "butterflies in formation."

Team goals create a foundation for action upon which the team will build its structure. Decisions on membership, leadership, meetings, core teams, extended teams, and subteams are all made based on developing the best structure to achieve the team goals. Without a strong foundation of clear goals, it is extremely difficult to build a high-performance team.

If the team does not agree on goals in the beginning, it will argue about them later on. Typically, these arguments take place during team meetings, in the heat of the moment, and at a time that destroys team productivity.

When possible, it will be more efficient for the team leader to start with the goals recommended in this section and lead a discussion on modifications or changes. The following process questions should be built into the discussion of each recommended goal.

◆ What about this goal do you agree with?
◆ What about this goal do you disagree with?
◆ Would you support the team adopting this goal?
◆ If not, what changes or alternatives would you suggest?

If the suggested goal is widely rejected, the leader can use the same process for developing each goal that was used to develop the team mission.

TASK 1: PLANNING GOALS   To accomplish this task, the team must achieve consensus on a goal that will result in the development and implementation of a planning process that organizes and coordinates all new product activities and actions across organizational boundaries. Without a plan, no one in the organization can see the big picture and how all of the pieces fit together. Typically, this will require a project plan with a critical path. In some instances, such as a line extension, the team may get by with a much simpler Gantt chart.

Project planning must be an ongoing process to assure that the plan contains the most accurate and current information available. Product development requires inventors to put creativity on a timeline. Therefore, plans must be regularly updated to capture and apply the learning that will take place as the project moves forward. Organizations are also dynamic, and circumstances can change over the life of a new product effort. Therefore, plans will also have to be updated to reflect changing circumstances and priorities. The level of project planning required varies greatly, depending on the complexity of the effort.

Recommended Planning Goal—New Product Development Team

Action Verb:	To establish
Subject:	A project planning process
Measurement:	That maintains an accurate and current project plan with a critical path.

TASK 2: IMPLEMENTATION GOALS   To accomplish this task, the team must achieve consensus on how it will assure the project plan is fully implemented and how it can protect the critical path. A project plan does not yield results until it is fully implemented. Implementation of the plan is the responsibility of functional management because they control resources and have the authority to get things done. In some cases, management may delegate this authority to the team or specific team members, but that authority can be revoked at management's discretion.

A major team responsibility in the implementation of the project plan is to hold management accountable for the commitments it made in the project planning process, and to resolve any issues—problems, decisions, opportunities—that may *delay, inhibit,* or, in the case of opportunities, *expedite* implementation. If management commits resources and does not deliver, or takes actions that negatively affect the project, or refuses to take action that will significantly enhance the effort, the team can hold management accountable by demonstrating the effect of the function's actions on the cost, quality, or cycle time of the project.

Recommended Implementation Goal–New Product Development Team

Action Verb:	To identify and resolve
Subject:	Critical issues
Measurement:	That affect the critical path or have a significant effect on cost, quality, or cycle time.

Note how this goal puts the team in the position of monitoring functional performance to identify any critical issues that may fall in the cracks or are ignored by the responsible function. This avoids a teams versus management relationship that pits teams against management for control of resources. Instead, it creates a situation where teams are focused on doing the right things for the business and functions are focused on doing things right the first time.

TASK 3: COMMUNICATION GOALS   To accomplish this task the team must achieve consensus on how it will communicate, both internally within the team and externally with the larger organization. Communication is the lifeblood of teamwork. Good communication can establish trust and build credibility. Poor communication can damage credibility, destroy trust, create conflict, and dramatically increase the team's workload.

Organizational Communications   High-performance teams gain access to a wide range of information. When they consolidate this information it creates a considerable amount of knowledge—the team will know more than anybody in the world about the product they are developing. This information and knowledge can be very valuable to others in the organization when they are making decisions and solving problems. By establishing a goal of sharing this information and knowledge, the team is improving organizational decision making and increasing the return on the organization's investment in the team.

Recommended Organizational Communications Goal

Action Verb:	To ensure that
Subject:	Key stakeholders receive information that could impact their actions and performance
Measurement:	In a timely fashion.

Internal Communications   Interpersonal communications is the action level of teamwork. How things get from our head to the flip chart is the process of interpersonal communications. Interpersonal communications is, therefore, the action level of teamwork.

Recommended Interpersonal Communications Goal

Action Verb:	To create shared understanding
Subject:	On complex and emotional issues
Measurement:	To ensure all perspectives on an issue are understood and valued.

These communications goals are clearly softer than the other goals. But establishing them up front empowers the team leader to hold team members accountable when their behaviors negatively affect team communications.

TASK 4: TEAM DYNAMICS GOALS   To accomplish this task the team must achieve consensus on goals for developing positive team dynamics. Establishing this process ensures the team makes the right investments, at the right time, and in the most efficient manner.

Strong team dynamics create the ability for teams to achieve high performance. A team with poor team dynamics is like a car with a bad engine. It is inefficient, it wastes a lot of energy, and it breaks down when it is put under stress. Therefore to maximize its performance and achieve success the team must make an investment in its team dynamics to create the ability of the team to perform.

Recommended Team Dynamics Goal

Action Verb:	To establish an on-going process
Subject:	Of team development
Measurement:	That results in positive personal, interpersonal and team dynamics.

TASK 5: OTHER GOALS   To achieve its mission, a team often will have to achieve goals beyond the recommended planning, implementation, communications, and team dynamics goals. To ensure that the team remains focused on high-priority activities, the team should establish specific goals for any additional actions it will undertake. This rigorous focus dramatically increases team productivity and performance.

For example, one new product team was asked to go back and determine why previous product development efforts on this product had failed. This was a significant undertaking that could consume a great deal of time and effort. Therefore, to control the level of effort it would make, the team set the following goal:

Action Verb:	To analyze
Subject:	Past new product development efforts
Measurement:	To identify technical reasons for failure of these efforts.

Notice how the goal measurement limited the effort to focus on technical reasons for failure. The team decided it did not want to get into analyzing non-technical issues that may have contributed to failure such as lack of priority, lack of resources, lack of market information because the organization was in a different spot and these issues no longer existed for this team.

To accomplish this objective, the leader will start by having team members brainstorm a list of potential goals and prioritize them in terms of action. Once a goal area has been identified, the team should use the same process for establishing the goal that was used to develop the team mission.

Milestone 3: Strategies

The third milestone in phase 2—organization is to establish specific strategies for each of the team's goals. The mission defines team success, the goals define what a team needs to do to achieve success, and strategies determine how the team will accomplish its goals (Figure 13-2).

The pressure to achieve results, combined with the action-oriented personalities of some team members, often causes the team to initiate action before reaching consensus on *how* it will take action. This typically results in false starts, as team members charge off in different directions and create unproductive conflict and blame is assessed for the false start. When this occurs, it reduces team productivity and wastes precious time and energy. Worse yet, the

---

Recommended Planning Strategies
New Product Development Team

Recommended Planning Goal

Action Verb:	To establish
Subject:	A project planning process
Measurement:	That maintains an accurate and current project plan with a critical path.

► Objective 1   **Plan**

Result:	Develop a project plan with a critical path
Measurement:	This is accurate and optimized
Deadline:	By March 1, 2006
Key Participants	(list functions and individuals)
Leader:	(determined by team)

► Objective 2   **Plan Revisions**

Result:	Provide major plan revisions
Measurement:	That incorporate learning and reflect changing circumstances
Deadline:	At beginning of a new phase or when learning and changing circumstances require
Key Participants	(list functions and individuals)
Leader:	(determined by team)

and

► Objective 3   **Review Plan Assumptions**

Result:	Review planning assumptions, concerns
Measurement:	To determine if plan is still valid
Deadline:	At least monthly
Key Participants	Core Team Members
Leader:	Core Team Leader

► Objective 4   **Progress Updates**

Result:	Provide progress updates to key stakeholders
Measurement:	To identify any slippage in tasks
Deadline:	At least monthly
Key Participants	Key Stakeholders
Leader:	Core Team Leader

FIGURE 13-2. Recommended planning strategies.

team may have to make additional expenditures of time and energy to repair damaged relationships.

Developing specific strategies to achieve its planning goals provides the team with several benefits:

◆ A well-defined strategy that establishes clear roles, responsibilities, and accountabilities is more likely to be efficient and successful.

◆ The consensus-seeking process avoids false starts by assuring that the team views possible actions from many different perspectives.

◆ The range of opinions expressed in a consensus-seeking process often results in innovative and creative approaches that increase team productivity and results.

◆ Achieving consensus on strategies through active participation increases team-member ownership for each strategy. Instead of doing what the team leader tells them to do, they are doing what they believe is the right thing to do.

◆ High levels of ownership translate into high levels of commitment to follow through—team members will tend to work longer and harder to achieve success.

◆ The process of developing strategies requires the team to establish a measurement for each objective that provides a tangible way to determine if the team succeeded in its efforts.

◆ Clear roles and responsibilities, combined with clear measure of success, allows the team to hold those responsible accountable for their performance.

A simple and very straightforward way to think about strategies is to break them into objectives that contain the following elements:

Result:	Develop a project plan with a critical path
Measurement:	That is accurate and optimized
Deadline:	By March 1
Key Participants:	Key resource allocators, major individual contributors, and those who are put at significant risk. (Team will list people and functions.)
Leader:	Name of team member who will lead the effort. This is not necessarily, and should not always be, the team leader.

The Core Team would have to reach out to the larger team before taking action on this objective. In essence, the team is creating a subteam composed of core team members and larger team members that will achieve this objective. Reaching outside the team to involve key participants increases ownership in the process and commitment to fully implement the results of the process.

To achieve this milestone the team must develop strategies for goals in each of the following areas:

Task 1:	Planning goals
Task 2:	Implementation goals
Task 3:	Communications goals
Task 4:	Team dynamics goals
Task 5:	Other goals

Figure 13-2 presents the planning strategies to achieve the previously recommended team-planning goal. Notice how the strategies essentially break the goal down into its components parts, and then assign specific responsibility to a leader and subset of team members to achieve the results. This division of labor allows teams to multitask and work on several goals and strategies in parallel. This dramatically increases team efficiency and productivity.

Milestone 4: Team Structure

The fourth milestone in phase 2 is to organize the team. With clarity on its mission, goals, and strategies, the team can now establish a team structure that will provide it with the best chance of success. This final step completes the transition from a collection of individuals to a high-performance team.

Up to this point, the team has been focused on effectiveness—doing the right thing. It has established the right mission, the right goals, and the right strategies. Now the team must focus on efficiency—doing things right the first time. To maximize its efficiency, the team must adjust its membership, establish team meetings, create a team structure, choose methodologies and processes, and define leadership roles.

There are no rules for organizing a high-performance team. The downside of this reality is that the team cannot simply copy what other teams have done. The upside is that the team has a great deal of flexibility and freedom to organize itself based on its specific mission, goals, and strategies. The answer to every question about how a team should be organized is the same—It depends! It will require a great deal of discussion for the team to identify all of the dependencies.

TASK 1: MEMBERSHIP To accomplish this task, the team must adjust its membership based on its mission, goals, and strategies. Consensus on the team mission, goals, and strategies defines the scope and action of the team. With this level of clarity, the team is now in a position to determine if it has the right team membership to achieve its mission and goals.

Often, teams will find they have redundant members (too many people from one function), or members that could just as easily contribute to the team effort without being a full-time member attending meetings. Remember, the team that is going to actually develop the new product is composed of many dozens of people (experts) from across the organization. The *new product team* is actually a core team of people who reach across organizational boundaries to assure that the right people are involved in the product development process at the right time, and in the most efficient way.

Reduction of team membership typically increases team productivity. With fewer members, it takes less time to discuss critical issues and less time to reach consensus on those issues. Since the team can gather input from people outside the team, removing a team member does not necessarily mean the team will lose access to his or her information and input.

When teams are able to reduce membership, without negatively affecting performance, it frees up people to take on other higher value activities within the organization. This contributes to overall organizational performance and resource utilization.

In some instances, the team may find it needs to add members to meet a specific need. Since the new member did not participate in the orienting and organizing phases, the team will need to establish a process that will allow new members to gain ownership and commitment through participation. A good way to accomplish this is to have the new members review the documentation (business analysis, organizational analysis, mission, goals, strategies) and interview team members, and then present their review in a team meeting. This helps the new members develop ownership and commitment through participation and may well identify issues the team has missed.

When adding members, the team must balance the resource needs of the functional organization with the resource needs of the team. Management has a responsibility to maximize its resource allocations to get the highest return on the organization's investment in its people. And it must also ensure that individuals do not become overloaded and cause a delay in any project. Allocating resources is critically important to both teams and management. Therefore, the team will need to work closely with management when making adjustments to team membership. The goal is not for teams or management to "win," but for teams and management to do the right thing for the overall business.

In most situations, there are good reasons why people are assigned to a team. But there are situations where this process does become politicized or distorted. For example, some teams will have several people from one function assigned to the team as a way of increasing that function's ability to control or "out vote" other functions. Sometimes a manager and his or her subordinate will both be on the team so that the manager can maintain control.

In such highly political situations, the team leader should first work behind the scenes to reach a resolution with those who are directly involved in the issue. The leader may want to seek the advice and support from a supervisor, a team sponsor, or others they trust. If issues cannot be resolved behind the scenes, the leader will need to bring the issue up to the larger team for discussion and action. Bringing the issue to the full team creates a great deal of visibility. In the case where the boss and subordinate are on the team, it will force the manager to defend the continued participation of both boss and subordinate on the team. This can create a great deal of peer pressure to do the right thing for the business. Situations such as these can have a very negative effect on team dynamics and may require extensive teambuilding to fix or to pick up the pieces after a resolution has been reached. But, if divisive issues are affecting performance, it is best if they are addressed sooner rather than later, when damage has been done and blame will be assessed.

TASK 2: MEETINGS By their very nature, teams are inefficient. Rules, roles, and structure must vary greatly between teams in order for teams to adjust to the specific needs and dynamics of each situation. In the beginning, this somewhat chaotic situation creates an advantage because it allows the team to avoid preconceived notions and determine right way to run this team. But once the team's mission and goals have been set, the team must now focus on creating the efficiency necessary to get things done.

In the search for efficiency, team meetings play a critical role in team success. Teams cannot achieve high performance when they run inefficient meetings that are unfocused and time consuming. Meetings of this type dramatically decrease team productivity and performance, while seriously eroding the ownership and commitment of team members, who begin to see team meetings as increasingly frustrating and a waste of time.

Establishing a formal agenda that reflects the team's mission and goals ensures that the team remains focused on what is important to the success of the team. It ensures that the time spent in team meetings advances the attainment of the team's goals and strategies and avoids the tendency of many teams to get sidetracked onto the latest hot issue.

High-performance teams establish a regular schedule for team meetings that goes out as far as possible. It is often difficult for team members to clear their schedules to attend team meetings when they are done on a month-to-month basis. Scheduling meetings out for one year or until the team's mission is completed—whichever comes first—makes it much easier for team members to work their schedules around team meetings. In the unlikely event that a team meeting is not needed, it is much easier for team members to use this gift of time to do work than it is to clear their calendars each month to schedule a meeting.

A monthly meeting schedule typically meets the needs of most new product teams. Keep in mind that there can be many meetings between monthly meetings—subgroup meetings, problem-solving meetings, decision meetings, etc.—that involve some or all of the team's members. The purpose of the regularly scheduled meeting is to keep the team focused on its mission, goals, and strategies, and to make changes in them based on learning and changing circumstances.

The final decision for a meeting schedule should be driven by the team's analysis in the orienting phase. A team that is critical, urgent, and faces a lot of organizational roadblocks may have to meet more often. A team that is less important and not urgent may find monthly meetings to be way too many.

Figure 13-3 presents a suggested agenda for routine team meetings. Notice how this agenda keeps the team directly focused on its mission, goals, and strategies. The team should schedule additional meetings to take specific actions that come up during team meetings, such as making a decision or solving a problem. This will assure the team completes its routine agenda and stays on course.

TASK 3: STRUCTURE Teams are a collection of individuals until they organize themselves around a team structure that gets the "butterflies in formation!"

Regular Team Meeting Agenda
1. Planning a. Review plan assumptions b. Review concerns
2. Implementation a. Identify and priortize critical issues b. Develop action plans c. Review current action plans d. Celebrate completed action plans
3. Other Goals a. Update by objective leader
4. Communications Goals a. Open Discussion b. Key Stakeholdeder communications
5. Other (specify):
6. Team Dynamics a. Team meeting evaluation

FIGURE 13-3. Regular team meeting agenda.

A defined team structure identifies and focuses on key integration points within the team that are critical to achieving success.

One of the strengths of a high-performance team is its extreme flexibility to organize itself in response to its task. Functional organizational units with job descriptions, chains of command, offices, and compensation plans are extremely cumbersome to reorganize. In contrast, the high-performance team should be like a chameleon that adjusts to the situation it finds itself in.

There is no right way to organize a team. How a team should be organized depends on the complexity of the project, the importance of the project, and the resources available to the project. Therefore, one of the major strengths of a high-performance team is its ability to consider these variables and create the right structure for its specific situation.

For example, establishing a sub-team for developing a new *liner* (the backing on a tape product) indicates that liner development is key to this team's success, and the development effort will require a great deal of focus and integration of activities to be accomplished. The increased focus and

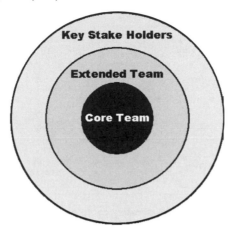

FIGURE 13-4. Simple team structure.

integration provided by a subteam will allow liner development to progress at a faster pace because subteam members are not constrained by having to wait for core team meetings, and they do not have to get consensus from core team members who add no value to liner development.

Simple Team Structure   Figure 13-4 provides an example of a team structure for a project that is limited in scope and complexity. This structure recognizes that the team that will develop this new product is made up of a great many contributors from across the organization. It then organizes this larger group into three smaller groups.

The core team comprises the full-time members who attend all meetings and whose responsibility it is to ensure that the right people are involved at the right time and in the most efficient manner. An ideal core team is six to eight people whose combined experience creates a broad understanding of the various aspects of the new product development effort.

In this structure, when a critical issue arises the core team's first action is to identify who from the extended team and/or key stakeholders needs to be involved in the issue. The core team will then organize a process for resolving the issue that actively involves nonteam members and may exclude some team members who will not add value to the process.

The commitment of the core team to involve the right people at the right time, in the most efficient manner, reduces the need for everyone involved in this effort to be part of the formal team. Keeping the core team down to a small number dramatically increases its efficiency, because key people are spending more time working on product development and less time in team meetings.

Moderately Complex Team Structure   Figure 13-5 presents an example of a structure for a team of moderate complexity. This team felt the project required action on several different efforts, each of which required high levels of focus

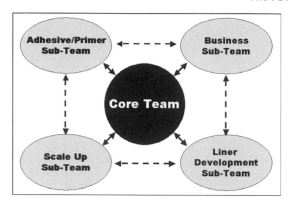

FIGURE 13-5. Moderately complex team structure.

and teamwork and basically involved different people. Therefore, the team established a structure where subteams had the freedom to actively pursue their areas of responsibility without the constraint of needing approval from everyone on the team.

This team also recognized that it was not part of the core business for this division, and therefore was not getting a lot of business direction. To address this situation, the team established a business subteam to provide additional focus and collaboration on business-related issues.

The core team was then composed of the leaders of each of the subteams and the team leader for the project. This small five-person team could efficiently integrate and coordinate the activities of the subteams with a minimum of effort.

Complex Team Structure   Figure 13-6 represents a complex team structure. Creating a complex team structure can clarify roles and responsibilities for teams that are working on more complex projects. For example, the team structure in Figure 13-6 was developed to recognize the complexity of the task and the many key integration points within the project. This team had to integrate market segments, technologies, existing products, functions, a business team, and several product development teams into a single effort. Members of this team can see at a glance where they fit into the larger project.

Since this project was implementing a technology platform, several new product development sub-teams were initiated within the large team framework. In many ways this complex team structure is like a mini-operating division within a larger organization. The flexibility provided by this team approach allowed the organization to develop this business without having to re-organize its basic functions until initial successes were achieved.

TASK 4: TEAM PROCESSES   To accomplish this task, the team must identify the specific tools and processes it will use for decision making and problem solving. Over the course of their existence, teams will make many decisions

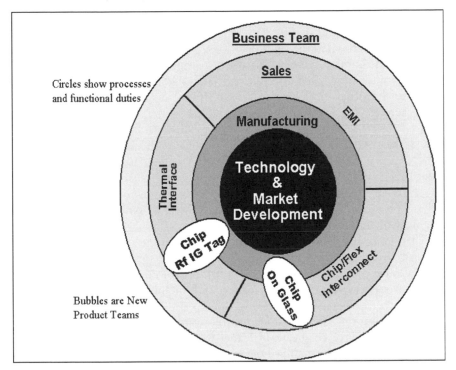

FIGURE 13-6. Complex team structure.

and solve many problems. Therefore, efficiency and effectiveness in team decision making and problem solving will have a significant effect on the team's overall performance and results. Failure to first define the process to be used will dramatically reduce team productivity and create unproductive conflict between team members.

There are many different methodologies for decision making and problem solving. The team will increase its efficiency if it selects a specific methodology for these processes. If the team does not do this, team members will tend to operate out of their past experiences, and since these experiences will be different, team members will be working different processes. If you ask team members, for example, how many steps there are in a decision making process you will get a range of answers. The team must agree on the steps first, or fight about them later as they are working on the process.

The team should also pick a common software package with collaboration tools to increase their performance. Web-based software tools increase team productivity by allowing it to operate asynchronously in cyberspace to gather and share information and thinking. Many of these tools also increase productivity by providing easy documentation and communication of the content of the decision making process. Once a decision is made, one simply clicks on an icon and the software creates a Word document or a PowerPoint presentation.

The team will have to modify the process for each decision and problem. Complex decisions that involve a great number of people may require a formal, detailed process to yield the right decisions. Simple decisions that involve a few people may be handled very informally in a single meeting. Most decisions will fall somewhere between these two extremes. Adapting the decision-making process to the situation helps ensure that the team spends the right amount of time to achieve success—no more, and no less.

TASK 5: LEADERSHIP To accomplish this task the team must define the leader's roles and responsibilities. Typically, management appoints the team leader based on its own criteria. But, since the team leader typically does not have authority over all team members, the team does not have to follow its leader. Therefore, the team needs to define the leadership role and legitimize the ability of the team leader to lead.

There may be a wide range of perspectives on the team leader's roles and responsibilities, resulting in a wide range of expectations on what the team leader should and should not do. Until the team comes to an agreement on what they expect from their team leader, each member will judge the leader on his or her own criteria.

By agreeing on basic team leader roles and responsibilities, the team will legitimize the ability of the team leader to lead and dramatically reduce the bid for power over leadership. For example, if the team decides it is the team leader's role and responsibility to keep the team focused on its agenda, it is legitimizing the team leader's authority to hold team members to the agenda. If a team member gets upset by this, the team leader can note that he or she is only doing what the team has already agreed needs to be done.

Agreement on the leader's roles and responsibilities also increases the ability of team members to hold the leader accountable for his or her performance. If a team leader is responsible for keeping team meetings focused on the agenda, and the team leader fails to do this, the team can hold the leader accountable for this failure in performance.

To complete this task, the team must define the team leader's role. This should include at least the following areas:

Facilitation

   ◆ What are the team leader's roles and responsibilities in facilitating team interactions in meetings and work sessions? Keeping the "butterflies in formation" during team meetings and processes often requires the leader to make decisions and confront disruptive behavior.

Decision Making

   ◆ What are the team leader's roles and responsibilities in making decisions and in breaking ties when the team cannot reach agreement?
   ◆ Does the leader make the decision or facilitate the process?

◆ When the team cannot achieve agreement, does the leader break the tie, or does the team use another approach such a majority rule?

Clarifying this role minimizes personality and politics and increases support for the final outcome.

Representation

◆ What are the team leader's roles and responsibilities in representing the team to the larger organization?
◆ How do you plan to avoid hurt feelings if the team leader's personal power increases because of the team's success?

The team leader is in a very visible position within the organization. This often leads team members to feel like the team leader is taking credit for the team's results and is using this access to increase his or her own personal power. To avoid this problem the team should define the leaders role in representing the team.

Bid for Power The team will also need to deal with bids for power over leadership among different elements within the team. For example, product champions, Six Sigma Black Belts, and people with strong personalities often vie for power and control over the team's actions. The team must agree on how the leader should handle these bids for power and control.

Strengths and Weaknesses Confidence in the team leader is critical to team performance. Typically, team members will have different perspectives on the leader's competence, based on their experiences and what they have heard from others. To maximize confidence in their leader, the team should objectively assess the leader's perceived strengths and weaknesses and initiate an ongoing process of feedback and support to ensure the leader's success.

PHASE 3: ACTION

In the first two phases of the Team Launch System, the team is making a major investment in getting oriented to the task and organized to take action. The return on this investment is the action phase:

1. Self-directed team members who are willing to step outside their functional roles and responsibilities in order to do the right things for the business.
2. An accurate assessment of the level of power, influence, and support that the team can expect within the organization
3. An accurate assessment of the time and effort it will require for this team to achieve success

4. Clear roles, rules, and structures for achieving maximum efficiency and effectiveness

5. A high degree of ownership and commitment on the part of team members to achieving the team's mission, goals, and strategies

6. A team that is very focused and well organized to achieve maximum performance and success

In phase 3, the team begins to take action focused on its goals and strategies—it begins to build the product. To maintain focus and maximize performance, the team does not take on any activities that do not directly relate to its mission, goals, and strategies. This will assure that the team does not chase every *hot issue* that comes up along the way.

But the team must also be responsive to learning and changing circumstances. This means the team may have to periodically return to the orienting and organizing stages when learning and circumstances require the team to make changes. A change in market conditions, for example, may require the team to update its business analysis, which in turn may cause the team to change its sense of urgency, which may, in turn, cause the team to change some of its goals and strategies. This ability to quickly adjust to learning and changing circumstances creates a more versatile and responsive organization.

Milestone 1: Team Norms

The need for versatility, flexibility, and responsiveness during the action phase means that the team cannot operate with one rigid set of rules, but instead must rely on establishing a set of *norms* that will allow it to run efficient and effective processes in a variety of situations.

TASK 1: TEAM MEETING NORMS   To accomplish this task, the team will need to establish norms for all team meetings. Team meetings are one of the most time-consuming elements in the team process. In addition to the direct costs of salaries and benefits that meetings consume, every minute spent in a team meeting is one less minute spent back on the job getting things done.

"Best Practice" Team Meeting Norms

1. *Stick to the agenda.* Every team meeting should follow an agenda. When it is necessary to adjust the agenda during a team meeting, the team must reach consensus before changing the agenda. Teams that fail to stay focused on their agenda are unproductive and become reactionary and crisis-driven over time. This creates a bid for power that can result in unproductive conflict that damage team dynamics.

2. *Complete the agenda for each meeting.* The team's agenda is established based on meeting the team's mission and goals. If the team consistently fails to complete its agenda, it increases the risk of not reaching its

mission and goals because critical issues will not be addressed in a timely fashion.

3. *Start and end on time.* When team meetings start late and run late, they waste resources, disrupt people's schedules, and punish those who show up on time.

4. *Have all team members in attendance.* Members that routinely fail to show up for team meetings deprive the team of the benefit of their participation, cause the team to have to revisit issues to bring the missing members up to speed, and often inhibit the team from taking timely action. If a team member cannot regularly attend team meetings, the team should strongly consider replacing him or her or find an efficient way of bringing the person up to speed that does not waste the time of all team members.

5. *Apply the team's learning and experience to continuously improve its performance in running efficient and effective meetings.* The complexity of the team situation makes it very unlikely that any team will start out with highly efficient and effective team meetings. Therefore, it is critical that the team establish a norm of continuous improvement in the efficiency and effectiveness of team meetings. This is especially true when the team starts to violate its norms of staying on and completing its agenda. These situations must be addressed and fixed before they can seriously affect team performance.

6. *Make frequent process checks throughout the meeting.* The team must regularly adjust its time frames during meetings to ensure that it spends meeting time on the most important issues. Since meetings tend to take as much time as they are allowed, the team should have a norm of establishing time limits for discussions and topics. Once these limits are reached, the team must either move on or agree that this discussion or topic is of sufficient importance and urgency that the team should stay with it for a longer period of time—and an additional time limit should be established.

TASK 2: TEAM PROCESS NORMS To accomplish this task the team must run highly participative processes that involve the right people, at the right time and in the most efficient manner. Team processes for planning, decision-making, problem solving, and conflict resolution dominate phase 3–action, and phase 4–results of the Team Launch System.

To maximize these efforts, the team must focus on team processes. Edward Deming, a legendary expert in statistical process control and Total Quality Management (TQM), points out that you do not "do" a result—you do a process that gets you a result. A decision-making process, for example, that starts with a preconceived outcome and selectively gathers information and criteria to support that preconceived outcome, cannot be trusted to have delivered the best result or do the right thing for the organization. The quality of the process will be determined by the quality of the result.

The credibility of the team and the trust that will be placed in its plans, decisions, solutions, and recommendations will be directly related to the quality of the process the team runs. But as with all other things in the team setting, there are no hard and fast rules on how to run a team process. Therefore, the team must again establish strong norms that guide the team's actions and individual behaviors when running any team process.

Best-Practice Team Process Norms

1. *Be highly collaborative.* Highly collaborative processes increase innovation and creativity and assure the team comes up with the best results. Processes should involve the right people at the right time, and in the most efficient manner.

2. *Be action-oriented.* Team processes should be action-oriented. A process that is dominated by discussion and inaction will achieve minimal results and is an extremely poor use of time and resources.

3. *Build ownership and commitment to the results through active participation.* Active participation by key stakeholders builds ownership for the results and willingness to follow through on commitments.

4. *Maximize information and analysis and minimize personality and politics.* Focusing on information and analysis dramatically reduces the level of personality and politics in the process. High levels of personality and politics tend to distort information and analysis, create winners and losers among key stakeholders and unnecessary conflict between organizational units (turf wars).

5. *Have a high degree of truth-telling.* Withholding, distorting, or selected use of information to support a particular position destroys the credibility of the process and trust in the results of the process. The computer metaphor of *garbage in–garbage out* holds true for team processes.

6. *Be open to a wide range of opinions and ideas.* Team processes that are open to a wide range of opinions and views find the most innovative and creative solutions to complex situations. A norm of openness minimizes the possibility of doing the wrong things and maximizes the possibility of finding innovative, breakthrough ideas.

7. *Manage costs and benefits to key stakeholders.* When costs and benefits flow unequally to key stakeholders, it creates win/lose situations. Those who win typically support the result of the process, and those who lose typically resist implementing the result of the process. By raising costs and benefits to a visible level and providing recognition and protection to those who will bear the cost of team actions, the team can dramatically reduce conflict, increase collaboration, and maximize support for its actions.

8. *Maximize the results of the process.* Team processes often consume a significant amount of time and energy to achieve results. The team can maximize the organization's investment in this process by fully implementing the results and capturing and sharing its learning.

TASK 3: TEAM LEADERSHIP NORMS   Team leaders play several different roles on a team. They play the facilitator role to ensure discussions are efficient and effective. They play a team member role when they contribute to team discussions. They play their functional role when they consider the effect of team actions on their function. They must also pay attention to the content the team is working on and the process the team is using. In this environment, expecting one person to see everything is asking for the impossible.

Team leaders often do not notice when team members are violating a team norm and are negatively affecting team performance as a result. If team members see this but do not speak up, the team will have missed an opportunity to increase team performance. Since there will be hundreds of these types of situations over the life of a team, this represents a tremendous opportunity for every team member to contribute to team performance and success.

Best-Practice Team Leadership Norms

1. *Delegate leadership of specific strategies, actions, and processes.* The team establishes shared leadership by assigning team members to lead specific team strategies. Sharing leadership responsibility allows the team to accomplish more and removes the team leader as a constraint in the amount of work the team can efficiently and effectively take on.

    Some teams delegate facilitation of team meetings to a team member who has excellent facilitation skills in leading meetings. This approach allows the team to take full advantage of the different skills sets of team members, and provides an opportunity for a team member to maximize his or her contribution to the success of the team.

2. *Enforce team norms.* When every team member takes responsibility for enforcing team norms, it creates a very strong group dynamic that will influence individual behaviors. Team members soon learn that if they violate a norm, they will be called on it. If they do not show up for meetings, for example, they know someone on the team will bring this violation of team norms to the attention of the team. Team members quickly learn there are consequences for violating team norms.

3. *Support the team leader's actions.* Shared leadership also ensures that when the team leader holds team members accountable for following team norms, the leader will have the support of team members. In fact, the team leader will be held accountable if he or she fails to hold people accountable for violation of team norms.

PHASE FOUR: MAXIMIZING RESULTS

Teams require a significant investment of organizational resources to achieve high performance and obtain results. When a team's plans, decisions, and solutions are not fully implemented, the organization gets a minimal return on its investment.

Milestone 1: Follow-Through

When key stakeholders fail to follow through on commitments, it is the team's responsibility to hold them accountable. There are many factors that cause key stakeholders to fail to implement their commitments, or fail to implement them in a timely manner. This is especially true when these actions are not central to the key stakeholders' priorities and must, therefore, compete with other commitments that are central to the key stakeholders' priorities and agenda. In these situations, the team may have to provide follow-up to keep the pressure on for results.

When consensus cannot be achieved or a key stakeholder refuses to implement a plan, decision, or solution, the team must escalate the issue to the appropriate management for resolution. For example, a function that is under pressure to reduce costs may back off on a resource commitment. In some cases, it may be best for the organization to reduce costs and not support the team. In other situations it may be best for the organization to absorb the costs and support the team. Balancing the needs of the team with the needs of functional organizations is a strategic decision that will have to be escalated to the manager who has the authority to make the decision.

The escalation process recognizes that there is a balance of power between teams and management that forces both sides to collaborate to do the right thing for the business. When an escalation occurs, the final outcome is less important to the team than getting a decision made. Whether the decision is in favor of the team's position is less important than getting a decision, and getting a decision that is best for the business.

Milestone 2: Documentation

High-performance teams document their results and the process used to obtain those results in order to create a transparent process that can be reviewed by others. This allows others to determine the validity of the results by looking at the quality of the process used to generate them. Documentation is also often required to satisfy legal or corporate policies.

Documentation creates a *public record* that increases the pressure on those who made commitments to follow through on them. It is not unusual in the hectic and fast-paced environment of business for a decision or solution to simply be forgotten or sometimes even ignored as memories fade and people change. Documentation and publication of results creates an incentive for key stakeholders to follow through on their commitments. If, after documentation and communication a key stakeholder does *not* follow through, the team will have to escalate to management for resolution.

The quality of the results is most often determined by the quality of the processes used to achieve those results. Documentation allows others to examine the quality of the process. A brief review of the documentation will show, for example, whether a team approached a decision with a preconceived solution and then gathered information to support only that conclusion. If this is the case, the results should be viewed with suspicion. By contrast, if the documentation shows a wide range of alternatives and criteria, the results will be much more credible.

Milestone 3: Learning

The third milestone of phase 4—results is to capture and share learning about the content and the process with those who can use this learning to increase their performance. By sharing learning, a team can also help create a higher-performing organizational culture and infrastructure. When an organization stops learning, its culture and infrastructure often become outdated and low performing.

Teams expand the organization's knowledge base and its intellectual properties when they capture and share what they have learned about the content (e.g., technology, processes, opportunities, shortcomings, etc.) of the team effort. Sharing this learning with the appropriate organizational units can increase both individual expertise and organizational core competencies.

Sharing learning on process avoids having to reinvent the wheel to discover the same learning on other teams. This learning often leads to the identification of systemic roadblocks in the organization's culture and infrastructure. For example, if teams identify a reward system that consistently inhibits team performance, the organization may decide to eliminate this roadblock or provide additional support when working through it.

Milestone 4: Recognition

The last milestone in phase 4—results provides recognition to those individuals and organizational units that made a significant contribution to the team's actions, performance, and results. Providing recognition to individuals and organizational units that make significant contributions to the team dramatically increases collaboration and trust. If individuals and organizational units know they will get credit for their ideas and contributions, they will be more willing to share them and make them. If the team fails to provide recognition for the contributions of others, whether intentionally or not, it will be seen as taking that credit for itself.

SUMMARY

The complexity of new product development teams means that the answer to every question about team development is the same—it depends. The Team Launch System (TLS) expedites the development of high-performance teams by defining the major dependencies that are critical to team success.

TLS organizes these dependencies into four phases of team development, with specific milestones and tasks for each phase. This provides a clear roadmap or checklist for developing a high-performance new product team.

The Team Launch System should be used as a guideline for team development. Therefore the team will need to determine the level of time and effort it should invest in each milestone and task. Teams that already share a common orientation to the business, for example, can simply check this milestone off and move to the next one.

Due to space limitations, this chapter did not address the personal and interpersonal dynamics that will affect the team as it moves through each phase of development. In many cases these dynamics will dramatically affect the ability of the team to move through these phases of development. Lack of trust, for example, may cause people to hold back information that is critical to team success.

14 Using a Rolling Wave for Fast and Flexible Development

Gregory D. Githens PMP, NPDP
Principal, CatalystPM

Innovation and product development projects—especially those that involve platform or radical innovations—have environments characterized by many unknowns and rapid change. Most innovation and product development success stories show that a handful of principles underlie their success; these include small organic teams of competent and motivated people, small batch sizes of information, iteration, and rapid feedback from the customer or user. More recently, an emerging set of practices termed *agile product development* has caused traditional orthodox product developers and projects managers to rethink their approach to creating fast and flexible projects. Agile organizations have the capability to quickly anticipate change as well as react to the unexpected. With this agile capability, these organizations are able to create value and do it with speed and flexibility.

The project management profession has used the term *rolling wave* for several years, using the expression "plan a little, do a little" to characterize its use of iteration. People now increasingly regard rolling wave as one of the newer agile tools that yield benefits of improved speed, flexibility, and customer value. Rolling wave is a robust, sophisticated way to manage the risks of innovation, to adapt to change, to align the organization, and to align the team toward breakthrough results.

Rolling wave—like other agile tools—offers clear benefits for complex, dynamic innovation efforts. Because rolling wave requires an open-minded adaptive work environment, it may not be suitable for incremental innovation in industries that have large capital commitments. Leaders in the product development profession are increasingly recognizing that traditional, orthodox project management tools are better for some organizations and applications whereas agile tools are more appropriate in other situations. Table 14-1 provides some elements of organizational readiness for agile methods.

This chapter describes the rolling wave approach in NPD. The chapter's focus is on program/project application at the project team level, but managers can also apply rolling wave to NPD portfolio management, as well as

TABLE 14-1.
Organizational Readiness for Agile Practices

Characteristics of Organizations Adopting Agile Methods	Characteristics of Organizations Using Traditional Methods
Understand that teams work collaboratively to create value, not just manage handoffs.	Compartmentalized and individualized. Development is by handoffs, and any "team" is a coordinating mechanism for handoffs.
People are willing and able to sense and respond to changes.	Culture values stability and consistency of methods.
Use integrated project management approach lead by "strong" project manager.	Rigid, highly proceduralized development environments. Task oriented.
Leadership style emphasizes learning and dialogue.	Leadership style emphasizes supervisory command and control.
Incorporates risk into schedules, and adjusts plans for risk.	Schedules by taking the assigned due date and working backward.
Strategy is to supply value to the customer, including accepting late changes.	Conformance/control, and minimization of costs are dominant features of project decisions.
Progress measured by uncertainty reduction.	Progress is measured by deliverables.
Recognize value and opportunity is in ambiguity.	Individuals avoid ambiguity and spend most of their time in personal comfort zones.
Investment objectives include preserving strategic options and minimizing regrets.	Investment objective is tactical efficiency.

strategic planning and budgeting. The chapter starts with a short examination of causes of brittle schedules, defined as a project schedule that breaks easily due to a change in assumptions. This understanding of brittle schedules will help the reader understand why they should consider this alternative approach. The chapter then presents a short case study of one organization's experience with the rolling wave technique, and then describes three agility principles that undergird rolling wave's effectiveness. Next, the chapter presents a six-step approach for applying rolling wave. The chapter's conclusion presents some of the reasons for rolling wave's effectiveness. Finally, there is an annotated bibliography and summary of key points at the end of the chapter.

UNDERSTANDING WHY BRITTLE SCHEDULES OCCUR

Rolling wave helps to overcome the problem of brittle schedules. Once a schedule is broken, individuals narrow their focus to their own subjective view of priorities. Simply put, once a brittle schedule has been broken, people lose the integrated perspective of the project and make local decisions that lead to frustration and suboptimization. As one observer accurately characterized product development projects, "The problem is not that we don't plan, it is that we don't believe our artifacts."

Understanding the practices that lead to brittle schedules can provide some insight. First, many people have a *mentality of scarcity*, meaning that they perceive they lack sufficient time and sufficient resources, rather than an *abundance mentality*, meaning that they regard limitations as an opportunity to creatively solve problems. Second, people tend to approach complex projects with an *all or nothing* approach to planning. The "nothing" approach is to skip any kind of planning. The "all" approach is to take the development team through a rigorous planning process for the purpose of developing a plan that tells management when the project would be delivered, the cost involved, the resource commitment needed. Teams have difficulty investing an appropriate amount of effort in planning. Third, participants tend to believe that the audience for planning artifacts is some bureaucrat; thus, people approach project planning as a compliance activity to "be gotten out of the way." More enlightened readers will recognize that "the plan" is better considered an asset that yields benefits. Fourth, people treat the planning process as if its purpose is to confirm a preconceived date, rather than manage the risks associated with the project. It is noteworthy that the common scheduling practice of picking the end date and working backwards is probably the number one reason for brittle schedules. Fifth, individual contributors tend to provide plans only for their own department and ignore interfaces, or assume that "someone else" will address those issues. Sixth, people plan as if they are able to predict everything in the project accurately and condense that knowledge in a spreadsheet. They oversimplify complex systems. Finally, during execution, people tend to ignore the plan in favor of *real* work. Real work is whatever they perceive as their area of competence. Since many individuals lack skill in good project planning, they don't perceive it as real work.

Brittle schedules reflect a mindset or culture: details are *knowable* and stable, the only important knowledge is documented, it is necessary to create a complete project plan, and then proceed to work the plan. More progressive product developers would argue that this mindset is outdated and ineffective for the complexity and change of innovation.

The alternative to brittle schedules is robust planning, defined as creating a project model that can withstand the stresses of change in the project, adapt to changes, and serve to focus the project on success. A robust plan is a useful tool for the project team to align and integrate its efforts to achieve a project characterized by speed, flexibility, and customer value added.

AN EXAMPLE OF THE ROLLING WAVE TECHNIQUE

One good case study of a project that overcame the problems of brittle schedules was reported in "Embracing Ambiguity: MDS Sciex Pilots Rolling Wave Project Management to Facilitate Fast and Flexible Product Development"

(Management Roundtable 2003). It is the story of how MDS Sciex, a Canadian company that provides products to pharmaceutical, diagnostic, and environmental companies, piloted a project to use the rolling wave technique. Sometimes past frustrations motivate an organization to try a new approach, and this was true for the Borg Project. A previous project had a 3,000-task project management schedule that took months to construct. When several changes in assumptions occurred, team members regarded it obsolete and followed their own instincts as what to work on. Thus, it became a brittle schedule.

In part due to the resulting frustration, managers were open to the benefits of the *plan-do-plan-do* approach for the Borg project, which involved the implementation of several new hardware and software technologies. Not surprisingly for high-technology innovation, there was considerable risk and uncertainty in the Borg project. The Borg Project did the initial planning in a week, and subsequent planning sessions took one day per month.

Figure 14-1 illustrates the rolling wave approach applied to a Gantt chart schedule for the Borg project, where the rolling wave is a sliding three months of planning and execution. Some people use the terms *window* or *planning horizon*—because the terms imply a defined field of vision—in place of the term *wave*. The top Gantt chart shows detailed tasks scheduled for the first three months of Borg. At t_a, the Borg project is only performing the detailed planning work for the current three-month wave, deferring the detailed planning of the distant future tasks until t_b, when the project better understands the uncertainty and risks associated with the future tasks. The project designated these tasks as ROM (Rough Order of Magnitude) to indicate their uncertain size. As the team moved into each wave, it decomposed activities and updated the planning baselines. The Borg team did not attempt to try to develop a detailed estimate of those activities that fell outside of the current rolling wave.

The Borg project encountered some initial resistance from senior managers, as well as project participants. Some key stakeholders were reluctant to accept initial project plans that had a low degree of accuracy. Most senior managers like to see finite targets with specific completion dates. Many people felt uncomfortable in guessing what was going to happen six to eight months from now and then basing commitments on the availability of a given resource. Eventually, functional managers developed faith in the project's short-term forecasts and more willingly committed resources to the project.

The rolling wave approach also allowed senior management and sponsors to have more realistic project cost and schedule to estimates at completion. The project manager said, "It took a little bit of selling to senior managers to get them to accept that the completion date is too far away for us to estimate it accurately." There was also some resistance from the project team members, who perceived the new initiative as an increased amount of work. Team members wanted to dispense with the often-tedious activity of planning so that they could get to the more satisfying development work.

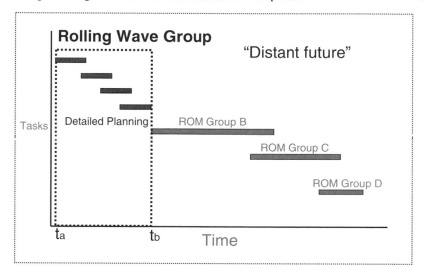

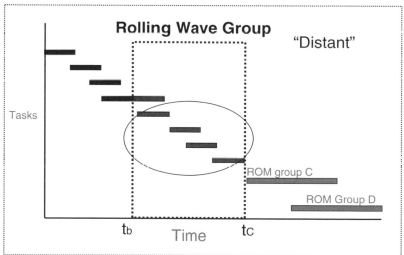

FIGURE 14-1. A simplified Gantt chart illustrating a rolling wave.

THREE PRINCIPLES THAT UNDERGIRD AGILITY

Principles are fundamental and apply across all situations. When practitioners understand the principles that underlie a tool or technique, they are more likely to reap the benefits of the tool or technique. The following paragraphs describe three principles that undergird agility:

1. Project and product architectures set the foundation for agility.
2. Strategic factors are embedded in uncertainty and ambiguity.
3. A functioning team with good leadership is essential.

Principle #1: Project and Product Architectures Set the Foundation for Agility

An architecture defines the basic structure of a system; it defines the chunks of product and project work. These chunks should not be too big, or too small. Good architectures allow the team to keep a big-picture perspective even as they drill into the details. Good architectures provide a basis for assessing the priorities and aligning activities. The result is improved development speed and flexibility. Product architecture and the project architecture are fundamental to formulating a rolling wave approach.

PRODUCT ARCHITECTURE   The product architecture is the strategy for partitioning the product subsystems and interfaces. Hardware, software, and service applications increasingly use a core unified technical architecture that allows for modular design, thus enabling better decisions about the evolution of individual product design decisions. Many projects can't provide every proposed feature, so the product architecture helps prioritize development priorities.

Here is an important insight that can help you better manage risks in the development project: *Interfaces are common failure points*. In fact, most product failure is at the interface, not within the subsystems. An early task in product architecting is to identify those interfaces and understand their structures, functions, and limitations. Manage the interface issues and you can avoid substantial frustration, rework, and delay.

PROJECT ARCHITECTURE   In this chapter, the phrase *project architecture* describes the project or program planning strategy: team composition, levels of authority, review and approval cycles, roles and responsibilities, risk and issues analysis approach, escalation strategy, etc.

Rolling wave is best suited for practitioners who have a basic understanding of standard program/project management concepts and principles. Rolling wave, like other good project management practices, is a process of improving the project-planning model, not just an exercise to develop a Gantt chart or other artifact. Other useful program/project management concepts that can help set the project architecture for rolling wave include: the project life cycle, organizational breakdown structure (OBS), product breakdown structure (PBS), and cost breakdown structure (CBS). The architectural perspective fosters an integrated appreciation of these structures. Validation of the project architecture is done bottoms up.

Plan a little, do a little works best when the team has a capability for seeing the big picture and the details so that the project can balance the top-down perspective with the bottom-up perspective. Figure 14-2 illustrates this shifting proportion of the emphasis from the top-down perspective to the bottom-up perspective. The proportion of top down thinking is greatest at the beginning, and then diminishes. Rolling wave achieves agility because it encourages both a strategic perspective and tactical control over the day-to-day work.

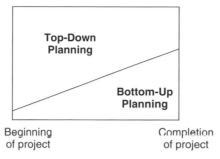

FIGURE 14-2. **Proportion of effort that is top down and bottom up shifts as project proceeds.**

Principle #2: Strategic Factors are Embedded in Uncertainty and Ambiguity

Innovation is typically fraught with unknowns and requires dialogue between different people with different expertise. Uncertainty (literally, without certainty) implies a lack of predictability of structure and of information. People may perceive this uncertainty either as risk or as opportunity. Rolling wave's focus is on uncertainty reduction, with questions like, What do we know? How good is what we know? What don't we know? How much have we progressed in evaluating what we don't know?

Research at Rensselaer Polytechnic Institute's Radical Innovation Program reveals four types of uncertainty in product development projects: technical uncertainty, resource uncertainty, market uncertainty, and organizational uncertainty (Leifer et al. 2000). *Technical uncertainty* has to do with scientific and engineering problem solving. Although most problems can be solved given sufficient time and money, it is difficult to estimate the amount of money and time is at early stages. *Resource uncertainty* has to do with the quantity and availability of key personnel and facilities and is closely associated with technical uncertainty. *Market uncertainty* has to do with the customer acceptance of the proposed functionality, feature set, and price. *Organizational uncertainty* refers to the stability of formal organizational relationships between different organizational units and individuals. For example, if the company is undergoing a major reorganization, people will tend to place attention on visible short-term results. Figure 14-3 illustrates these four types of uncertainty

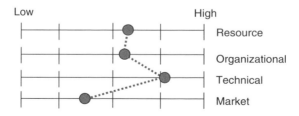

FIGURE 14-3. **Characterize the uncertainty in the project.**

for a codevelopment project involving technologies and NPD resources from two firms.

Radical innovations are particularly noteworthy for the indeterminate nature of the uncertainty associated with the project. The greater each of these types of product development uncertainty, the more the program needs to use techniques that foster agility, such as rolling wave.

Individuals generally attempt to avoid ambiguity. They dive into the details pertinent to their subject matter expertise and start solving the most comfortable problems. However, the easy (or familiar) problems to solve may not be the most important problems to solve. Leaders must balance individuals' tendency to work on the most familiar, comfortable problems with the need to tackle the messy, abstract, ambiguous problems that will ultimately be the most strategic.

Principle #3: A Functioning Team with Good Leadership Is Essential to Communication and Decision Making

Experts in new product development have long regarded personal leadership and effective teamwork as key success factors. This is because the rapid exchange of tacitly held knowledge allows for better communication and decisions. For example, the *scrum model* of a team has the idea that people are self-organizing for empowered decision making. It is the job of top managers to remove barriers from the team.

Increasingly for business-to-business offerings that are adopting agile tools, the development teams are including a knowledgeable representative of the customer or of the user as a member of the development team. This provides relevant feedback on the feature and functions that produce the most value. Because customers can and do change their priorities, flexible approaches take that dynamic as a given and work to react quickly and effectively.

ROLLING WAVE STEPS

Figure 14-4 illustrates the six steps of the rolling wave approach. Note that Figure 14-4 builds on Figure 14-2, which was the transition of the team's attention from top down to bottom up. Along the bottom of Figure 14-4 is a listing of six steps, described in this section. The curved arrows on Figure 14-4 represent the plan a little activity of rolling wave. Notice that the planning activity becomes more bottoms up as the project proceeds.

In implementing the steps, keep in mind the problem of *brittle schedules* described earlier in this chapter. Brittle schedules are easily broken, and when they break the project manager has lost control of the project. Recall that rolling wave results in a robust schedule that will withstand the stresses of change in the project and will serve to focus the project on success.

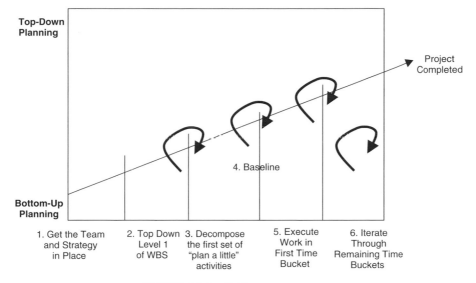

FIGURE 14-4. Rolling wave process.

Also, keep in mind the previous section of this chapter, "Three Principles that Undergird Agility" that describes fundamentals for planning a little, and then doing a little. Product and project architectures and the team's management of uncertainty drive the process of partitioning project work. Leadership and team functioning are the glue that holds the process together.

Step One: Get the Team and Strategy in Place

"Plan a little, do a little" is dynamic, and it is this dynamism that leads to speed and flexibility. The need for an integrated, top-down perspective is greatest early in a rolling wave project.

The goal is to get the team and strategy right in order to achieve and maintain a robust approach. The quality of communications in the project team largely drives project speed and flexibility. Thus, Step 1 of the rolling wave approach is basic: Ensure that you have a charter, an approach to capture and manage requirements (see the *PDMA ToolBook 1*, Chapter 12), the right people committed to the team (see the *PDMA ToolBook 2*, Chapter 6), and a good project vision. The project charter is the formal authorization from management that gives the team resources and the authority to use those resources. The project vision is a description of the expected outcome in terms of customer benefits, form of the product, and function of the product. For more on common tools for the front end of projects, see Githens (January 2004, April 2005).

Table 14-1 presented earlier in this chapter can be considered an assessment tool for rolling wave. Using it as a checklist, go through the table and make sure you can confidently say yes to six of the nine characteristics in the ready column. If your organization is not ready, you need to invest some effort in organizational development or rethink your strategic intentions in light of your organizational capabilities.

Step Two: Perform Top-Down Planning, Starting with the Level 1 of the Work Breakdown Structure

Step 2 establishes the work scope management strategy for the project. The common tool for organizing and managing the project's work scope is the work breakdown structure (WBS). Most expert project managers believe that the work breakdown structure is the singlemost important tool of project management. Not all product developers understand this important tool, so a brief explanation follows. The reader can find more on the topic in Rosenau and Githens (2004).

The work breakdown structure is a hierarchical listing of all the work of the project. Through it, the project can establish its control approach for the project scope, including cost, risk, responsibilities, and so on. Figure 14-5 illustrates a generic work breakdown structure showing the project level, the Level 1, and Level 2. Later discussion in the chapter will speak of Level 2 as work packages and will discuss defining the work packages as black boxes.

The waves become the planning horizons in which the project will plan a little and do a little. Three months seems to be a common time horizon in NPD. Project team members should be able to estimate the close-in work with a higher degree of accuracy (plus or minus 10 percent). The team decomposes in detail the work within this near-term three-month wave. It estimates longer-term activities in the range of plus or minus 30 percent.

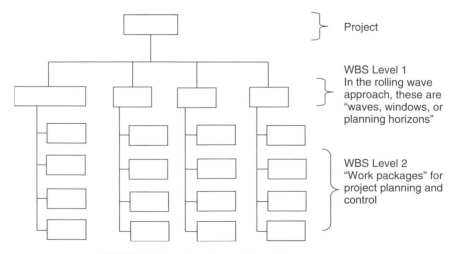

FIGURE 14-5. Generic work breakdown structure.

Step Three: Decompose the First Set of ``Plan a Little'' Activities

In this step, the team details individual work packages. Also called a *task*, a *work package* is a defined unit of work with a budget of duration and required resources. The work package is that element that the project scheduler places into the schedule for control purposes. Since rolling wave uses a "plan a little, do a little" approach, the team defines the work packages for the first time horizon.

PARTITIONING WORK INTO BLACK BOXES Innovation and development work is difficult to define. NPD practitioners overcome this difficulty through the defined architectures and uncertainty management. Imagine, if you will, the product and the project as lumps of clay that the program partitions and shapes to the environment and the organization's strategies. The Borg project described earlier used the term *rough order of magnitude* (ROM) to show the partitioning.

The *black boxes* analogy can be helpful. Black boxes have defined boundaries, but no effort is made to describe the internal detail, as it is poorly understood work. Thus, the rolling wave work breakdown structure captures two kinds of work: the work contained in not-totally-defined black boxes and detailed work. The work package is simply the unit of work that the project is planning and doing.

Figure 14-6 illustrates a black box and the work inside it. Black Box A would represent a large chunk of development work, such as designing a major subsystem. Inside that subsystem are numerous tasks and activities, which are noted as B, C, D, and E. Note that B has the potential for further decomposition. The skeptic might ask, "Why not just go ahead and detail those work packages?" The answer is: "If you can accurately and confidently define the work activity, go ahead, but if you are making unfounded guesses, you might be better off continuing to treat the work as a black box."

Now, inspect the work breakdown structure shown in Figure 14-7. Note the placement of A, B, C, and so forth. The left-most branch of the work breakdown structure is labeled Rolling Wave #1, and contains the Work

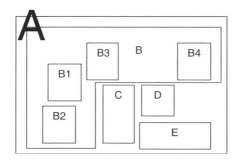

FIGURE 14-6. Work package.

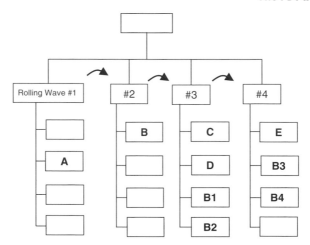

FIGURE 14-7. Planning is progressively elaborated.

Package denoted as A. The work breakdown structure should recognize the work that might occur in the later horizons. Next, compare Figure 14-6 to Figure 14-7, find Work Package B in Rolling Wave #2, and then identify Work Packages B1, B2, B3, and B4 in future waves.

Hopefully, the reader will see the progressive elaboration of the black box work and its placement on the work breakdown structure. For the first wave, it is not necessary to have extensive detail; it is only necessary to have a general and conceptual understanding of the work involved in A. The project team will decompose A into detailed work packages in future waves. The goal is to detail the right work at the right time. Remember that the project advances in cycles of *plan a little, and do a little.*

RANGE ESTIMATING   Range estimating is the practice of developing estimates in ranges rather than single point values. For example, rather than saying a year is 365 days long, you would say it is between 364 and 367 days long (which is more accurate when you consider that once every four years there is a leap year, so, really 365 days is a value derived by rounding down). Instead of committing to a single value, the team generates a range that implies its confidence. In innovation environments, the goal is to understand and bound the amount of error that is natural to each estimate.

Figure 14-8 illustrates how initial ranges for project cost (left Y axis) and duration (right Y axis) are converged toward a true value. Notice that the expected range for the project duration at the first wave is 0.6x to 1.5x. Thus, if $X = 12$ months, the range is reported as 7 months to 18 months. Note that the range narrows considerably from the first wave to the second; that narrowing describes the expectation that the project will remove considerable uncertainty during the *do a little* portion of the first wave.

Most people remark that estimating a broad range like 7 months to 18 months would make their management uncomfortable because there is

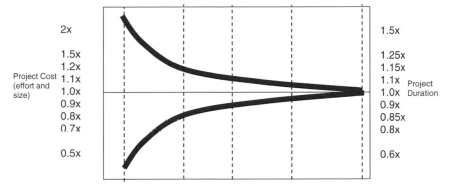

FIGURE 14-8. Convergence chart for range estimating.

too much *trust me* involved. Perhaps, but project leaders need to show some courage and manage expectations. The benefits of improved speed and flexibility are possible for anyone, but one big barrier to overcome is people's mental models on how best to plan, execute, and control a project.

Why does range estimating offer benefits? All estimates have error, and single-point estimates create an unwarranted illusion of confidence. Range estimating frees people from wasting time trying to refine an estimate to perfection. Creating the *perfect estimate* requires much energy, but adds little value in the way of useful knowledge. Range estimating encourages the team to focus on the strategic issues. By reviewing and testing estimating assumptions, the team continually narrows the range as it learns more about the work of the project.

RISK ANALYSIS AND UNCERTAINTY REDUCTION   Keep in mind the principle that strategic factors are embedded in uncertainty and ambiguity. The most important work of the project team is uncertainty reduction. As part of a preliminary risk analysis, examine, validate, and document system-level assumptions, then subsystem-level assumptions, and then component-level assumptions. Emphasize the capture and use of knowledge. Ask these questions: Are our current assumptions still valid? What new information do we need? What are the warning signals of potential project problems, and are they present? (see also the *PDMA ToolBook 1*, Chapter 8).

COMPLETION CRITERIA   Note the phrase *Project Completed* on the right-hand side of Figure 14-4. This is a good time to define and validate the project's processes for verifying that the design and deployment meet customer requirements and administrative closure (the organizational policy and practices for ending a project—for example, lessons learned). There is an old saying, "Never start a project that you don't know how to finish!" You do not want to fall into the trap of having the project team retain responsibility for the product after the project closes. Do not conclude this first pass at planning without discussing the launch and transition plan.

DOCUMENT   As part of Step 3 you should also document your project planning artifacts and distribute your report to the appropriate stakeholders.

Step Four: A Commitment to Proceed: The Baseline

The term *baseline* is a basic project management concept discussed in most project management texts: it is the committed scope, schedule, budget, risk, and so on against which the project team will monitor its actual performance. The project manager establishes the baseline by securing the appropriate approvals from executives, sponsors, users, project managers, and participants. By subtracting the actual performance from the baseline expectation, the project can determine variances from the baseline. If the variance is excessive, the project can take corrective actions to nudge the project back toward the baseline.

In the newer agile product development environment, there is a shift of values from documentation and enforcement to one of adaptation and responsiveness. Thus, it is not common to develop baseline budgets and measure variances. Instead, it is a continuous flow of prioritizing and reprioritizing work. Whether or not you decide to baseline, perform a risk analysis and develop risk responses. Also, calculate and be prepared to defend a risk reserve (also called a contingency reserve). You should also revisit your change management strategy and determine how you will use the work breakdown structure to track the new work packages that will emerge as you decompose the black boxes.

Step Five: Execute the Planned Work in the First Time Wave

Inside the waves, work is straightforward: Plan your work, and work your plan. This list of questions provides a useful structure for information sharing in the rolling wave project meeting:

- ◆ Are we in agreement on the vision for this project?
- ◆ What scheduled tasks were and were not completed in the last reporting period?
- ◆ What are the scheduled tasks for the next reporting period?
- ◆ Have any recognized risks appeared? What is our risk response plan?
- ◆ Are there any new risks?
- ◆ Are there any issues that can be closed out?
- ◆ Do we have any new issues?

Because top management wants status information, it is useful to set expectations on variance data. For example, you might set a trigger level of 20 percent, and if the project deviates by more than plus or minus 20 percent, the project manager would explain the reasons for the variance to senior management. Otherwise, top management can assume that the project is on course. Obviously, this requires trust in the sense that management must believe that competent people are on the project, and that they are competently applying their knowledge and skills. Management must also give them the authority to independently make decisions.

The iterative nature of *plan a little, do a little* fosters knowledge capture for application in future phases. This learning allows the team to anticipate and avoid future problems, or to react quickly to the risks that the team decides to accept.

Step Six: Iterate through the Planning Horizons and Close the Project

This last step involves the continuing iteration of the *plan-a-little-do-a-little* approach and ends with launch- and post-launch project activities. The important elements include the following:

◆ Assess the team's learning, the needed work, and forward plan the next horizon of the project (return to Step 3). As the project completes work in the current rolling wave, it increases the attention paid to work in the next phase. The successive waves of *planning a little and then doing a little*, roll through the entire project. One key to the designing the rolling wave approach is that the waves overlap, so that people are always looking a little further ahead. You want to assure that all members of the project team maintain some awareness of the strategic, top-down future perspective, even though they are embroiled in the tactical details of their current work.

◆ Further, decompose the black boxes and modify the work breakdown structure to reflect the added detail as time and cost baselines become refined.

MOVING THE ROLLING WAVE FORWARD Inter-phase planning involves the transition from closing processes (of the prior phase) to the initiating process (of the following phase). There are many forces that the project needs to manage, including execution inertia, high emphasis on detail, functional and personal conflict, creeping elegance, and listening closely to customer requests. Here are three questions to ask to help manage the interphase transition:

1. *"Do we have enough information to move to the next gate?"* This question is a good criterion for helping the team to decide when to start and stop planning.

2. *"Is that still our vision?"* Often, vision and objectives change. This question encourages the development and validation of a vision/completion statement that can serve as a touchstone. A good project manager will recognize the dynamics and continue the efforts in uncertainty reduction. A mediocre project manager has tunnel vision and will get lost in technical details and problem solving.

3. *"What changes need to be made to our work breakdown structure?"* You might not be able to do everything that you thought you could at the start. You could consider *descoping*—a process of making time and scope tradeoffs, generally inferring that features will be eliminated from the launch or features will be delayed to later releases. The complexity and dynamism of a development project is no excuse for not practicing change management.

CLOSING THE PROJECT Projects are completed when the product meets the customer's requirements and when it has completed the necessary administrative processes demanded by organizational policy. Refer back to Step 3 for a review of completion criteria. Keep focus on how the project will validate and verify requirements.

The previous paragraphs described rolling wave from initiation to closure. There is a lot of activity, and many decisions are being made. There are also several pitfalls to avoid. One pitfall to avoid is using the rolling wave method on routine NPD projects. In routine projects, there are few unknowns and planners can describe done for the project and the plan that will get them to that endstate. Another pitfall is that many people will confuse *agile* with *ad hoc*; that is, they will act as if they can do whatever they want when they want. Instead, they need self-discipline and a commitment to communicating with their teammates. A third pitfall is that the project will succumb to the temptation to capture all activities in the form of process documentation and run the project to a script. Although consistency is important, people often subordinate the goals of high value-add, speed, and flexibility to it. Balance is essential.

The most important success key is recruiting good people to your project; they need technical skills, business acumen, a commitment to their team, willingness to experiment and take risks, and self-discipline. Since organizations have a limited amount of these people, their assignment sends a clear signal about the priority of the project in the portfolio of all development projects. With good people, the team leaders' role evolves into one of removing barriers and setting high expectations for communication, decisions, and results.

CONCLUSION

What are the underlying reasons for improved project performance when using rolling wave?

◆ It shortens the duration of front-end planning and transitions the project into implementation sooner. It gets the individual work teams started with work sooner, because the best way to know what is coming is to actually face it.

◆ It improves the sense of ownership and accountability among team members. People who have used it report that they feel more in tune with each other and the needed work. They learn to regard the plan as a resource for the project rather than some bureaucratic document that is filed out of sight and forgotten.

◆ It promotes an open-minded and flexible project environment. It helps the project meet the challenge of keeping energy focused on both short-term work with an eye to the long term. It yields a balanced approach for control and flexibility.

◆ It opens people up to the idea that it's acceptable to say, "I don't know," or "I can find out," or "The data may be unclear."

◆ It facilitates better estimating. There is a more realistic view of the near-term resource needs, leading to timely commitments from managers to assign the needed staff to the project. It allows the team to combine the advantages of both top down (for the tasks that are far into the future) and bottom up (for the near term tasks) estimating.

◆ It gives management a modicum of control and of predictability in the project plans without a fantasy that the project plan can know and describe all project details. It facilitates aligning work with product development strategies.

◆ It gives the team a tool to manage ambiguity, which helps to effectively address risk, respond to opportunity, and be creative.

Plan a little, do a little is a response to uncertainty: "The further out in time we go, the more uncertain the future is, and the harder it is to make accurate plans." The effectiveness of rolling wave rests on the users' ability to develop a balance between top down and bottom up; between strategic thinking and tactical doing; between emphasis on the individual and emphasis on the team; and so forth.

Agility is not haste; it is discipline. If people want flexibility, speed, and performance, they have to do the work. Although some of the needed hard work is learning the tool, the tool can't work unless mental models support the agile principles that the tool is based on. Ultimately, the adoption of agile techniques is one of individual and organizational motivation to overcome the inertia of established values and behaviors.

REFERENCES

Augustine. S. 2005. *Managing Agile Projects*. Upper Saddle River NJ: Prentice Hall Provides complementary information on use of agile techniques for software development.

Belliveau. P., Griffin, A. and Somermeyer, S. eds. 2002. *The PDMA ToolBook 1 for New Product Development*. New York: John Wiley & Sons. Four chapters are particularly recommended: Chapter 6, "Managing Product Development Teams Effectively," supports Step 1 of the rolling wave technique; Chapter 8, "How to Assess and Manage Risk in NPD Programs: A Team-Based Risk Approach," provides a complementary structure for organizing a project and for managing risks and issues; Chapter 11, "Technology Stage Gate: A Structured Process for Managing High Risk New Technology Projects;" and Chapter 15 on "Risk Management: The Program Manager's Perspective."

Belliveau. P., Griffin, A. Somermeyer, S. eds. 2004. *The PDMA ToolBook 2 for New Product Development*. Hoboken, NJ: John Wiley & Sons. Chapter 12: "Integrating a Requirements Process into New Product Development."

Githens, G. D. 1999. "Building Schedule Discipline." PDMA *Visions* magazine (July and October). The reader wanting to further explore appropriate project scheduling technique should review this two-article series. The articles can help explain more about techniques that lead to brittle schedules, as well as development of robust schedules. *http://www.pdma.org/visions/jul99/githens.html* and *http://www.pdma.org/visions/oct99/discipline.html*.

Githens, G. D. 2004. "You Can't Be Done If You Don't Know What Done Looks Like!" PDMA *Visions* magazine (January). The article describes tools such as completion criteria, success criteria, requirements and metrics. *http://www.pdma.org/visions/jan04/focus.html*.

Githens, G. D. 2005. "Tools for Aligning the Front End of Innovation." PDMA *Visions* magazine (April). The article describes tools such as the elevator test, use case, scope exclusion, and product innovation charter. *http://www.pdma.org/visions/apr05/focus.html*.

Githens, G. D. 2005. "Agile and Lean Development: Old Wine in New Jugs." PDMA *Visions* magazine (October). An overview of the emerging practices of agile product development describing essential principles and selected terms. *http://www.pdma.org/visions/october05/in-focus.php*.

Hohmann, L. 2003. *Beyond Software Architecture: Creating and Sustaining Winning Solutions*. Boston: Addison-Wesley. Software people use the term *release management* to describe elements of the *plan-a-little-do-a-little* technique. This book provides tools and insights for relating any product architecture into useful product offering.

Liefer, R., C. M. McDermott. G. C. O'Connor. L. S. Peters, M. Rice, and R. Veryzer. 2000. *Radical Innovation: How Mature Companies Can Outsmart Upstarts*. Cambridge, MA: Harvard Business School Press. Source of the four types of uncertainty described in this chapter that were mapped onto Figure 3.

Reinertsen, D. G. 1997. *Managing the Design Factory: A Product Developer's Toolkit*. New York: The Free Press. This book provides useful and complementary insights on partitioning work, product architecture, and getting the team right.

Rosenau, Jr., M. D., and G. D. Githens. 2004. *Successful Project Management: A Step-by-Step Approach with Practical Examples, Fourth Edition*. Hoboken, NJ: John Wiley & Sons. This elaborates on basic project management practices such as work breakdown structures, work packages, baselines, Gantt charts, requirements, and project charters.

The Management Roundtable. 2003. *Product Development Best Practices Report.* Waltham MA: The Management Roundtable (May). "Embracing Ambiguity: MDS Sciex Pilots Rolling Wave Project Management to Facilitate Fast and Flexible Product Development." The article provides a more detailed description of the Borg Project and the use of the rolling wave technique. *http://fasttrack.roundtable.com/app/content/knowledgesource/item/158.*

15

Gaining Competitive Advantage by Leveraging Lessons Learned

Ken Bruss, Ed.D.
Global Learning and Development Consultant PAREXEL,
Int'l.

> *"If H-P knew what H-P knows, we would be three times as profitable."*
>
> *—Lew Platt, former CEO of Hewlett-Packard*

Confronted with the accelerated rate of technology development, shortened product life cycles, and global competitive pressures to reduce costs, new product development (NPD) organizations are compelled to optimize every aspect of their process. Many best-practice companies have gained benefit from implementing such improvement initiatives as *phase gates, core teams*, and *just-in-time* manufacturing. As these processes have matured, they are increasingly only yielding incremental gains. In the quest to maintain market leadership many of these companies are critically examining the benefits they're securing from their knowledge assets. By strategically managing the manner in which knowledge is collected and shared, these companies are reaping rich benefits from a previously underutilized resource.

The chapter will examine how to leverage knowledge assets from a strategic as well as a tactical level. In particular, this chapter will explore the *after action review* (AAR), a powerful, low-tech tool that the reader can immediately utilize. Organizations as diverse as British Petroleum, Harley Davidson, and Sprint have formally implemented AAR as a standard component of their work. NPD project leaders likely will find the AAR process most useful. However, throughout the discussion, the chapter will highlight the critical role development managers and senior company management play to ensure that their organization secures full economic benefit from their knowledge assets.

417

LEVERAGING KNOWLEDGE ASSETS FOR COMPETITIVE ADVANTAGE

Knowledge Asset Management Defined

Knowledge has been defined as "information in action; it is what people need to know to do their jobs" (O'Dell and Grayson 1998). The term Knowledge Asset Management introduces the value proposition that an organization's business and technical knowledge should be viewed as a valuable asset, which when appropriately managed can provide a sustainable source of competitive advantage. Successfully managing organizational knowledge requires efficient processes to ensure that organizational information effectively flows to the right people; when, where and how they need it. Research demonstrates that companies that effectively leverage their knowledge assets in pursuit of business objectives will consistently achieve stronger results. Such companies:

◆ Make better product selection decisions, aligned with market needs and organizational capabilities.

◆ Have a stronger up-front understanding of project risks and necessary mitigation strategies.

◆ More effectively monitor development efforts, responding to problems and opportunities.

◆ Successfully reduce cycle time and development expense through technology re-use.

Knowledge Asset Management Strategic Overview

All companies to some extent engage in creating and sharing organizational knowledge. The problem is that most companies don't do it well, and don't do it consistently. They typically engage in these activities in an ad hoc, informal manner. In some cases, companies launch a KAM initiative, complete with banners and T-shirts. Frequently, these initiatives coincide with significant investment in web-enabled information management technology. These types of initiatives tend to lose sight of the business problems they were intended to address. Rather than being seen as concretely related to daily work, employees perceive them to be stand-alone added work.

Instead of proposing development of a KAM strategy per se, a more effective approach is to incorporate KAM within existing business strategies. As part of strategy development, most companies assess organizational strengths and weaknesses. Incorporating KAM into this process requires asking a few additional questions:

◆ Are practices that contributed to project success being successfully shared within the organization and, as appropriate, incorporated into standard work processes.

◆ Are painful lessons learned being effectively shared, enabling other projects to avoid making the same mistake, and/or are improvements being made to the process?

◆ In making business decisions, are all pertinent data easily accessible?

Common to all of these questions is the aim of consistently identifying opportunities to effectively leverage organizational knowledge during standard daily work.

Embedding KAM within the NPD Phase Gate Process

How knowledge can be incorporated into standard workflow may be demonstrated by looking at the front-end of new product development (NPD). Many companies utilize a phase gate process to manage NPD activity. *Phase gate processes* are conceptual and operational road maps for successfully moving new product ideas from concept to launch. Product development teams complete a prescribed set of cross-functional tasks in each stage prior to obtaining management approval to proceed to the next stage of product development.

As Figure 15-1 demonstrates from a KAM perspective, each phase gate also represents a critical learning opportunity. To facilitate accomplishment of the stage's specific tasks, successful teams leverage existing organizational knowledge. Along the way, these very same teams may create new knowledge that, if shared, can provide the organization with a unique source of competitive advantage. Throughout this process, successful managers check to ensure that the team is effectively leveraging the organization's knowledge base.

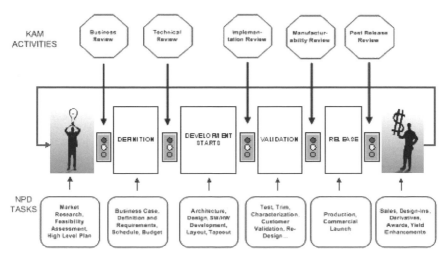

FIGURE 15-1. KAM activities in the NPD process.

Typically, at the beginning of a development effort some form of preliminary business and technical review is conducted, culminating in submission of a high level proposal. Some companies have a formal concept phase whose goal is to determine if the opportunity is sufficiently promising to warrant additional investment of time and money to conduct a more in-depth feasibility analysis.

Applying KAM to the Business Review

Among the most widespread KAM activities is incorporating formal processes to leverage in-house business and market intelligence to effectively evaluate the proposed business opportunity. The sales and marketing organization has a wealth of up-to-date, relevant knowledge about customers, markets, and competition. A common practice at many companies is for account managers, field application engineers, and salespeople to be included in a preliminary product definition meeting. Frequently sales engineers are aware of organizational and market changes months before the data becomes public. Companies that are able to effectively act on this market intelligence can make the proposed product more competitive. Some larger organizations such as AMD and Gateway supplement these more ad hoc efforts by more formally and routinely capturing and sharing sales and marketing information by using Web-enabled databases such as *salesforce.com*.

Another valuable internal source of customer and market information resides within the customer service organization. This information may be collected via a corporate Web site or by a customer service representative. Many R&D groups routinely review records of customer complaints, questions, and requests as part of their business review discussions. Some company databases are programmed to automatically forward this information, either on an incident basis or at regular intervals, to R&D. These organizations are then able to introduce product enhancements and derivatives that directly incorporate customer feedback.

Common to these examples is the idea that companies that are able to effectively leverage the latest customer and market intelligence stand to gain significant competitive advantage. In some cases, team members actually accompany sales and marketing staff on customer visits so that they can more directly hear the *Voice of the Customer* (see the *PDMA ToolBook 2*, Chapter 7). By actually experiencing the working conditions in which customers are using their products, many design engineers have developed profitable product enhancements, which they would not have considered if not for their out-of-lab experience.

The quality of the business review can be further enhanced by formally incorporating information from externally focused search engines into the business review process. In addition to collecting pertinent customer and market data, the team can effectively engage in patent searches, both to assess the competitive landscape and to identify innovative ideas that should be protected.

Time and again market research has shown that the quality of front-end activities is a key differentiator between successful and unsuccessful new products. To be successful, new products must be in tune with a complex array of market forces. It is during the business review that product decisions are made to ensure this market alignment. Effectively utilizing KAM methods to collect and share pertinent business data residing within the organization can significantly enhance the quality of the decision-making process. Many companies informally tap pockets of business knowledge residing within the organization, but best-in-class organizations are utilizing KAM on a systematic basis as a key enabler of their business reviews.

Applying KAM to the Technical Review

Equally important early on is tapping the technical knowledge within the organization. Particularly for those companies pushing the proverbial innovation envelope, technical risks are quite high. To reduce these risks, many companies require technical reviews at various stages of the development process. Best-in-class companies formally integrate KAM methods into these technical review activities, thereby ensuring that knowledge sharing becomes part of the standard workflow.

At the beginning of the project, a key goal of technical reviews is to ensure the team members accurately understand project risks and plan accordingly (e.g., develop a realistic schedule and resource plan, develop appropriate mitigation and/or contingency measures, etc.). From a knowledge asset management perspective, this process can be significantly strengthened by ensuring that the team has access to relevant experts and organization history.

The idea is to leverage the full breadth of the organization's knowledge assets, as opposed to being limited to the expertise within the individual's own business unit. In many cases, by leveraging expertise outside of their immediate group, the team will learn about alternative approaches and/or reuse opportunities, which can reduce costs and risks. At some companies, lead designers are provided incentives to consult on new projects reusing their designs. Mentoring junior engineers, in many instances, has resulted in early identification of valuable intellectual property (IP) that was proactively patented, providing the company future competitive advantages.

Many companies have formal initiatives to facilitate technology reuse. Geographically dispersed companies are increasingly deploying Web-enabled content-management systems that integrate with other core business applications (e.g., cell libraries) to increase technology reuse. For example, Analog Devices recognizes that its expertise in the areas of analog and digital IC design is one of its core competencies. To effectively leverage this expertise, Analog Devices has developed several proprietary databases where reusable designs are securely catalogued and easily accessed by company design engineers.

Management's Role with Front-End KAM Activities

Projects have a much greater chance of success if teams have the most accurate, up-to-date business and technical information throughout the project. Risk also can be reduced by systematically learning from the experience of similar projects. In planning a new project, there is always the risk that the team isn't aware of what it doesn't know.

Creating a culture where learning occurs *before doing* requires active management support. As part of the phase gate process, in addition to the technical and business assessment questions asked, successful managers constructively challenge the team to ensure that its plans reflect lessons learned elsewhere within the organization. Phase gate questions may include the following:

- ◆ Does the plan reflect business, market and technical lessons learned from past projects?
- ◆ What problems did similar projects encounter, and how does this new team plan to address these issues?
- ◆ How do the task estimates in their schedule compare to actual company history with similar projects?
- ◆ How have they incorporated best practices from other projects?

If the team comes to the phase gate meeting without having sufficiently done its homework, successful managers will delay project approval. The team will be instructed to come back when it can effectively answer these key questions. This concrete demonstration of management commitment to *learning before doing* is much more effective than any set of policy guidelines.

LEVERAGING LESSONS LEARNED

In addition to leveraging an organization's recognized technical experts, an equally important source of knowledge is created by individuals and teams engaged in daily work. As this example demonstrates, companies can secure significant financial benefits by effectively acting on *lessons learned* in real time.

LSC is a leading-edge technology company. One of its teams engaged in developing a product using a newly qualified package in order the meet its customers' aggressive power and cost requirements (a *package* houses a chip or other discrete electrical device, electrically connecting the chip with other circuitry and providing physical and chemical protection to the chip). When parts came back from the factory, the test engineer discovered a serious performance problem. Testing the part revealed that the package did not fully meet the specs promised in the data sheet. To correct this flaw, the team had to modify the design and produce a second round of silicon. This activity easily cost $100,000, factoring in the expense of materials and engineering time. This unplanned design iteration added to project cost and pushed out delivery of samples to the customer.

The team's sponsor required that the team conduct an after action review (AAR), both to ensure that all flaws had been addressed and to identify implications outside of the team. Fortunately, the AAR revealed that the team had corrected all problems. Three action items were generated by the AAR:

1. To prevent other projects from experiencing this problem, the manufacturing representative on the team was tasked with identifying other company teams using this package. Four other new product development efforts, scattered across the globe, were identified. Benefiting from the learnings generated in the AAR, four other teams made the necessary design modifications. To varying degrees all four teams were required to engage in unplanned design iterations, but by making these changes prior to sending the part to manufacturing, all four projects were spared the significant costs and schedule delays associated with a second round of silicon.

2. The design engineer was tasked with contacting the vice president of R&D to revise the design rules to reflect the actual package characteristics, thereby preventing future teams from encountering this problem.

3. The manufacturing representative of the team was tasked with speaking with the VP of manufacturing regarding how to address this package problem with the supplier.

For many readers, this example is reminiscent of the postmortem many companies typically conduct at the end of the project. However, rather than waiting for project completion, operating from a KAM perspective companies are able to achieve the greatest competitive advantage by engaging in learning and knowledge sharing activities *throughout* the product development life cycle. In the case of the LSC example, if the first team had delayed in sharing its knowledge of package flaws until completion of the project, four other teams would have experienced the same costly and time-consuming problems.

The idea of actively learning and sharing knowledge throughout NPD is captured in the concept of *The Learning Cycle* (Collison and Parcell 2004). Fundamental to the learning cycle is the idea that teams will benefit from engaging in learning activities, *before, during*, and *after* a project. Each of these learning points represents an opportunity to both leverage the experience of others, as well as to contribute to the organization's knowledge base:

◆ *Before* starting a new project, teams are encouraged to ask: Who has done this type of activity before, and what can we learn from their experience? Are there opportunities for technology reuse, potentially reducing risk, cost, and development time?

♦ *During* the project, teams are encouraged to stop periodically and reflect on their progress. What has our team learned that we want to apply going forward (e.g., successful practices we wish to standardize, mistakes to avoid repeating)? In addition, the team is encouraged to ask who else in the organization would benefit from this new knowledge.

♦ *After* the project has been completed, the team is encouraged to more formally reflect on what it has learned, how as individuals they'll apply this learning to future projects, and how this information can be shared within the broader organization.

Of the various KAM methods successfully employed by NPD organizations, After Action Review (AAR) is unique due to its ease of use and immediacy of impact. The remainder of this chapter will be devoted to a discussion of AAR, both the theory and hands-on instructions for its application.

USING AAR TO CAPTURE AND SHARE NEW KNOWLEDGE CREATED DURING NPD

AAR Overview

As the LSC example illustrated, AAR is a very effective mechanism for capturing and leveraging lessons learned. *After action review* was originally developed by the U.S. Army to rapidly prepare troops for foreign missions. Used both on the battlefield and in training situations, the military's goal was to rapidly collect data from troops on the ground, share it with other troops in real-time, and then use it to revise military processes and procedures. The key focus was collecting actionable knowledge that can be immediately applied to the next battle or similar situation. AAR now is widely used by such diverse companies as British Petroleum, Ford, Harley Davidson and Sprint. By conducting AARs, their teams are able to tease out tacit knowledge into actionable *lessons learned* and then explicitly document this information to be stored and shared with the wider organization.

An AAR compares what actually occurred to what was planned. AARs strive to capture the sequence of events, and gain understanding into the thinking behind the participants' actions. AARs identify what went well that future teams may wish to replicate, as well as mistakes to be avoided and/or process weaknesses that must be corrected.

AAR Format

An after action review consists of several specific sets of questions, followed by action planning to define how to act on lessons learned:

1. What were the project objectives?
2. What were the actual results?

3. What accounted for the team achieving, exceeding or missing their stated objectives?

4. What have we learned?

5. How will we act on this learning?

The following example contains a portion of the AAR described previously.

LSC 93887 AAR: March 5, 2006   **Participants:** Steve Bourque, Elaine Davis, Chris O'Farrell, Maureen Nee, Max Patel, Misha Rubinovich, Sandra Yee, Ken Bruss Facilitator

I. Objectives	II. Results
A. Send customer samples—12/18/05	A. Missed customer samples date
B. Achieve revised samples date—2/23/06	B. Achieved revised samples date
C. Reduce test time from 1 minute to 35 seconds	C. Successfully reduced test time from 1 minute to 35 seconds

III. Reasons for Variances:

A Why was original release date missed?

  1. New XLB package proved unable to hit performance requirements described in product spec

    ◆ XLB is a new, recently approved package, but the approval process did not test the part at the limits to which we would be subjecting it.

    ◆ When the part came back from the factory it failed on the test bench. Further testing identified package flaws.

    ◆ Six weeks were spent redesigning the part, and then having it re-manufactured.

  2. We underestimated the scope of hardware and software changes necessary to reduce test time and costs.

    ◆ . . . .

B What enabled the team to reduce test time?

  1. . . . .

IV. Lessons Learned?

A Due to the uniqueness of our products, we need to exert an extra degree of caution when accepting generically approved new packages and processes

B Be cautious when labeling a project a derivative. While this was a derivative design, the test effort was much more complex, which was not adequately reflected in the project plan.

C . . . .

V. Next Steps		
Action Items	Action Owner	Due Date
1. Identify other projects using XLB package and inform them of the parts inability to achieve performance specs. 2.	Steve	March 10th

A successful AAR involves three sets of activities: *preparation, conducting the AAR, and then leveraging lessons learned*. The next sections of this chapter cover the specific actions required to effectively complete each phase of the AAR process.

After Action Review Preparation

PLANNING An AAR can be as simple as two individuals conducting a 10-minute assessment at the end of a customer visit or as complex as a day-long event held upon completion of a large project. Organizations receive the greatest benefit when AARs are conducted on a regular basis (e.g., after completion of major activities or at scheduled milestones), thus becoming an ongoing continuous improvement activity. This approach enables the knowledge to be captured before the team disbands, and, as the LSC example demonstrated, when it can provide greatest benefit to the organization.

Some organizations use an AAR inspired template for monitoring project status. In the case of one company, twice a quarter teams will conduct a brief AAR. As demonstrated in the LP332 example in Figure 15-2, teams are given 15 minutes to discuss results achieved over the past six weeks, areas of risk and contingency plans, and status of project spending and plans for the next six weeks. In addition to providing management a concise status report on the project, it is an opportunity for teams to request management support (e.g., additional resources, advocacy, etc.). The engineering manager is tasked with monitoring trends. Themes that surface in multiple AARs may suggest process weaknesses to be addressed, as well as best practices to standardize.

In a similar vein, another company uses an AAR-inspired format to review all new products 6 to 12 months after release. This postrelease AAR does not address issues that surfaced during development. These issues will have already been discussed in a prerelease AAR. The focus of the postrelease AAR is exclusively on the extent to which the new product did or did not achieve the targets established when the product was released. Using this format to review released projects, Product Line management is able to:

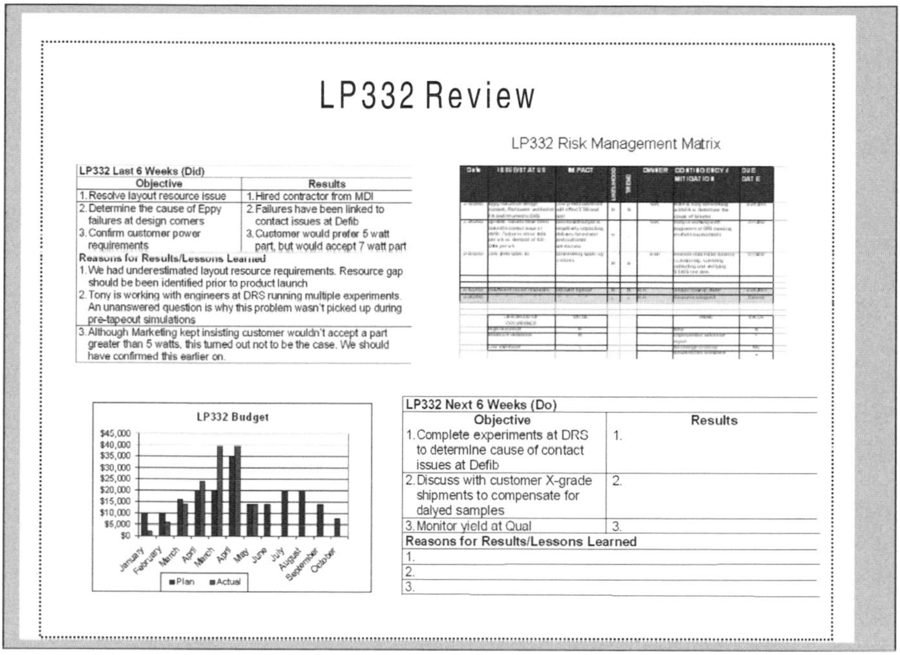

FIGURE 15-2. Example: AAR inspired project monitoring template.

◆ Identify actions that can be taken to boost revenue for poor performing products.

◆ Utilize fresh market data to evaluate products currently in development.

◆ Identify opportunities for new products and/or line extensions.

In all of these examples a strategic decision was made to integrate AAR into the NPD process. These businesses are consistently collecting lessons learned, and requiring project planning teams to incorporate lessons learned from previous projects.

PARTICIPANT SELECTION   A key question in planning in AAR is: Who should participate? A rule of thumb is to invite everyone who has direct experience with the project and to avoid outsiders, as they may inhibit conversation. A potential obstacle is that sometimes this results in participants with a diverse level of project experience (e.g., core team members bring hands-on functional experience, whereas others individuals such as marketing and sales staff may come from a 30,000-foot view). What works best in these situations is to conduct two or more parallel AARs, and then bring the diverse subgroups together to discuss their learnings. In this case, the first AAR would involve those individuals who had hands-on technical involvement with the development effort. The second AAR would involve individuals, who were involved with account management aspects of the project, possibly including members of the customer organization. To ensure continuity between the two AAR's it is recommended that at least one individual (e.g., the team leader) participate in both AARs, and that the same individual facilitate both AARs.

In determining the participant list for an AAR, another issue to be considered is how best to create and maintain a climate of openness and trust in which participants are able to critically examine issues with the focus on learning, not finger-pointing or blame. There are no hierarchies in AARs—everyone is regarded as an equal participant. To maintain this sense of equality, ground rules may need to be established. Additional preparation will likely be required when several layers of management are participating in the same AAR.

THE FACILITATOR'S ROLE   While an informal AAR (e.g., held upon completion of a customer visit) does not require a facilitator, in most cases recruiting a facilitator is a critical requirement for success. The facilitator's role is to ensure that an appropriate climate is maintained—a climate in which all team members feel comfortable expressing their opinions honestly. The facilitator keeps the discussion on track, only entering the discussion when necessary.

An AAR does not require a professional facilitator, but the individual assuming this role should be experienced in the AAR process, and should be effective at meeting management. The facilitator should not be an actual participant in the AAR, nor should the facilitator be the manager of an AAR participant, or anyone else who has a vested interest in the outcome of the AAR. The facilitator should possess familiarity, but not necessarily expertise in the issue being addressed. This profile ensures that the facilitator possesses the ability to understand the issues being discussed, while still maintaining objectivity. This person should be at the appropriate level in the organization to have the respect of the participants. A common practice is that a team leader from one project will facilitate AARs for another team, and the second team's leader will reciprocate.

THE SCRIBE'S ROLE   A key output from an AAR is the write-up. Taking good notes while facilitating an AAR is an extremely difficult task. Unless the facilitator has a lot of experience juggling these two tasks, it is recommended that a separate scribe should be recruited for the AAR. The facilitator criteria also apply in recruiting a scribe. The scribe's primary role is to take detailed notes of what has been discussed, not to editorialize or correct what might be perceived as incorrect comments made by a participant. The scribe may edit the text to make the AAR write-up easier to read, but may not change the intent of the participants' answers or insert their own views.

During the meeting, scribes can use flipcharts, white boards, and/or computers (preferably connected to a laptop projector). They should type up their notes as close to the meeting's conclusion as possible. Afterward, they should distribute their draft write-up to participants for input.

Conducting an AAR

STEPS 1 AND 2: DEFINING OBJECTIVES AND RESULTS   An AAR begins with the team defining the activity's goal and results. Teams are encouraged to be as specific as possible. Sample probe questions to identify objectives include:

- ◆ What was the project goal?
- ◆ Were there any metrics (e.g., time, cost, defects, etc)? For new product development AARs, this type of information can often be taken directly from phase gate documents.
- ◆ If there weren't metrics, how would management assess whether the project was successful?
- ◆ Does everyone agree that these were the objectives?

Typically, this is a quick process. If the team is having difficulty gaining consensus as to the project's objectives, this is a definite red flag.

Sample probe questions for defining results include:

- ◆ Were the objectives achieved?
- ◆ If the team either exceeded or did not achieve stated objectives, can the results be described quantitatively (e.g., planned release date was October 15; actual release date was December 3)?
- ◆ Were there any unintended results?
- ◆ Does everyone agree with this description of results?

When defining objectives and results, teams have a tendency to want to begin explaining the reasons why various results were achieved. It is important to keep the discussion focused just on intent and actual outcomes, and not allow the team to discuss reasons for variance until consensus is reached on objectives and results.

When facilitating an AAR on a project covering a lengthy period of time, it is often helpful to hold a quick planning meeting that only addresses these first two topics. This meeting identifies the key issues that will be covered in the AAR, enabling participants to prepare (e.g., review their project notes, confirm schedule dates, etc.). With lengthy projects when a preparatory meeting hasn't been held, team members often disrupt the meeting by walking in and out of the meeting to collect pertinent information they have back at their desk.

STEP 3: DETERMINING REASONS FOR SUCCESS OR VARIANCE   The brunt of an AAR is aimed at determining what enabled the team to successfully attain its goal or what caused variances, positive or negative. Here are typical probe questions:

◆ What contributed to successful execution of the project?

◆ What enabled you to save time and/or reduce cost?

◆ What difficulties or unpleasant surprises did you encounter?

A variety of discussion methods can be used to collect this information. Three popular approaches follow:

1. *Chronological order of events.* A logical, straightforward approach often helps participants recall the sequence of events. The challenge with this approach is that multiple events often happen in parallel, which are not always visible to all of the participants. There is also a tendency to get bogged down in too low a level of detail.

2. *Key events.* This approach focuses on critical events that directly support objectives. This technique is particularly effective when time is limited. The challenge is to maintain a tight focus on critical events or else the discussion will become sidetracked by less important issues

3. *Bucketing issues.* The team identifies broad categories and then fills in the details. In a potentially contentious AAR this is particularly effective approach. By quickly capturing and acknowledging hot button issues to be examined in-depth later in the meeting, the team is able move forward and identify other issues. The team can use tools such as an *affinity diagram* or a *fishbone diagram* to rapidly collect and sort data. Figure 15-3 contains a portion of the fishbone diagram used in the previously described LSC AAR.

Regardless of which discussion method is selected, the focus of the discussion is on gaining understanding in order to improve, not to assign blame.

STEP 4: IDENTIFYING LESSONS LEARNED   The next phase of an AAR involves identifying lessons learned. This portion of the meeting represents a high-level summation of the previous discussion, with the aim of creating new knowledge for future application. With lessons learned, the AAR discussion addresses two groups of knowledge users—the team itself, as well as future users. Unlike when a postmortem is held upon completion of a project, when a team conducts periodic AARs during the project it is able to apply this new knowledge as part of their own continuous improvement process. Speaking to this point, one AAR expert observed:

Many people believe that the main purpose of AARs is to capture lessons for the benefit of other teams. But our belief is that the team itself is the first, best customer for what is learned, and the best time to apply 'lessons learned' is in the current project itself. What a shame to wait until the end of a project to hold an AAR and gain an insight that might have helped improve the results of that project! (Darling, 2005)

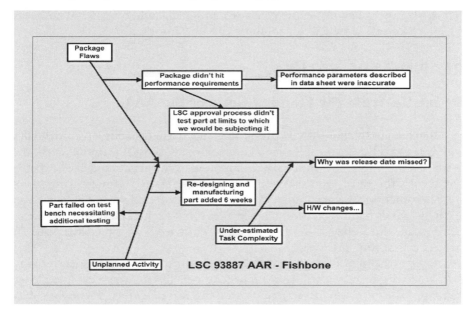

FIGURE 15-3. Example fishbone diagram.

Sample probe questions include:

◆ What has worked really well on this project that you want to continue doing and/or replicate with other projects?

◆ What mistakes should you stop making or avoid making in the future?

Oftentimes both lessons learned and action items logically emerge during the course of the variances discussion. Writing them down in real time is a useful way to manage the discussion and move on to a new issue.

STEP 5: ACTION PLANNING This portion of the discussion is aimed at multiple audiences—the team itself, current and future teams who would benefit from this new knowledge, as well as other segments of the organization and/or external parties who may be assigned improvement tasks.

During this phase of the AAR it is the facilitator's responsibility to ensure that specific items are assigned, with designated action owners and due dates. In some cases it's helpful to include a formal planning table as part of the AAR write-up. Sample probe questions include:

◆ As a team, where do we need to improve? What behaviors/processes do we want to change? What actions do we want to approach differently?

◆ Are there successful practices we wish to standardize?

◆ Did we encounter problems requiring attention that are beyond the scope of this team? If yes, who needs to be contacted? Who owns the action of communicating this information?

◆ Are there lessons learned that would benefit other teams? Who needs to know? How will we communicate this information?

REAPING THE VALUE

Define Up-Front the Learning Goals of the AAR

A common complaint with AARs is that the write-up ends up in a binder somewhere collecting dust. Prior to conducting an AAR, it is useful to ask the team to think about how the results will be used. Returning to the AAR expert's comments, the most immediate audience for the AAR learnings is the team members themselves. Accordingly, a useful final question when facilitating an AAR is to ask team members what key learning they have gained from this project, and how they will apply this new knowledge to either their current project or a future one.

During the action planning phase, a critical question is to identify who else in the organization will gain immediate benefit from the new knowledge generated during the AAR. In many cases, the team members themselves may not know the answer to this question, and based on the issue identified may need management support to ensure that the knowledge is shared with appropriate parties. In the case of LSC example, the manufacturing representative was tasked with checking the projects in the development database to identify other projects that were using the new package, and then sharing the knowledge with them. The manufacturing representative was also tasked with speaking with the VP of manufacturing to determine who should deal with the vendor regarding the package problem.

Sharing Lessons Learned

A common practice is for teams to share their findings at a *brown-bag seminar*. Typically, brown-bag seminars are voluntary forums held on company time. At best-practice companies, presenting at such an event is considered prestigious, and employees are encouraged to attend. Sometimes these meetings are held at lunchtime. If the business doesn't provide lunch, it at least provides beverages and dessert.

A benefit of using brown-bag seminars to share AAR learnings is that since participation is voluntary, attendance is motivated by interest in the topic, and attendees are more likely to act on the learning. Presenting at a brown-bag seminar also gives the team recognition and concretely demonstrates usage of the tool. Senior managers can role model commitment to knowledge sharing by regularly attending these meetings. Awards, whether financial or otherwise, are also useful methods to acknowledge teams that contribute to the organization's knowledge base by sharing lessons learned.

One challenge with knowledge sharing events is that employees who may need this knowledge at a future date may not attend the brown-bag seminars

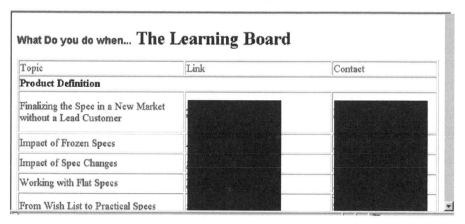

FIGURE 15-4. Web-based example of AAR knowledge capture.

due to lack of current need. It is for this reason that it is so important to effectively capture the knowledge in a format that's easily available to employees at a future time. For small companies this may be as simple as a loose-leaf binder kept in a central location. For larger, multisite organizations, a searchable Web site is desirable. Figure 15-4 is a portion of *The Knowledge Board* utilized by one multi-national company. (Both product and contact names were blacked out to maintain confidentiality.) Key learnings were organized into searchable categories. Prospective users can skim through the functionally arranged categories on the Web site or use the search engine. Clicking on the link, the user can read the full AAR. Because the organization realizes that a write-up can never fully capture the tacit knowledge contained in an AAR, The Knowledge Board also includes the name of a contact person who would be available to answer further questions.

Posting AARs on a searchable Web site makes them easily accessible to users. The main reason, however, for the success of this site was not technology, but rather, management commitment. Teams knew that when they appeared before product line managers at a phase gate meeting they would be asked if and how they had leveraged relevant lessons learned. Capturing lessons learned and reviewing this information during project planning had become part of the standard workflow.

FINAL THOUGHTS

Knowledge Asset Management fundamentally involves the realization that an organization's knowledge assets are a valuable resource, which when effectively managed can produce a sustainable source of competitive advantage. Best-in-class organizations systematically collect, share, and apply lessons learned. Rather than being seen as added work, KAM activities are considered the normal manner in which work gets done.

Promoting and sustaining culture change is difficult. Although successful KAM initiatives require strong management support, the goal is to secure commitment, not compliance. Some managers may complain they do not have the time to engage in these types of activities. The appropriate response is to ask them whether in today's increasingly competitive marketplace, they can afford *not* to support KAM initiatives. One of the key metrics in NPD is cost. In the case of the LSC example, timely sharing of information resulted in a savings of at least $400,000 of rework expense across four projects. This figure does not take into account missed-opportunity costs (e.g., projects delayed since resources committed to new projects are engaged in rework, customer frustration due to missed milestones, etc.). Although change is never easy, change efforts that produce measurable benefits have an easier time gaining mass acceptance.

REFERENCES

AQPC. *Measuring the Impact of Knowledge*. 2003. Houston.

Collision, C., and G. Parcell. 2004. *Learning to Fly*. West Sussex: Capstone Publishing.

Cooper, R.G. 1993. *Winning at New Products*. Reading, MA: Perseus Books.

Cooper, R.G. 2003. "Making Your New Product Process Really Work". *inKNOW-vations* (December).

Darling, M., 2005, "Getting Better at Getting Better—How the After Action Review *Really* Works". *Leverage Points*, Issue 61.

De, A., and R. Sathyavgeeswaran. 2003. "KM at Hughes Software Systems: Certification, Collaboration, Metrics." *Leading with Knowledge*. New Delhi: Tata McGraw-Hill.

Falvey, D., and O. Nagel. 2000. "Implementation of the After Action Review at Sprint." Presentation to the APQC's Fifth Knowledge Management Conference, December.

Maybury, M. 2003. "Knowledge Management at the MITRE Corporation: Partnership, Excellence and Outcomes." *Leading with Knowledge*. New Delhi: Tata McGraw-Hill.

O'Dell, C., and C. Grayson. 1998. *If Only We Knew What We Know*. New York: The Free Press.

Reinertsen, D. G. 1997. *Managing the Design Factory*. New York:, The Free Press.

16 Metrics that Matter to New Product Development: Measuring Actions, Getting Results

Wayne Mackey
Product Development Consulting, Inc.

Almost every company attempts to measure its new product development (NPD) efforts in some way, yet industry research shows that very few are satisfied that they are measuring the right things (Suomala 2001). Further, product development practitioners often find metrics burdensome, disconnected from their real work, or an infringement on what they view as an inherently creative craft. This chapter is written for both product development practitioners and leaders to help them form a common language around measuring the difference between successful and unsuccessful product development. We begin by offering definitions of some metrics terms to provide a common language of discussion, then go on to define six keys to metrics success along with the associated missteps that development organizations often make. We introduce the concept of a *metrics tree* and explain how using it as a tool for implementing metrics offers an array of valuable benefits. We conclude with a brief case study. Putting this all together allows you to set metrics that really matter to product development.

One obstacle to the use of metrics is the fact that product development leaders often request that practitioners evaluate activities that don't necessarily indicate the success or failure of NPD—for example, hours spent working on a project or the number of changes to a design drawing. Developers get cranky about tallying up things that don't seem to matter, and their bosses are frustrated that nothing seems to change as a result of all the data they collect. Both expend energy and time, and neither gets what he wants.

These issues most often are rooted in a basic misunderstanding of metrics by both groups of people. Metrics don't fix problems—ever. Instead, the power of metrics is in accurately highlighting situations and issues that can, if handled properly, make a difference in the outcome. The goal of any metrics implementation should be allowing a capable product development or leadership team to make early, informed decisions without expending extra effort on setting up or using the system.

Finally, most companies focus on measuring the wrong things. Like a dieter who gets on the scale every day only to find the number going down a little one day and up the next, companies that measure *outcomes* rather than *causes* end up with the same poor results they've always had. To lose weight successfully, a dieter needs to focus on *food in* and *energy out*, not on the number on the scale. To achieve its product development goals, an organization needs to focus on whatever *actions* will achieve the desired result—such as an early customer-driven design specification, a properly resourced development team or fully simulated and bench-tested prototypes—not on the result itself. This chapter shows how this key shift in approach makes metrics much more useful and relevant to product development.

WHAT IS A METRIC?

The American Heritage Dictionary defines a metric as "A standard of measurement." This seems simple enough; however, there is no universal understanding of how to put this simple definition into practice in a product development environment. Table 16-1 provides a definition of a metric, together with definitions of related concepts. This table provides a common framework for this chapter's discussions.

The field of metrics is a bit like parenting. Few people go through formal training before becoming parents. Likewise, although most product development people use metrics, few have actually received training in the subject. There is no Metrics 101 class in college to give people an idea about what makes for successful versus unsuccessful use of metrics before they are charged with measuring performance. There are few standards of what to measure, especially when measuring difficult things. All this explains why there is so much confusion about what metrics are and how to use them, and why almost nobody in the product development community is satisfied with the metrics they do use.

CHARACTERISTICS OF GOOD METRICS

So what does a good metric look like? Ideally, a metric should be simple and unambiguous and should tell exactly what is being measured (e.g., tolerance, software code defects, prototype test time). Explicit in the metric should be the units (millimeters, number of occurrences, minutes or hours). For example, a designer of ergonomic desk chairs should not use "comfort" as a metric, since comfort depends on the individual judgment of the person sitting in the chair and does not tell what to measure. Instead, the designer might use the pliancy of the foam (measured in centimeters of pliancy per pound of applied pressure) and the number of combinations of seat back/seat bottom positions.

TABLE 16-1.
Definition of a Metric

Metrics Term	Definition
Metric	The description of the standard of measurement used to assess an element or key characteristic of a product development effort to monitor its progress toward a goal.
Goal	The desired final state representing completion of a product development effort or task.
Result metric	The description of a standard of measurement used to assess end results. Examples include design cycle time, number of drawing errors or software code bugs.
Causal action	A specific event or subprocess that, if successful, contributes directly to the achievement of a goal. An example is simulation testing of tolerance margins between two moving parts in a mechanical design.
Process or predictive metric	The standard of measurement used to assess a causal action. An example is the number of hours to complete an intermediate task such as prototype testing of a new technology.
Measurement	The execution of a metric resulting in data at any point in time. A metric is not a number, while a measurement is. For example, a metric might be the tolerance stack-up of a mechanical subsystem. The corresponding measurement could be 0.1 millimeters.

Good metrics have the characteristic of being applied at the proper level. A CEO should not be measuring the heat absorption of a silicon wafer, nor should a semiconductor designer measure shareholder value. Even a good metric can be bad if applied at the wrong level.

Good metrics also have the property of being time-based. Without a time frame, metrics cannot be meaningful. For example, suppose the desired outcome is the creation of a new handheld high-definition DVD player. Whether the product has to be ready in time for this Christmas buying season in 2 months or for introduction at a consumer electronics show 18 months from now dictates completely different approaches to achieving the goal.

Although most people haven't been trained in metrics, almost all employees have a sense of whether things are going well or poorly with their jobs. The key to developing good metrics is capturing the essence of what experienced professionals already know in their heads about what makes any product development project successful or unsuccessful. Even an organization whose business is innovation can apply metrics to projects. The challenge for a team tasked with doing things no one has ever done before is to figure out the ways in which something new makes sense. If the team's business is innovation, team members will have an idea about what made them successful in the past and can use this knowledge as the basis of metrics development.

Best practices in metrics involve simple, clear-cut statements of desired progress toward a goal. They cover an area that's problematic or challenging.

Worst practices involve ambiguous statements that focus on a process or output the company already executes well. Using the right metrics can make a good management team look great: They always seem to be ahead of problems, fixing minor glitches before they become disasters. Using poor metrics makes even a great management team look only fair to middling: They are always one step behind major problems, operating over budget and behind schedule.

SIX KEYS TO METRICS SUCCESS

Key 1: Measure Only Those Things the Company Does Poorly

One of the problems with metrics is that most companies measure what they're doing well. They do so because it's easy and it feels good. The boss gets a glowing report; the team gets a pat on the back. But such metrics are largely meaningless and can even thwart the true purpose and value of metrics, which is to give people the information they need to make critical decisions.

In a high-tech business, for example, it's common to develop a rainbow of metrics around the technology challenges that engineers face, such as faster clock speeds than currently available products, better heat dissipation with greater miniaturization, or more languages for a selectable use interface. But often such technical challenges are the very ones the company knows it can conquer, because it has achieved similar successes on many other products. And while the firm expends (wastes) time and resources focusing on technology metrics, it neglects to allocate sufficient resources to the fledgling project. It falls short of resource targets, extending the schedule and missing the market opportunity. Instead of measuring technology, the company should have focused on its resource plan versus actual project staffing.

Key 2: Understand Exactly What Metrics Can—and Can't—Do

To use metrics effectively, companies must understand what metrics can and can't do. Companies often use metrics as police officers. Yet just as a city wouldn't use a crime report to keep the community safe, companies can't use metrics to solve problems. Actions must come from people, not numbers. People, armed with the information acquired through wise use of metrics, diagnose and solve problems. Better still, metrics help people keep problems from developing in the first place.

Often, the challenge is that the executives requiring teams to use metrics don't understand what metrics can and can't do. The value of metrics comes in highlighting problems early and accurately so decision makers can make good decisions to fix problems early when there are just a few wisps of smoke instead of a raging fire. For example, a company that's developing a new transducer for consumer audio products knows there's a potential risk of distortion. If the team chooses the *amount of distortion* as a metric, it must wait until the product is ready for final testing before receiving any feedback on the potential for distortion. If instead it conducts early bench tests of prototype materials to understand the interaction of transducer materials (e.g., pliancy, tensile strength, and surface deformation), it will have an early indicator of the *sources of distortion* prior to building the final product. This will allow the team to modify the specifications for the materials long before testing the first unit.

Often, the product development executive's dilemma isn't that product development cycles are too long, but that each product has to go through the cycle multiple times. When problems don't surface early, developers must do a lot of reworking, essentially going back to the drawing board to fix problems discovered late in the cycle. Understanding problems when they can be fixed with relatively little pain drastically reduces cycle time.

Key 3: Measuring the Right Number of Things

Usually, companies measure *too many* things. They end up with too many metrics because they measure the *wrong* things. Since their initial measurements don't yield useful results, they measure more wrong things, and end up in an escalating spiral of metrics, clogging the pipes of the enterprise with the resulting numbers and wasting time that could more profitably be spent developing products. One of the most extreme examples of this was a company that had an entire department devoted to nothing but metrics. Not only did this result in too many metrics, it also divorced the people developing the metrics from the processes they were measuring.

People tend to focus on metrics that are not the critical few because they naturally are drawn to *good news metrics* that measure company strengths. With good news metrics, everybody knows the reports will be positive, employees will feel good, and the boss will hand out awards. The key to reducing the number of metrics is to stop measuring what the company is good at and start measuring things it's bad at—the processes that are challenging and that will require the early and accurate attention of decision makers.

Product development organizations also need to avoid the *tyranny of importance*. People often believe they *must* measure something because it's *important*. There's an analogy here to individual health metrics. It's true that blood pressure is an important metric, and that a sudden sharp fluctuation in blood pressure can signal a life-threatening medical emergency. Yet, in the absence of other symptoms, most of us are content to monitor our blood pressure at an annual physical. The cost—in time and information overload—of continual blood-pressure monitoring for most people far outweighs any predictive or problem-solving benefit.

Similarly, product development organizations must decide what metrics are the critical few and measure those, not necessarily the seemingly important ones. For example, an engineering firm's CAD (computer aided design) system is extremely important. It must work smoothly, allowing designers to easily enter, store, and access drawings. However, the product development staff should not measure aspects of the CAD system such as downtime on a daily basis. Once a year—the equivalent of the annual health checkup—it can conduct an overall review of whether the system provides the proper reliability or burdens staff with excessive downtime. Only if downtime fluctuates from the norm or indicates unacceptable data problems would the company need to change the frequency of measurement.

Key 4: Align Metrics with Corporate Goals

The very first question to ask when introducing metrics to an organization is: *What's the business or customer goal?* Metrics shouldn't exist in a vacuum, but should be tied to the larger picture. If an organization can't or doesn't state a clear goal, it ends up with a metric in search of a goal, which, like an answer in search of a question, is usually a colossal waste of time. Team members with a positive perspective on their work decide beforehand that they'll succeed, while those with more pessimistic views reach the foregone conclusion that they will fail. And nobody will be able to distinguish the truth, because nobody knew where they were trying to go in the first place.

To reinforce how important it is to start metrics with goals, consider what happens when a company creates a metric not tied to a goal. A common metric in engineering organizations is the number of changes in engineering drawings. Organizations then aim to minimize that number because they believe that "too many changes are bad." But what happens if the initial design is poor and the team is constrained by its metric to keep drawing changes to a minimum? It ends up with a lousy design!

Another example of a misguided use of the engineering drawing metric is when an organization decides to use this metric to try to force engineering changes early in development, when presumably they will be less expensive to implement. That approach, however, might result in engineers (who may enjoy *designing* and *improving* as opposed to *completing*) making lots of changes early on as allowed by the metric, in the process destroying what started out as a simple but effective design. The point is that the number of engineering drawings generated does not relate in a meaningful way to product development success factors such as product quality or value to the customer.

Working with a metrics tree (discussed in more detail in the next section) is a simple, explicit way of tying metrics to multiple levels of goals: corporate, executive, business unit, director, manager, and team. The tree lets organization members understand why they're doing what they're doing.

Key 5: Involve the People Who Are Responsible for What's Being Measured in Defining the Metrics

The people closest to whatever process is being measured must participate in the development of metrics. The product development leader's job is to pass the goals down, helping people understand the bigger picture and how their activities fit into that picture, then ask development team members what causes each goal to happen. Human nature dictates that people will cooperate more fully and be more productive when they're not told what to do but, rather, are invited to participate in the design of a solution. By approaching metrics in this way, the people closest to the process determine what needs to be measured and how to measure it to meet the team, manager, director, executive, and, ultimately, corporate goals.

One of biggest challenges in implementing metrics is the *hand of God* phenomenon. An executive—let's call her Vice President Smith—spends the weekend playing golf in a group with Vice President Jones from a competing company, who goes on and on about how his company's CAD terminal usage and prototype test expenditures have improved. Early Monday morning, VP Smith decides that her company should start measuring things. The edict comes from on high: *Put a report on my desk every two weeks showing metrics on development hours per CAD terminal and testing hours per prototype printed circuit board*. The problem is that VP Smith is not involved in the day-to-day operations of the team. CAD terminal use and PC board prototype testing may be completely irrelevant to how the team accomplishes its work. Team members and managers have not bought into the metrics. They now experience them as an added burden required by an executive who has no sense of the actual problems of their department or project. When higher-ups require teams to track and monitor unimportant activities, team members find themselves doing their real work and *also* tracking metrics that are disconnected from their real work. When metrics are not connected to people's real jobs, they are perceived as an extra burden that serves no useful purpose.

Key 6: Monitor and Immediately Act on Metrics

The need for governance ties together the previous three areas: getting to the critical few metrics, aligning them with organizational goals, and instilling ownership at the proper level. A metric should be simple enough for people to understand without training—and it will be if the people doing the work have developed the metric. It should be as easy to measure as possible. And it should be measured *regularly* and *locally*.

Regular review helps maintain focus. But take care not to let the review process itself lead everyone astray. Implementing metrics requires two kinds of reviews: a process review, which occurs often, and a more formal results review, which occurs less often. This chapter covers both in more detail later. Any metric that isn't both *watched regularly* and *acted upon immediately* is a waste of time. Period. Remember, the purpose of metrics is to provide *early* and *accurate* data so *people* can make better decisions.

As Figure 16-1 illustrates, a good metric is one that is aligned with and connected to corporate goals, is owned by a specific person responsible for

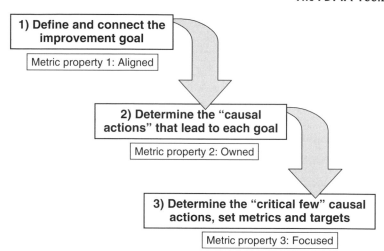

FIGURE 16-1. Steps to metrics success.

the actions that lead to fulfilling each goal, and is focused on the critical few processes that need attention, not those that the company already executes well.

CREATING A METRICS TREE

Once a company understands the common problems around metrics, what steps does it take to make sure that its metrics implementation is successful? Creating a metrics tree is a simple and relatively painless way to tie metrics to corporate goals and propagate them throughout the organization. A metrics tree is built from the top down, starting with high-level product development goals and asking, "What causes these goals to happen?" The answers to this question are the causal actions that become the next level down in the tree. Each of these causal actions becomes a goal of the next level down in the organization, which in turn looks at what causal actions lead to fulfillment of those goals. Those causal actions then become the goals of the next level, and so on. The metrics tree proceeds from top to bottom in the same goal/causal action relationship, from the top of the organization down to the individual product developer. At each level of the tree, the causal actions can be converted to actual metrics by simply asking the question, *how much/how many?*, *how good?*, or *when?*

Start with the Goals

Every discussion of metrics should begin with the question: "What's the goal?" This ensures that every metric will support the company's or customers' goals rather than producing metrics in search of a goal. So how, exactly, does a

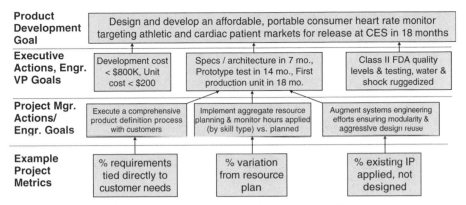

FIGURE 16-2. Metrics tree: Schedule branch detail for portable heart rate monitor project.

firm tie metrics to specific and meaningful goals? First, as discussed earlier, goals must be set at an appropriate level. An example of a high-level product development goal is: *Design and develop an affordable portable consumer heart rate monitor targeting athletic and cardiac patient markets for release at the Consumer Electronics Show in 18 months* (see Figure 16-2).

Beginning with the product development goal, ask, what are the causal actions for the engineering vice president that will make this happen? Taking a simple cost/schedule/quality approach, we could come up with (1) Cost: development cost less than $800,000, unit production cost less than $200; (2) Schedule: specifications and architecture in 7 months, prototype tests in 14 months, first production unit in 18 months; and (3) Quality: Class II FDA quality levels and testing, water and shock ruggedized. Beginning with the top-most product development goals, *each causal action becomes a goal of the next level.*

Proceeding to the project manager level, what causal actions support the schedule goals as outlined? The project manager, based on his or her experience, might define actions around executing a comprehensive product definition process, implementing aggregate resource planning, and encouraging a high level of design reuse. Those actions will become the schedule goals of the project team. The metrics tree develops more and more branches at each level, starting at the top and flowing seamlessly through the organization.

At the project level, the metrics associated with each causal action are set by simply asking *how much/how many?*, *how good?*, or *when? How good* is the execution of the product definition process with customers? The metric is the percentage of total product requirements tied directly to customer needs. *How many* hours, by skill type, have been applied to the project versus the hours originally planned? The metric is the percentage variation from planned hours. *How much* design reuse is taking place? The metric is the percentage of intellectual property applied but not designed. Good causal actions make for good metrics.

A welcome byproduct of working this way is *automatic strategy deployment*. Because each level in the organization bases its goals on those of the level above it and supports those goals through its actions and associated metrics, a high-level product development strategy flows through the organization without the need to set up additional mechanisms for propagating the strategy. Furthermore, when everybody works on metrics with an understanding of how the metrics relate to individual goals, and how individual and departmental goals relate to one another, there is improved buy-in for the larger goals of the organization. Remember, though, that while the entire organization may be working on hundreds of metrics, each individual is concerned only with the critical few that relate to his or her goal.

Defining Causal Actions

The previous example of dieting helps clarify causal actions and their relationship to goals and metrics. Suppose a person set a goal of losing 10 pounds. The usual approach to dieting is to use weight as a metric. Every day, the dieter gets on the scale to see whether his or her weight has gone down, up, or stayed the same. But by measuring weight, the dieter achieves little more than informed fluctuation. What he or she should measure instead are the *actions* that *cause a person to gain or lose weight*: calories consumed in food or burned through exercise.

Causal actions form the basis of the predictive nature of metrics. In the diet example, measuring calories daily or weekly gives an idea very early on whether the dieter will be losing or gaining weight, whereas measuring pounds gained or lost reveals success or failure *only after the fact*. In NPD, a common goal is not to lose weight but is instead to shave time off development cycles. If the development team working on the portable heart rate monitor wants to shorten its development cycle time by 10 months, measuring time leads only to informed fluctuation: The team will know that it's a month behind schedule, but this won't contribute to fixing the problem. Instead, the development project team should measure things that *cause* cycle time to be reduced, such as whether product definition is complete prior to the start of design, whether it is executing the resource plan, and the amount of design reuse. By keeping an eye on those key items, the team has a much better chance of achieving its goal of reducing cycle time by 10 months. The metrics tree in Figure 16-2 shows that monitoring resources applied versus resources planned provide a predictive indicator of whether or not the schedule will slip.

Once the goal is defined, stop focusing on it and focus instead on the causal actions: what it takes to make the goal happen. This may seem counterintuitive, but it's key to the proper alignment of metrics with goals. Using this approach in an organization leads to the creation of the metrics tree. Relying on the people who are actually doing the work to decide what actions are necessary to achieve a goal ensures that responsibility for developing metrics lies at the right level. Metrics are not handed down from on high, so you avoid the lack of ownership that often causes metrics to fail. Finally, focusing on causal actions results in metrics that are predictive in nature and can give early, regular feedback about progress toward a goal.

Limiting the Key Causal Actions

As mentioned earlier, one of the reasons metrics fail is that organizations become overloaded and clogged with too many metrics. To overcome this, we suggest using only a critical few metrics (as illustrated in step 3 of Figure 16-1) chosen by examining only the causal actions the product development team does poorly. People undertake hundreds of activities in the course of doing their jobs that need not be measured because they are part of basic expectations of competency. Businesses don't use "health and hygiene" metrics to measure whether people bathe or dress appropriately for work, and yet certainly they expect employees to come to work not smelling offensive and looking professional. Metrics cannot be used to fill a void in competence—that's the role of training, mentoring, or appropriate assignments of tasks. This leads back to the edict: "Don't measure what you're already doing well."

People inevitably ask, "What's the *right number* of causal actions or metrics to examine?" The simple answer for any organization is *four to eight*. The more complex answer is that there are many metrics that the team or individual can watch regularly *and* act upon immediately. For some, this might be as many as 20, for others as few as 3. Getting rid of the health and hygiene metrics makes room for the more relevant ones and lets team members pay attention to the causal actions that will truly advance them toward their goals.

BEST METRICS FOR NEW PRODUCT DEVELOPMENT AND RELATED AREAS

So what *should* we focus on in measuring product development? The usual suspects of Stage-Gate™ criteria, cycle time, cost, and quality, pop up at most companies. More insightful companies focus additionally on design reuse, staffing, knowledge management, waste reduction, and risk mitigation.

But different industries also have different needs in metrics. In medical industries, regulatory approval drives cycle time and cost, so metrics *supporting* cycle time and cost (such as patent protection, testing, and trials) are more prevalent than measurements of cycle time and cost themselves. In consumer industries, time to market and cost play a bigger role, with many companies additionally driven by trade show deadlines. Metrics in consumer industries tend to focus more directly on schedule reduction, design for manufacturability, product differentiation, and outsourcing. High-tech industries push the leading edge of what can be done and still be understood, so they focus more on patents, computer simulation, and customer needs mapping. Aerospace companies generally have product development cycles spanning many years, and therefore have longer schedules and less extreme product cost constraints, so their product development metrics focus more on return on assets, partnering, and quality.

Innovation—a huge and important concern for many companies—is related to, but not the same as, product development. Thus, the metrics used to measure innovation are different from NPD metrics. Product development has a known end point: shipping a product. Innovation, by contrast, is itself about discovering the end point. The metrics that help monitor progress toward a product development goal, such as the achievement of prototype milestones, simply won't work in the context of innovation.

Innovation is about diving into the unknown, then charting the potential for diffusing an idea into successful products. Thus, innovation metrics focus less on schedule and more around the environment that allows ideas to grow and the diffusion of the ideas into multiple and high-margin opportunities. A company also can create metrics around the two distinct phases of innovation, the first being the idea phase (an inspiration that comes in the shower) and the second being the phase that brings the idea to a point when the innovator understands it well enough to pass it on to product development. At that point, metrics shift to assessing those things that will contribute to the product being delivered.

Let's say a company has an idea for a new process to gold-coat the leads on semiconductor chips. The goal is not to produce something at this point, but rather to understand the innovation. Can it be applied to all semiconductors, or only certain types? Will it work in very small architectures? Is this an innovation that will affect one product or product line, or the entire industry? To measure innovation, the firm needs to look at what *actions* lead to deep understanding about the innovation's potential application. This requires understanding the materials science that underlies the innovation by studying the process at the molecular or atomic level. Only then will the company be able to gauge how big the idea is. The metrics for this effort would include percent coverage of testing of the key characteristics of gold lead bonding, percent of test results supporting applications to variable sizes of silicon dies, and patent diffusion rates to existing product lines. The success of an innovation product manager is measured not by schedule and cost but by how well he or she has understood the technology and mitigated the risks of applying it to specific product development efforts.

Finally, portfolio management, also related to NPD, is a beast unto itself, one famous for having both the most comprehensive and the least useful metrics. Portfolio management decisions about which products to fund usually are made on the basis of financial projections. The problem is, every product manager has his or her own story to tell, accompanied by charts showing hockey-stick growth curves and limitless potential with a small up-front investment. Of course, it would be wonderful to have completely accurate financial projections. The problem is that nobody has found a reliable way to make such projections early enough in the development cycle to be useful.

So, rather than continuing to rely on inaccurate financial data as metrics to assess new product potential, it's more sensible to use things we really *can* measure successfully, such as customer need and customer value. (Contrary to popular belief, it is possible to measure customer value. There are even methods to uncover and quantify *latent* or *unexpressed* customer needs. See also *Toolbook 1*, Chapters 2 and 10; *Toolbook 2*, Chapters 7, 8, and 9). Measuring customer value—while also measuring how well a new product idea aligns with business strategies and the cost to develop a product to a point where its potential can be more accurately assessed—provides a much more accurate assessment of a product's potential as an addition to the portfolio.

IMPLEMENTING A BETTER WAY: ALIGNMENT, OWNERSHIP, AND FOCUS

So what does it take to actually make this happen? The approach to metrics described so far is not only simple—it also functions well in the real world of product development. A company can begin implementing it at *any* level of the organization, from the very top to the project level. There is no need to have complete buy-in from everyone in the product development organization before getting started. In fact, the example presented later in this chapter began as a pilot project in one part of the company tangential to product development, and then spread organizationwide after achieving success. Successful implementation of metrics demands *alignment*, *ownership*, and *focus*. This means bringing metrics in line with corporate goals, clearly identifying who is responsible for executing the metric, and maintaining focus on watching and regularly acting on the metric.

Regardless of what part of the organization begins the process, the first step is to set goals or make sure that existing goals are well stated. A goal like "Increase shareholder value by raising the quality of our product" is simply too generic. It could apply equally to a ketchup manufacturer, a chip producer for home electronics products, or a multinational petroleum company. Contrast this generic goal with FedEx's original goal statement: "Everything, every time, delivered by 10 a.m. the next day." This goal is specific to the company and forms a unifying force that people at *that particular company* can rally around. It's a high-level goal, but it is specific to overnight delivery and is understandable by everyone. At whatever level goal setting begins, consider the goals one level up in the organization, and then develop actions and goals that relate to those goals. To ensure that a goal is specific, that its timing is correct, and that it's appropriate for the level of the organization, use the following checklist:

Goal Checklist

◆ Does it clearly define the desired result or end state? (Could you explain it to a teenager?)

◆ Is the path from this goal to the bigger picture goal(s) explicit? (Provides or supports customer/business value)

◆ Is it specific and relevant to the organization responsible to meet it? (Is it realistic?)

◆ Does capability exist to meet goal? (Marketplace demands this performance level)

Use the checklists after the following sections to further test how goals relate to causal actions and to the people within the organization charged with fulfilling the goals.

Capturing Causal Actions

The next step is to capture the causal actions that will lead to achieving the goals. Product development leaders do this not by sitting alone in their offices, but by going to the people who are actually doing the work, at whatever level, and asking them what factors account for the difference between succeeding and failing at the goals. The beauty of this approach is twofold: first, it captures latent expertise within the organization. Professionals usually have a pretty good idea of what makes things work—or not. Second, working this way generates a powerful byproduct: buy-in. When the people doing the work participate in developing the measurements for success, they're invested. They can't come back later and say, "This didn't work because you gave us the wrong metrics."

This is potent indeed, since one of the most common reasons metrics fail is that people don't pay attention to them because they are not invested in them. Even in cases where the product is *innovation*, and the group is being asked to measure something that has never been done before, those involved with innovating for a living usually have a good sense of what will work and can provide useful information that can become an early indicator for success or failure.

Once a product development team has created good causal actions, turning them into metrics is an almost trivial matter of asking *how much/how many*, *how good*, or *when*? Having decided that calorie consumption and exercise are the causal actions for losing weight, the dieter then asks, "How many calories should I consume each day, and how much exercise should I get?" The questions become metrics and the answers become targets. Similarly, having decided that proper resources applied at the proper points in the development cycle lead to reducing cycle time, an NPD manager would ask, "How many resources, and of what type, should I apply, and when?" An appropriate metric would be the percentage by which actual resources applied during development vary from the resource plan.

Causal Actions Checklist

◆ Does your experience show a cause/effect relationship to the goal?

◆ Is each action different from failed approaches aimed at the same goal in the past?

◆ Is each action distinctly different from the goal?

◆ Do the actions occur early and often relative to the goal?

◆ Can each action be objectively evaluated for success?

◆ Are actions resistant to causing inappropriate behavior? (Doesn't pit one organization against another, isn't easily gamed)

Sorting the Critical Few

The beginning of this chapter discussed the importance of winnowing causal actions to the critical few to avoid overloading the organization with meaningless metrics. The process for doing this involves applying the rather counterintuitive approach of ignoring the things the company is already doing well, *no matter how important they may be.* Probably the largest class of activities that fall in this category is technical capabilities. Companies know how important it is to overcome technical hurdles, so they focus lots of energy on measuring and studying technical capabilities. However, most competent development organizations *will* figure out the technical problems. These are often not the problems that stand in the way of completing projects on time and successfully. Instead, companies should measure such things as resources, whether they have assigned the right people to the project, and whether they have the right suppliers lined up. Bottom line: Get rid of causal actions that may be important but that the company already consistently achieves.

Simple, Effective Governance

Metrics are useless unless they are reviewed early and often. This implies that some sort of governance process must be in place. In general, the simpler the governance process, the more effective it will be. The first step to effective governance is to place responsibility for it in the hands of the right people. Only people who have the ability to affect the outcome of a process should be in charge of the metric for that process. If they can't affect the outcome, shift responsibility to another part of the organization or have responsibility shared by someone who can.

One productive way to keep focus through process reviews is to have a regular stand-up meeting. Every week, the people involved in the process or product being measured gather in a room *without chairs.* This keeps the meeting short and focused. Review each metric. If someone needs help, they ask for it; if not, move on. The only preparation required is to make one person responsible for reporting on each metric. They are not necessarily the ones responsible for attaining the numbers, simply the ones charged with bringing the numbers to the meeting. These meetings look at the early indicators for potential problems. For example, one person may say, "I'm having trouble getting the two ASIC designers I had in my resource plan working on the project full time." Someone else could say, "Oh, we can give you two half-time ASIC designers through next month because our chips are out at the foundry being manufactured until then." The problem—insufficient design resources—never sees the light of day because it's solved before it turns into something major. Everyone on the team knows where everyone else stands.

Once every two months, which translates into approximately one meeting for every eight process reviews, conduct a more formal, in-depth, results review to cover such questions as these: Are we making progress toward our goals? Are there major problems? Do we have to adjust metrics? Are we over- or underachieving? The purpose of this review is not to bang on the table and complain. Rather, it's to refine, change, or jettison process metrics if they aren't helping reach the goal. This meeting includes more people—not only the group in charge of metrics, but bosses and other stakeholders—and should take place in a room that has tables and chairs.

The final step is to set targets for each metric. Usually, this is done annually, but could be more or less frequent depending on the industry. Basically, this involves defining the shape of the curve for progress toward the goal. It may look different for each metric—some may be a line, with slow, steady progress throughout the time period; others may begin rapidly and taper off; others may show no activity for several months and then take off sharply. Whatever the shape of the curve, the idea is to reach 100 percent of the goal at the end of the time period. The team doesn't have to become bogged down in setting targets exactly and tracking minute changes. But if the team should be 25 percent of the way toward the goal and is at only 5 percent, trouble is brewing.

Frame Actions—Then Look Outside the Frame

What happens when something outside of an individual's or team's control impinges on a carefully constructed metrics tree? The product development department, for example, may rely on the quality assurance (QA) department for test results—but product developers don't work for QA and can't directly influence their activities.

In the real world, we're all subject to forces we can't control. The dieter, for example, may also be the person who cooks for a large family of growing kids or may frequently be invited to business lunches and cocktail parties, where lots of forbidden foods will be available. It's no different when implementing product development metrics. To deal with this, *frame* the team's activities: Imagine an empty picture frame. Put inside the frame everything the product development organization can do to affect the attainment of its goals. The product development group controls these actions directly. Now place outside the frame all other organizations, whether internal to the company (different departments) or external (suppliers).

Let's return to the example of the company developing portable heart rate monitors. At the product definition stage, the marketing department is involved with research to determine the need for a proposed new heart rate monitor: Who is the target consumer? What factors are important in terms of appearance, function, and price? When is the optimal time to launch the product? Certainly, the answers to these questions are important to the product development team, but the marketing department owns the activities that lead to the answers; the product development team doesn't have direct control over these activities. Some causal actions related to product development's goals fall outside of its control. Since the actions of the marketing department affect the product development organization's metrics, product development metrics associated with product definition need to be tied explicitly to marketing metrics. This way, the product development organization can receive updates and notifications of customer visits and perhaps even attend pertinent process and results reviews. Likewise, marketing should tie its metrics to the product development organization to avoid last-minute surprises across organizational barriers. Executing the schedule can't be done in a vacuum. This highlights how crucial it is that the two teams—product development and marketing—work closely to gather data, and that they understand the trade-offs involved in certain product feature choices and other product definition activities.

Although you can't exert direct control over many of the processes or activities that influence achievement of your department's or project's goals, you can take some steps to be sure that what's outside the frame doesn't take you by surprise. First, list all the organizations and activities that substantially affect the organization either positively or negatively. While an individual can't control the actions or behaviors of others, he or she can monitor changes in how the other organization is conducting business and can make adjustments early.

For the heart rate monitor example, an engineering director might assign a technical manager to attend the marketing organization's weekly status meetings. The technical manager might assign a member of the technical staff to get on marketing's distribution list to review updates pertinent to the project. Additionally, the engineering director could take things up the chain, asking the vice presidents of engineering and marketing to be cognizant of how any executive-level strategy actions might affect the heart rate monitor project's goals. Similar framing actions should be established with the quality, supply chain and manufacturing organizations. *Framing* adds an incremental complexity to NPD metrics but results in a substantial improvement in real-world metrics performance.

To ensure that metrics are tightly focused, governance responsibilities are fully defined, and that interorganization dependencies are accounted for, use this checklist:

Simplicity Checklist

◆ Would an *otherwise competent organization* succeed if it succeeded at these actions?

◆ Are the metrics prioritized so only four to eight metrics are assigned to any one individual?

◆ Is an organization or individual responsibility set for each process metric?

◆ Is a target set for each metric for the foreseeable future?

◆ Is there a governance plan to watch each metric regularly and act on the data immediately?

◆ Are frames defined to monitor affecting organizations, people, and situations?

ACTION PLANNING: A REAL-WORLD METRICS PROJECT PLAN APPROACH

The project plan in Figure 16-3 shows the steps taken by a medical devices company on a metrics implementation over 18 months. The impetus for the project came from the company's supply chain organization, which wanted to transform itself from a transaction processing organization to a more strategic product development force within the company, scouting and selecting design partners, assessing intellectual property sharing contracts, and improving design partner performance. The organization originally was measuring things like transaction times and cost reduction for component parts. Transformation began with development of a metrics tree (see Figure 16-4), starting with the product development goals: to determine which projects could best leverage outside companies' design capabilities, to find and then get the right suppliers on board, and to develop both internal and external design and intellectual property management capabilities. What actions would lead to these goals? It became immediately obvious that the supply chain organization needed more engineering talent. So recruiting more engineers became a causal action for achieving the group's goals. Another causal action was to have supply chain staff involved in more engineering and product development activities.

These discoveries led the group to set up a completely different set of metrics. Instead of discussing in weekly status review meetings how quickly it could complete a certain transaction, the group began looking at whether it was being represented at engineering planning meetings. The executive who initiated the process later reported that one of the most valuable things to emerge from the process was that it completely changed the focus and discussions in his weekly

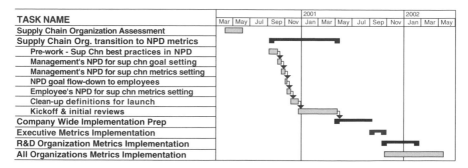

FIGURE 16-3. Real-world implementation.

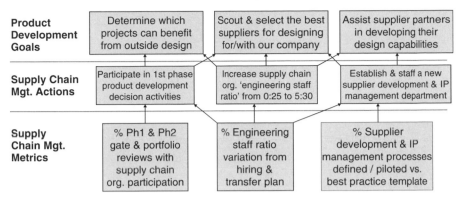

Product Development Goals	Determine which projects can benefit from outside design	Scout & select the best suppliers for designing for/with our company	Assist supplier partners in developing their design capabilities
Supply Chain Mgt. Actions	Participate in 1st phase product development decision activities	Increase supply chain org. 'engineering staff ratio' from 0:25 to 5:30	Establish & staff a new supplier development & IP management department
Supply Chain Mgt. Metrics	% Ph1 & Ph2 gate & portfolio reviews with supply chain org. participation	% Engineering staff ratio variation from hiring & transfer plan	% Supplier development & IP management processes defined / piloted vs. best practice template

FIGURE 16-4. Metrics tree: Supply chain organization transition to new product development focus.

meetings. Now that the group was regularly discussing how to get more engineering talent into its organization, the issue remained top-of-mind for everyone involved, contributing to actually finding and integrating engineers into the staff.

Once the metrics implementation had been up and running for a few months in the supply chain organization, the company's CEO saw the difference it had made in product development partnering and decided to roll it out through the whole organization. He implemented the process again for the entire organization, starting at the very top with goals for the CEO and his direct reports, then rolling down through the various departments.

The ultimate reward for the company was higher sales, higher profits, and a major industry award. The system has since evolved. Although it's not 100 percent perfect, the company still realizes benefits more than five years after its initial rollout. Instead of many disconnected, disorganized metrics, it has a critical few that are tied to corporate goals. It staff—from the CEO to engineers and support staff—continue to focus on the things that matter, and sales, profits, and recognition have followed as a natural consequence.

SUMMARY

The level of satisfaction with NPD metrics is very low. Often, this dissatisfaction has at its root a misperception of the use and value of metrics. By refocusing metrics around those things that really matter—getting away from measuring what the company already does well, paring down the number of metrics to a critical few, aligning metrics with corporate goals, applying simple tools like a metrics tree, involving the people closest to the process in setting the metrics, and establishing a simple but effective governance system—companies can create and implement a metrics system that fits into the mainstream of the NPD process. NPD metrics will vary by industry and by application, but the principles outlined here will work in any type of organization.

Appendix I

The PDMA's Body of Knowledge

Gerry Katz
Executive Vice President, Applied Marketing Science, Inc.

INTRODUCTION

Purpose and Background

Product and service development is part of virtually every organization, and now it is the subject of a great deal of study in many universities. However, due to the cross-functional nature of new product development, it has never fit very easily into the existing organizational structure of most business schools. Likewise, relatively few corporations to this day have a VP or C-level executive in charge of new product development. Thus, the status of product development and management as a recognized profession varies considerably. If product development is ever going to become a mature functional area in its own right (similar to marketing, finance, and operations management), consensus on a core body of knowledge is a crucial first step.

PDMA, through the New Product Development Professional certification program, serves to improve the level of professional practice. Recognizing a core body of knowledge supports the certification of professionals and is central to furthering practice.

Achieving consensus by the profession on a core body of knowledge is a strategic pillar identified by the PDMA board of directors as crucial for the evolution of product development and management toward professional status.

What is the PDMA Body of Knowledge, and What are its Benefits?

Initiated in 2003 by PDMA past president Mark Deck, the PDMA Body of Knowledge (PDMA-BOK) organizes, distills, and provides ready access to the continuously evolving core knowledge needed and used by product development and management professionals and their organizations. The body of knowledge starts with the basics and is expected to grow over time as the PDMA community at large adds to it.

The BOK provides the following important benefits to the PDMA community:

◆ Generally agreed-to definitions and summaries of important concepts, tools, methodologies, processes, and so on to promote better use and more widespread adoption of practices that improve product development effectiveness.

◆ Access to best-of related reference information around key knowledge areas to minimize the time needed to search for deeper information and to further promote continuous learning.

Linkage to the latest writing, presentations, and discussion forums provides a way to learn about leading edge concepts and innovations in the application of current and new practices.

What is the Scope of the Body of Knowledge?

The body of knowledge covers all aspects of product development and management across the entire lifecycle—from opportunity generation and strategy through product launch and on to product iteration and renewal.

◆ Product development management—strategy setting, planning, organizing, resourcing, prioritizing, researching, scheduling, renewing, recycling, retiring, enabling, supporting, measuring, improving, collaborating—for products and services, portfolios, and a company's entire product development program.

◆ Product development operations—generating, defining, designing, testing, validating, prototyping, modeling, building, developing, provisioning, sourcing, maintaining, changing, and so forth.

It does *not* include those areas of product and operations management that are directed primarily to existing products such as the management of ongoing promotion, advertising, branding, pricing, distribution, and customer support (but does address these topics as related to new product development).

Specifically, What Can Be Found?

The body of knowledge is a growing, dynamic work. The most recent release already includes extensive glossary definitions, processes, organizational approaches, practices, tools, methodologies, academic research, and metrics, along with many helpful references for obtaining further information. Future releases are expected to include case examples from conferences and discussion forums, as well.

Where Can It Be Found?

The body of knowledge currently exists in the form of a Web site. It can be accessed through the PDMA's main Web site (*www.pdma.org*) or directly through its own URL (*www.pdmabok.org*). The Web site contains an easy-to-use visual navigation system based on the matrix shown in Exhibit 1. Many of the topics and references contain hyperlinks to important content found elsewhere in the BOK as well as in other important related Web sites, including sites for the *Journal of Product Innovation Management* (JPIM—the PDMA's highly acclaimed academic journal), *Visions* (the PDMA's award-winning practitioner's magazine), and the PDMA/*Amazon.com* bookstore. While all of the current content in the BOK is fully reviewed and edited by an editorial board, the BOK team is also exploring the inclusion of a parallel *wiki* area in which any author can post content without official editorial approval, and only subject to peer review by other BOK users.

Most of the higher-level content in the BOK is fully accessible to anyone, so long as they register on the Web site. However, only PDMA members have access to the more detailed content that lies further down within the matrix—an important benefit of PDMA membership!

Who Is Developing the Body of Knowledge and How Can I Participate?

The PDMA body of knowledge team has included more than 40 volunteer product development professionals (NPDPs) from a broad cross-section of companies. We are continuously soliciting additions and expansions from the PDMA community at large. The body of knowledge Web site offers full attribution to contributors. Authors' names and contact information are included for each of the knowledge areas in order to promote feedback, to suggest new content, and to seek volunteers to develop additional topics.

BODY OF KNOWLEDGE ARCHITECTURE

The architecture or organization scheme for the PDMA body of knowledge is displayed pictorially as a matrix consisting of six rows and three columns (see Exhibit 1). This architecture provides a foundation upon which to capture and organize knowledge about new product and service development. It encompasses three macro *phases* of the product development life cycle as shown in the columns of the matrix: the front end (Discovery), the middle (Development), and the back end (Commercialization). Within each phase, knowledge is organized around six key *knowledge areas*, as shown in the rows of the matrix. Finally, within each "cell" of this matrix, the content can, in many cases, be further broken down into topics that apply to individual products or projects versus those that apply to an entire business or portfolio of products. Each of these major *phases* and *knowledge areas* are described next.

THE PHASES

Discovery Phase

This first phase is primarily about discovery. It covers the entire process of searching for and identifying opportunities—whether market-based or technology-based—and all of the planning and strategy to accomplish this. It requires the identification of customer needs, problems, and benefits, and development of the conceptual features that are envisioned for the products it wishes to build. It ends with the publication of a formal product specifications document. This phase is sometimes referred to as the *fuzzy front end* or the *front end of innovation.*

Development Phase

This second phase is primarily about realization. It covers the entire process of converting specifications into specific features, designs, and platforms— whether for an individual product or a complete portfolio of products—and all of the tasks necessary to accomplish this. It usually requires detailed resource management, creative engineering and process design capabilities, and sophisticated information technology. It ends when the products or services achieve their first commercial availability.

Commercialization Phase

This third phase is primarily about fulfillment. It covers the entire process of new product introduction and the organization's management of its product and service portfolio as it attempts to fulfill its financial potential. It ends when the products or services have reached the end of their useful lifecycle and are to be considered as candidates for retirement, renewal, and regeneration. At this stage, the process begins anew with the undertaking of a new product development initiative, and a return to the Discovery Phase.

THE KNOWLEDGE AREAS

Customer and Market Research

This knowledge area includes anything having to do with bringing external insight into product innovation, development, and growth—especially insight about customers (both buyers and end users), but also information about channels, competitors, markets, alternatives, and the overall environment. This includes gathering and scanning for this information as well as processing, analysis, storage, and use. It includes going to outsiders for insight, validation, confirmation, and feedback. It includes the gleaning of insights from primary as well as secondary sources.

Technology and Intellectual Property

Anything having to do primarily with the invention, development, acquisition, licensing, and management of the technologies and intellectual property (IP) that enable and become part of products is included in this topic area. To the extent that commercializing a technology means creating new products, then that would be dealt with by the larger BOK. But topics such as planning for technology commercialization would be in this row.

Strategy, Planning, and Decision Making

This topic includes anything having to do primarily with strategies, plans, and decision making around product innovation, development, and growth. These would include strategies, plans, and decision making at the business level (as it relates to product innovation, development, and growth), as well as for platforms, product lines or product families, and individual products. The focus is on business, platform, and product planning as opposed to planning for marketing, operations, or customer support, which would be captured in process, execution & metrics. This row would include strategy and planning around resource capacity and throughput management across multiple projects (pipeline resource management), since resourcing is part of decision making. Strategy, planning, and decision making for technologies would be included in the technology and IP knowledge area.

People, Teams, and Culture

Anything having to do with the people side of product development across the NPD lifecycle—including organization/team structures, people management, skills development, culture, organization change management, and human interaction—is included in this knowledge area.

Codevelopment and Alliances

This knowledge category includes anything having to do primarily with innovation, development, and growth activities that take place in unison with external partners of all sorts including customers, suppliers, service providers, and channels. This would include codevelopment or development chain strategy, partner management, codevelopment execution processes, and codevelopment teams, for example. Even if the topic might fall into another row, if the focus is on this activity in a codevelopment context, then information on it would be found in this row primarily.

Process, Execution, and Metrics

Anything having to do with the operational dimension of product innovation, development, and growth will be found in this knowledge topic. This row does not include processes and tools for customer/market input, technology/IP management, strategy and planning, people management, and codevelopment. It does include processes and tools for requirements development and management, design, manufacturing, supply chain, (engineering) change management, channel management, pricing, positioning, promotion, financial management, and customer support. This row would include process and performance management in general, including metrics and benchmarks for product innovation, development, and growth.

THE CELLS

For each of the 18 cells in the BOK matrix, there is an overall cell description, along with a list of references and subtopics (these subtopics are referred to as *In-Depth Knowledge* or IDKs). Following is an example of one of the 18 cell descriptions:

EXAMPLE: CUSTOMER AND MARKET RESEARCH / DISCOVERY PHASE

In the discovery phase, the primary focus of customer and market research is to identify customer needs. These needs, which may or may not have already been addressed, and which the customer may or may not even be aware of, help define the goals and performance parameters, as well as the target market, for a potential new product.

There are many processes and market research tools available for identifying customer needs, ranging from the very general such as focus groups and surveys, which can be used for a wide variety of objectives, to highly specific techniques such as Voice of the Customer, Lead User Analysis, Attribute Testing, and Conjoint Analysis, which address more targeted objectives. Market research methods can be categorized as qualitative or quantitative in terms of the types of data collected and analytical techniques required.

Key decisions must be made in preparation for any market research initiative:

◆ *Who to study: designing a sample.*
◆ *How to study them: in groups vs. individually.*

- ◆ *Where to study them: in their normal environment or in a structured setting.*
- ◆ *How many to study: a small sample studied in depth vs. a large sample that allows for the generalization of the results to the entire population of interest with statistical confidence.*
- ◆ *What the research objectives are: to help guide the research design and/or questionnaire development.*
- ◆ *How to capture the information: recording and coding of data.*
- ◆ *How to analyze the information: compiling, extracting, organizing, drawing conclusions from the data, and presenting them.*

In-Depth Knowledge Topics

- − Conjoint analysis
- − Ethnography
- − Individual depth interviews
- − Lead user research
- − Quantitative market research
- − Shadowing
- − Voice of the Customer: needs identification

REFERENCES

Concept Testing

Developing an Effective Concept Testing Program for Consumer Durables, JPIM 9:267–277 (1992), Page, A. L. and Rosenbaum, H. F.

- ◆ Provides a step-by-step description of the development of a concept test and its use.

Anschuetz, Ned F. 1996. ``Evaluating Ideas and Concepts for New Consumer Products.'' In *The PDMA Handbook of New Product Development*. Ed. M.D. Rosenau, New York: John Wiley & Sons, pp. 195–206.

- ◆ A good compact description of how to conduct Concept Testing for consumer products.

Paul, Ronald N. 1996. ``Evaluating Ideas and Concepts for New Business-to-Business Products.'' In *The PDMA Handbook of New Product Development*. Ed. M.D. Rosenau. New York: John Wiley & Sons, pp. 207–216.

- ◆ A good compact description of how to conduct Concept Testing for business-to-business products.

Market Research Tools

Cohen, Steven H. 1996. ``Tools for Quantitative Market Research.'' In *The PDMA Handbook of New Product Development*. Ed. M.D. Rosenau. New York: John Wiley & Sons, pp. 253–267.

- ◆ A good overview of a number of quantitative market research tools and where they fit into the product development process.

Davis, R. E. 1993. ``From Experience: The Role of Market Research in the Development of New Consumer Products.'' *JPIM* 10: 309–317.

♦ A practitioner's overview of where and how market research fits into the development process for consumer products.

Redesigning Product Lines with Conjoint Analysis: How Sunbeam Does it, JPIM 4:120-137 (June 1987), Page, A.L. and Rosenbaum, H.F.

♦ Provides an example of how a conjoint analysis technique is applied to a real product development challenge.

Voice of the Customer

Burchill, Gary, and Brodie, Christina Hepner. 1997. *Voices into Choices*. Madison, WI: Joiner Publications.

♦ A complete ``how-to'' guide of the process developed at the Center for Quality Management in the early 1990s.

Cohen, Lou. 1995. *Quality Function Deployment: How to Make QFD Work for You*. Reading, MA: Addison-Wesley Publishing.

♦ While the main thrust of this book is on QFD, it contains many insightful suggestions on how to gather and use Voice of the Customer data.

Griffin, Abbie, and Hauser, John.1993. The Voice of the Customer. *Marketing Science* 12(1): (Winter): 1–27.

♦ The first truly empirical study of Voice of the Customer, the goal of which was to identify best practices.

Katz, Gerald. 2001. ``The `One Right Way' to Gather the Voice of the Customer.'' *PDMA Visions*, 25 (2) (October).

♦ Examines all of the various trade-offs in how to go about gathering Voice of the Customer information, with the conclusion that there is no one right way.

McQuarrie, Edward F. 1998. *Customer Visits*. Thousand Oaks, CA: Sage Publications.

♦ A comprehensive description of this most common way of gathering customer wants and needs.

Ulwick, Anthony. 2002. ``Turn Customer Input into Innovation.'' *Harvard Business Review*, 80 (1) (January).

♦ A paper that describes the Voice of the Customer process in more managerial language (rather than product developer or market researcher language).

THE IN-DEPTH KNOWLEDGE AREAS (IDKS)

Finally, within each cell of the matrix, there are many pages of rich content about a myriad of in-depth knowledge (IDK) areas. Each of these includes a definition, a description, and a list of detailed references. Following is an example of an IDK:

EXAMPLE: VOICE OF THE CUSTOMER: NEEDS IDENTIFICATION

DEFINITION Voice of the Customer is a detailed set of customer wants and needs, organized into a hierarchical structure, and then prioritized in terms of relative importance and satisfaction with current alternatives.

DESCRIPTION The Voice of the Customer is a market research technique consisting of both qualitative and quantitative research steps. It is generally carried out at the start of any new product, process, or service design initiative in order to better understand the customer's wants and needs, and as the key input for new product definition, quality function deployment (QFD), and the setting of detailed design specifications.

Much has been written about this process, and there are many possible ways to gather the information—focus groups, individual interviews, contextual inquiry, ethnographic techniques, and so on. But all involve a series of structured in-depth interviews that focus on the customers' experiences with current products or alternatives within the category under consideration. Needs statements are then extracted, organized into a more usable hierarchy, and then prioritized by the customers.

It is critical that the product development core team own and be highly involved in this process. They must be the ones who take the lead in defining the topic, designing the sample (i.e., the types of customers to include), generating the questions for the discussion guide, either conducting or observing and analyzing the interviews, and extracting and processing the needs statements.

A good Voice of the Customer study provides (1) a detailed understanding of the customer's requirements; (2) a common language for the team going forward; (3) key input for the setting of appropriate design specifications for the new product or service; and (4) a highly useful springboard for product innovation.

REFERENCES

Burchill, Gary, and, Christina Hepner Brodie. 1997. *Voices into Choices*. Madison, WI: Joiner Publications.

Griffin, Abbie, and John Hauser.1993. "The Voice of the Customer." *Marketing Science*, *12* (1): 1–27 (Winter).

Katz, Gerald, (2001). The "One Right Way" to Gather the Voice of the Customer. PDMA Visions, *25*(2) (October).

McQuarrie, Edward F. 1998. *Customer Visits*. Thousand Oaks, CA: Sage Publications.

Ulwick, Anthony. 2002. "Turn Customer Input into Innovation." *Harvard Business Review*, *80* (1) (January).

Appendix II

The PDMA Glossary for New Product Development

Accidental Discovery: New designs, ideas, and developments resulting from unexpected insight, which can be obtained either internal or external to the organization.

Adoption Curve: The phases through which consumers or a market proceed in deciding to adopt a new product or technology. At the individual level, each consumer must move from a cognitive state (becoming aware of and knowledgeable about), to an emotional state (liking and then preferring the product) and into a behavioral state (deciding and then purchasing the product). At the market level, the new product is first purchased by the innovators in the marketplace, which are generally thought to constitute about 2.5 percent of the market. Early adopters (13.5 percent of the market) are the next to purchase, followed by the early majority (34 percent), late majority (34 percent), and finally, the laggards (16 percent).

Affinity Charting: A bottom-up technique for discovering connections between pieces of data. Individuals or groups start with one piece of data (say, a customer need). They then look through the rest of the data they have (say, statements of other customer needs) to find other data (needs) similar to the first, and place it in the same group. As they come across pieces of data that differ from those in the first group, they create a new category. The end result is a set of groups where the data contained within a category are similar, and the groups all differ in some way. See also *Qualitative Cluster Analysis*.

Alliance: Formal arrangement with a separate company for purposes of development, and involving exchange of information, hardware, intellectual property, or enabling technology. Alliances involve shared risk and reward (e.g., co-development projects). (See also Chapter 11 of *The PDMA HandBook, 2ⁿᵈ ed.*)

Alpha Test: Preproduction product testing to find and eliminate the most obvious design defects or deficiencies, usually in a laboratory setting or in some part of the developing firm's regular operations, although in some cases it may be done in controlled settings with lead customers. See also *beta test* and *gamma test*.

Alpha Testing: A crucial first look at the initial design, usually done in-house. The results of the alpha test either confirm that the product performs according to its specifications or uncovers areas where the product is deficient. The testing environment should try to simulate the conditions under which the product will actually be used as closely as possible. The alpha test should not be performed by the same people who are doing the development work. Since this is the first flight for the new product, basic questions of fit and function should be evaluated. Any suggested modifications or revisions to the specifications should be solicited from all parties involved in the evaluation and should be considered for inclusion. Since the testing is done in-house, special care must be taken to remain as objective as possible.

Analytical Hierarchy Process (AHP): A decision-making tool for complex, multicriteria problems where both qualitative and quantitative aspects of a problem need to be incorporated. AHP clusters decision elements according to their common characteristics into a hierarchical structure similar to a family tree or affinity chart. The AHP process was designed by T. L. Saaty.

Analyzer: A firm that follows an imitative innovation strategy, where the goal is to get to market with an equivalent or slightly better product very quickly once someone else opens up the market, rather than to be first to market with new products or technologies. Sometimes called an imitator or a fast follower.

Anticipatory Failure Determination (AFD): A failure analysis method. In this process, developers start from a particular failure of interest as the intended consequence and try to devise ways to assure that the failure always happens reliably. Then the developers use that information to develop ways to better identify steps to avoid the failure.

Applications Development: The iterative process through which software is designed and written to meet the needs and requirements of the user base or the process of enhancing or developing new products.

Architecture: See *product architecture*.

As-Is Map: A version of a process map depicting how an existing process actually operates. This may differ substantially from documented guidelines.

Asynchronous Groupware: Software used to help people work as groups, but not requiring those people to work at the same time.

Attribute Testing: A quantitative market research technique in which respondents are asked to rate a detailed list of product or category attributes on one or more types of scales such as relative importance, current performance, current satisfaction with a particular product or service, for the purpose of ascertaining customer preferences for some attributes over others, to help guide the design and development process. Great care and rigor should be taken in the development of the list of attributes, and it must be neither too long for the respondent to answer comfortably or too short such that it lumps too many ideas together at too high a level.

Audit: When applied to new product development, an audit is an appraisal of the effectiveness of the processes by which the new product was developed and brought to market. (See Chapter 14 of *The PDMA ToolBook 1.*)

Augmented Product: The core product, plus all other sources of product benefits, such as service, warranty, and image.

Autonomous Team: A completely self-sufficient project team with very little, if any, link to the funding organization. Frequently used as an organizational model to bring a radical innovation to the marketplace. Sometimes called a *tiger* team.

Awareness: A measure of the percent of target customers who are aware that the new product exists. Awareness is variously defined, including recall of brand, recognition of brand, recall of key features or positioning.

Back-up: A project that moves forward, either in synchrony or with a moderate time lag, and for the same marketplace, as the lead project to provide an alternative asset should the lead project fail in development. A back-up has essentially the same mechanism of action performance as the lead project. Normally a company would not advance both the lead and the back-up project through to the marketplace, since they would compete directly with each other.

Balanced Scorecard: A comprehensive performance measurement technique that balances four performance dimensions: (1) Customer perceptions of how we are performing; (2) Internal perceptions of how we are doing at what we must excel at; (3) Innovation and learning performance; (4) Financial performance.

Baton-Passing Process: See *relay-race process.*

Benchmarking: A process of collecting process performance data, generally in a confidential, blinded fashion, from a number of organizations to allow them to assess their performance individually and as a whole.

Benefit: A product attribute expressed in terms of what the user gets from the product rather than its physical characteristics or features. Benefits are often paired with specific features, but they need not be.

Best Practice: Methods, tools, or techniques that are associated with improved performance. In new product development, no one tool or technique assures success; however a number of them are associated with higher probabilities of achieving success. Best practices likely are at least somewhat context specific. Sometimes called *effective practice.*

Best Practice Study: A process of studying successful organizations and selecting the best of their actions or processes for emulation. In new product development it means finding the best process practices, adapting them and adopting them for internal use. (See Chapter 36, in the *PDMA HandBook 2nd ed.,* Griffin, "PDMA Research on New Product Development Practices:

Updating Trends and Benchmarking Best Practices," *JPIM*, 14:6, 429-458, November, 1997, and "Drivers of NPD Success: The 1997 PDMA Report," PDMA, October, 1997.)

Beta Test: An external test of preproduction products. The purpose is to test the product for all functions in a breadth of field situations to find those system faults that are more likely to show in actual use than in the firm's more controlled in-house tests before sale to the general market. See also *field test*.

Beta Testing: A more extensive test than the alpha, performed by real users and customers. The purpose of beta testing is to determine how the product performs in an actual user environment. It is critical that real customers perform this evaluation, not the firm developing the product or a contracted testing company. As with the alpha test, results of the beta test should be carefully evaluated with an eye toward any needed modifications or corrections.

Bill of Materials (BOM): A listing of all subassemblies, intermediate parts, and raw materials that go into a parent assembly, showing the quantity of each required to make an assembly.

Bowling Alley: An early growth stage strategy that emphasizes focusing on specific niche markets, building a strong position in those markets by delivering clearly differentiated *whole products* and using that niche market strength as a leverage point for conquering conceptually neighboring niche markets. Success in the bowling alley is predicated on building product leadership via customer intimacy.

Brainstorming: A group method of creative problem solving frequently used in product concept generation. There are many modifications in format, each variation with its own name. The basis of all of these methods uses a group of people to creatively generate a list of ideas related to a particular topic. As many ideas as possible are listed before any critical evaluation is performed. (See Chapters 16 and 17 in *The PDMA HandBook 2nd ed.*)

Brand: A name, term, design, symbol, or any other feature that identifies one seller's good or service as distinct from those of other sellers. The legal term for brand is *trademark*. A brand may identify one item, a family of items, or all items of that seller.

Brand Development Index (BDI): A measure of the relative strength of a brand's sales in a geographic area. Computationally, BDI is the percent of total national brand sales that occur in an area divided by the percent of U.S. households that reside in that area.

Breadboard: A proof-of-concept modeling technique that represents how a product will work, but not how a product will look.

Break-even Point: The point in the commercial life of a product when cumulative development costs are recovered through accrued profits from sales.

Business Analysis: An analysis of the business situation surrounding a proposed project. Usually includes financial forecasts in terms of discounted cash flows, net present values or internal rates of returns.

Business Case: The results of the market, technical and financial analyses, or up-front homework. Ideally defined just prior to the *go to development* decision (gate), the case defines the product and project, including the project justification and the action or business plan. (See Chapter 21 of *The PDMA HandBook 2ⁿᵈ ed.*)

Business Management Team: Top functional managers and business unit head who work together throughout the design of the decision-flow component of a stage-gate process.

Business-to-Business: Transactions with nonconsumer purchasers such as manufacturers, resellers (distributors, wholesalers, jobbers and retailers, for example) institutional, professional and governmental organizations. Frequently referred to as *industrial* businesses in the past.

Buyer: The purchaser of a product, whether or not he or she will be the ultimate user. Especially in business-to-business markets, a purchasing agent may contract for the actual purchase of a good or service, yet never benefit from the function(s) purchased.

Buyer Concentration: The degree to which purchasing power is held by a relatively small percentage of the total number of buyers in the market.

Cannibalization: That portion of the demand for a new product that comes from the erosion of the demand for (sales of) a current product the firm markets. (See Chapter 34 in *The PDMA HandBook 2ⁿᵈ ed.*)

Capacity Planning: A forward-looking activity that monitors the skill sets and effective resource capacity of the organization. For product development, the objective is to manage the flow of projects through development such that none of the functions (skill sets) creates a bottleneck to timely completion. Necessary in optimizing the project portfolio.

Category Development Index (CDI): A measure of the relative strength of a category's sales in a geographic area. Computationally, it is the percent of total national category sales that occur in an area divided by the percent of U.S. households in that area.

Centers of Excellence: A geographic or organizational group with an acknowledged technical, business, or competitive competency.

Certification: A process for formally acknowledging that someone has mastered a body of knowledge on a subject. In new product development, the PDMA has created and manages a certification process to become a New Product Development Professional (NPDP). See *http://www.pdma.org/certification/* for additional information.

Champion: A person who takes a passionate interest in seeing that a particular process or product is fully developed and marketed. This informal role varies from situations calling for little more than stimulating awareness of the opportunity to extreme cases where the champion tries to force a project past the strongly entrenched internal resistance of company policy or that of objecting parties. (See Chapter 5 in *The PDMA ToolBook 1*.)

Change Equilibrium: A balance of organizational forces that either drives or impedes change.

Charter: A project team document defining the context, specific details, and plans of a project. It includes the initial business case, problem and goal statements, constraints and assumptions, and preliminary plan and scope. Periodic reviews with the sponsor ensure alignment with business strategies. (See also *Product Innovation Charter*.)

Checklist: A list of items used to remind an analyst to think of all relevant aspects. It finds frequent use as a tool of creativity in concept generation, as a factor consideration list in concept screening, and to ensure that all appropriate tasks have been completed in any stage of the product development process.

Chunks: The building blocks of product architecture. They are made up of inseparable physical elements. Other terms for chunks may be modules or major subassemblies.

Classification: A systematic arrangement into groups or classes based on natural relationships.

Clockspeed: The evolution rate of different industries. High clockspeed industries, like electronics, see multiple generations of products within short time periods, perhaps even within 12 months. In low clockspeed industries, like the chemical industry, a generation of products may last as long as 5 or even 10 years. It is believed that high clockspeed industries can be used to understand the dynamics of change that will in the long run affect all industries, much like fruit flies are used to understand the dynamics of genetic change in a speeded-up genetic environment, due to their short life spans.

Cognitive Modeling: A method for producing a computational model for how individuals solve problems and perform tasks, which is based on psychological principles. The modeling process outlines the steps a person goes through in solving a particular problem or completing a task, which allows one to predict the time it will take or the types of errors an individual may make. Cognitive models are frequently used to determine ways to improve a user interface to minimize interaction errors or time by anticipating user behavior.

Cognitive Walkthrough: Once a model of the steps or tasks a person must go through to complete a task is constructed, an expert can role play the part of a user to cognitively walk through the user's expected experience. Results from this walk-through can help make human-product interfaces more intuitive and increase product usability.

Collaborative Product Development: When two firms work together to develop and commercialize a specialized product. The smaller firm may contribute technical or creative expertise, while the larger firm may be more likely to contribute capital, marketing, and distribution capabilities. When two firms of more equal size collaborate, they may each bring some specialized technology capability to the table in developing some highly complex product or system requiring expertise in both technologies. Collaborative product development has several variations. In customer collaboration, a supplier reaches out and partners with a key or lead customer. In supplier collaboration, a company partners with the provider(s) of technologies, components, or services to create an integrated solution. In collaborative contract manufacturing, a company contracts with a manufacturing partner to produce the intended product. Collaborative development (also known as *codevelopment*) differs from simple outsourcing in its levels of depth of partnership in that the collaborative firms are linked in the process of delivering the final solution to the intended customer.

Co-location: Physically locating project personnel in one area, enabling more rapid and frequent decision making and communication among them.

Commercialization: The process of taking a new product from development to market. It generally includes production launch and ramp-up, marketing materials and program development, supply chain development, sales channel development, training development, training, and service and support development. (See Chapter 30 of *The PDMA HandBook 2nd ed.*)

Competitive Intelligence: Methods and activities for transforming disaggregated public competitor information into relevant and strategic knowledge about competitors' positions, size, efforts, and trends. The term refers to the broad practice of collecting, analyzing, and communicating the best available information on competitive trends occurring outside one's own company.

Computer-Aided Engineering (CAE): Using computers in designing, analyzing, and manufacturing a product or process. Sometimes refers more narrowly to using computers just at the engineering analysis stage.

Computer-Aided Design (CAD): A technology that allows designers and engineers to use computers for their design work. Early programs enabled two-dimensional (2-D) design. Current programs allow designers to work in 3-D (three dimensions), and in either wire or solid models.

Computer-Enhanced Creativity: Using specially designed computer software that aids in the process of recording, recalling and reconstructing ideas to speed up the new product development process.

Concept: A clearly written and possibly visual description of the new product idea that includes its primary features and consumer benefits, combined with a broad understanding of the technology needed.

Concept Generation: The processes by which new concepts, or product ideas, are generated. Sometimes also called *idea generation* or *ideation*. (See Chapters 15 and 17 in *The PDMA HandBook 2nd ed.*, Chapter 9 in the *PDMA ToolBook 2 for New Product Development*, and Chapters 1, 4, 5, and 6 in the *PDMA ToolBook 3 for New Product Development*.)

Concept Optimization: A research approach that evaluates how specific product benefits or features contribute to a concept's overall appeal to consumers. Results are used to select from the options investigated to construct the most appealing concept from the consumer's perspective.

Concept Statement: A verbal or pictorial statement of a concept that is prepared for presentation to consumers to get their reaction prior to development.

Concept Study Activity: The set of product development tasks in which a concept is given enough examination to determine if there are substantial unknowns about the market, technology, or production process.

Concept Screening: The evaluation of potential new product concepts during the discovery phase of a product development project. Potential concepts are evaluated for their fit with business strategy, technical feasibility, manufacturability, and potential for financial success.

Concept Testing: The process by which a concept statement is presented to consumers for their reactions. These reactions can either be used to permit the developer to estimate the sales value of the concept or to make changes to the concept to enhance its potential sales value. (See Chapter 6 in *The PDMA HandBook 2nd ed.*)

Concurrency: Carrying out separate activities of the product development process at the same time rather than sequentially.

Concurrent Engineering (CE): When product design and manufacturing process development occur concurrently in an integrated fashion, using a cross-functional team, rather than sequentially by separate functions. CE is intended to cause the development team to consider all elements of the product life cycle from conception through disposal, including quality, cost, and maintenance, from the project's outset. Also called simultaneous engineering. (See Chapter 30 of *The PDMA HandBook 1st ed.*)

Conjoint Analysis: Conjoint analysis is a market research technique in which respondents are systematically presented with a rotating set of product descriptions, each of which contains a rotating set of attributes and levels of those attributes. By asking respondents to choose their preferred product and/or to indicate their degree of preference from within each set of options, conjoint analysis can determine the relative contribution to overall preference of each variable and each level. The two key advantages of conjoint analysis over other methods of determining importance are: (1) the variables and levels can be either continuous (e.g., weight) or discrete (e.g., color), and (2) it is just about the only valid market research method for evaluating the role of price, i.e., how much someone would pay for a given feature. (See Chapter 18 of *The PDMA HandBook 2nd ed.* and Chapter 3 of the *PDMA ToolBook 3 for New Product Development*.)

Consumer: The most generic and all-encompassing term for a firm's targets. The term is used in either the business-to-business or household context and may refer to the firm's current customers, competitors' customers, or current nonpurchasers with similar needs or demographic characteristics. The term does not differentiate between whether the person is a buyer or a user target. Only a fraction of consumers will become customers.

Consumer Market: The purchasing of goods and services by individuals and for household use (rather than for use in business settings). Consumer purchases are generally made by individual decision makers, either for themselves or others in the family.

Consumer Need: A problem the consumer would like to have solved. What a consumer would like a product to do for them.

Consumer Panels: Specially recruited groups of consumers whose longitudinal category purchases are recorded via the scanner systems at stores.

Contextual Inquiry: A structured qualitative market research method that uses a combination of techniques from anthropology and journalism. Contextual inquiry is a customer needs discovery process that observes and interviews users of products in their actual environment.

Contingency Plan: A plan to cope with events whose occurrence, timing, and severity cannot be predicted.

Continuous Improvement: The review, analysis, and rework directed at incrementally improving practices and processes. Also called *Kaizen.*

Continuous Innovation: A product alteration that allows improved performance and benefits without changing either consumption patterns or behavior. The product's general appearance and basic performance do not functionally change. Examples include fluoride toothpaste and higher computer speeds.

Continuous Learning Activity: The set of activities involving an objective examination of how a product development project is progressing or how it was carried out to permit process changes to simplify its remaining steps or improve the product being developed or its schedule. (See also *Learning Organization.*)

Contract Developer: An external provider of product development services.

Controlled Store Testing: A method of test marketing where specialized companies are employed to handle product distribution and auditing rather than using the company's normal sales force.

Convergent Thinking: A technique generally performed late in the initial phase of idea generation to help funnel the high volume of ideas created through divergent thinking into a small group or single idea on which more effort and analysis will be focused.

Cooperation (Team Cooperation): The extent to which team members actively work together in reaching team level objectives.

Coordination Matrix: A summary chart that identifies the key stages of a development project, the goals, and key activities within each stage, and who (what function) is responsible for each.

Core Benefit Proposition (CBP): The central benefit or purpose for which a consumer buys a product. The CBP may come either from the physical good or service, or it may come from augmented dimensions of the product. (See also *Value Proposition.*) (See Chapter 3 of *The PDMA ToolBook 1.*)

Core Competence: That capability at which a company does better than other firms, which provides them with a distinctive competitive advantage and contributes to acquiring and retaining customers. Something that a firm does better than other firms. Can include technical, organizational, supply chain, operational, financial, marketing, partnership, or other capabilities. The purest definition adds *and is also the lowest cost provider.*

Corporate Culture: The *feel* of an organization. Culture arises from the belief system through which an organization operates. Corporate cultures are variously described as being authoritative, bureaucratic, and entrepreneurial. The firm's culture frequently impacts the organizational appropriateness for getting things done. (See Chapter 1 of the *PDMA ToolBook 2 for New Product Development.*)

Cost of Goods Sold (COGS or CGS): The direct costs (labor and materials) associated with producing a product and delivering it to the marketplace.

Creativity: "An arbitrary harmony, an expected astonishment, a habitual revelation, a familiar surprise, a generous selfishness, an unexpected certainty, a formable stubbornness, a vital triviality, a disciplined freedom, an intoxicating steadiness, a repeated initiation, a difficult delight, a predictable gamble, an ephemeral solidity, a unifying difference, a demanding satisfier, a miraculous expectation, and accustomed amazement." (George M. Prince, *The Practice of Creativity*, 1970). Creativity is the ability to produce work that is both novel and appropriate.

Criteria: Statements of standards used by decision makers at decision gates. The dimensions of performance necessary to achieve or surpass for product development projects to continue in development. In the aggregate, these criteria reflect a business unit's new product strategy. (See Chapters 21 and 29 of *The PDMA ToolBook 2.*)

Critical Assumption: An explicit or implicit assumption in the new product business case that, if wrong, could undermine the viability of the opportunity.

Critical Path: The set of interrelated activities that must be completed for the project to be finished successfully can be mapped into a chart showing how long each task takes, and which tasks cannot be started before which other tasks are completed. The critical path is the set of linkages through the chart that is the longest. It determines how long a project will take.

Critical Path Scheduling: A project management technique, frequently incorporated into various software programs, which puts all important steps of a given new product project into a sequential network based on task interdependencies.

Critical Success Factors: Those critical few factors that are necessary for, but don't guarantee, commercial success. (See Chapter 1 of *The PDMA HandBook 2nd ed.*)

Cross-Functional Team: A team consisting of representatives from the various functions involved in product development, usually including members from all key functions required to deliver a successful product, typically including marketing, engineering, manufacturing/operations, finance, purchasing, customer support, and quality. The team is empowered by the departments to represent each function's perspective in the development process. (See Chapters 9 and 10 in *The PDMA HandBook 2nd* ed. and Chapter 6 in *The PDMA ToolBook 1*, Chapter 5 in *The PDMA ToolBook 2* and Chapter 13 in the *PDMA ToolBook 3 for New Product Development*.)

Crossing the Chasm: Making the transition to a mainstream market from an early market dominated by a few visionary customers (sometimes also called innovators or lead adopters). This concept typically applies to the adoption of new, market creating technology-based products and services. (See Chapters 2 and 3 in *The PDMA ToolBook 2 for New Product Development*.)

Cross-sections: An explanation of a part that is referenced by slicing through the area that needs to be explained.

Customer: One who purchases or uses your firm's products or services.

Customer-based Success: The extent to which a new product is accepted by customers and the trade.

Customer Needs: Problems to be solved. These needs, either expressed or yet to be articulated, provide new product development opportunities for the firm. (See Chapter 14 in *The PDMA HandBook 2nd* ed.)

Customer Perceived Value (CPV): The result of the customer's evaluation of all the benefits and all the costs of an offering, as compared to that customer's perceived alternative. It is the basis on which customers decide to buy things. (See Chapter 4 of *The PDMA ToolBook 1 for New Product Development*.)

Customer Site Visits: A qualitative market research technique for uncovering customer needs. The method involves going to a customer's work site, watching as a person performs functions associated with the customer needs your firm wants to solve, and then debriefing that person about what they did, why they did those things, the problems encountered as they were trying to perform the function, and what worked well. (See Chapters 15 and 16 of *The PDMA HandBook 2nd* ed. and Chapter 5 of *The PDMA ToolBook 3 for New Product Development*.)

Customer Value Added Ratio: The ratio of WWPF (worth what paid for) for your products to WWPF for your competitors' products. A ratio above 1 indicates superior value compared to your competitors.

Cycle Time: The length of time for any operation, from start to completion. In the new product development sense, it is the length of time to develop a new product from an early initial idea for a new product to initial market sales. Precise definitions of the start and end point vary from one company to another, and may vary from one project to another within the company. (See Chapter 12 of *The PDMA HandBook 2nd* ed.)

Dashboard: A typically colored graphical presentation of a project's status or a portfolio's status by project resembling a vehicle's dashboard. Typically, red is used to flag urgent problems, yellow to flag impending problems, and green to signal on projects on track.

Data: Measurements taken at the source of a business process.

Database: An electronic gathering of information organized in some way to make it easy to search, discover, analyze, and manipulate.

Decision Screens: Sets of criteria that are applied as checklists or screens at new product decision points. The criteria may vary by stage in the process. (See Chapter 7 in *The PDMA ToolBook 1* and Chapter 21 of *The PDMA HandBook 2ⁿᵈ ed.*)

Decision Tree: A diagram used for making decisions in business or computer programming. The *branches* of the tree diagram represent choices with associated risks, costs, results, and outcome probabilities. By calculating outcomes (profits) for each of the branches, the best decision for the firm can be determined.

Decline Stage: The fourth and last stage of the product life cycle. Entry into this stage is generally caused by technology advancements, consumer or user preference changes, global competition or environmental or regulatory changes. (See Chapter 34 of *The PDMA HandBook 2ⁿᵈ ed.*)

Defenders: Firms that stake out a product turf and protect it by whatever means, not necessarily through developing new products.

Deliverable: The output (such as test reports, regulatory approvals, working prototypes or marketing research reports) that shows a project has achieved a result. Deliverables may be specified for the commercial launch of the product or at the end of a development stage.

Delphi Processes: A technique that uses iterative rounds of consensus development across a group of experts to arrive at a forecast of the most probable outcome for some future state.

Demographic: The statistical description of a human population. Characteristics included in the description may include gender, age, education level, and marital status, as well as various behavioral and psychological characteristics.

Derivative Product: A new product based on changes to an existing product that modifies, refines, or improves some product features without affecting the basic product architecture or platform.

Design for the Environment (DFE): The systematic consideration of environmental safety and health issues over the product's projected life cycle in the design and development process.

Design for Excellence (DFX): The systematic consideration of *all* relevant life cycle factors, such as manufacturability, reliability, maintainability, affordability, testability, etc., in the design and development process.

Design for Maintainability (DFMt): The systematic consideration of maintainability issues over the product's projected life cycle in the design and development process.

Design for Manufacturability (DFM): The systematic consideration of manufacturing issues in the design and development process, facilitating the fabrication of the product's components and their assembly into the overall product.

Design of Experiments (DOE): A statistical method for evaluating multiple product and process design parameters simultaneously rather than one parameter at a time.

Design to Cost: A development methodology that treats costs as an independent design parameter, rather than an outcome. Cost objectives are established based on customer affordability and competitive constraints.

Design Validation: Product tests to ensure that the product or service conforms to defined user needs and requirements. These may be performed on working prototypes or using computer simulations of the finished product.

Development: The functional part of the organization responsible for converting product requirements into a working product. Also, a phrase in the overall concept to market cycle where the new product or service is developed for the first time.

Development Change Order (DCO): A document used to implement changes during product development. It spells out the desired change, the reason for the change, and the consequences to time to market, development cost, and to the cost of producing the final product. It gets attached to the project's charter as an addendum.

Development Teams: Teams formed to take one or more new products from concept through development, testing and launch. (See Chapters 9 and 10 in *The PDMA HandBook* 2$_{nd}$ ed. and Chapter 6 in *The PDMA ToolBook1*, Chapter 5 in *The PDMA ToolBook 2* and Chapter 13 in the *PDMA ToolBook 3 for New Product Development*.)

Digital Mock-Up: An electronic model of the product created with a solids modeling program. Mockups can be used to check for interface interferences and component incompatibilities. Using a digital mock-up can be less expensive than building physical prototypes.

Discontinuous Innovation: Previously unknown products that establish new consumption patterns and behavior changes. Examples include microwave ovens and cellular phones.

Discounted Cash-Flow (DCF) Analysis: One method for providing an estimate of the current value of future incomes and expenses projected for a project. Future cash flows for a number of years are estimated for the project, and then discounted back to the present using forecast interest rates.

Discrete Choice Experiment: A quantitative market research tool used to model and predict customer buying decisions.

Dispersed Teams: Product development teams that have members working at different locations, across time zones, and perhaps even in different countries. (See Chapter 5 in *The PDMA ToolBook 2 for New Product Development*.)

Distribution: The method and partners used to get the product (or service) from where it is produced to where the end user can buy it.

Divergent Thinking: Technique performed early in the initial phase of idea generation that expands thinking processes to generate, record and recall a high volume of new or interesting ideas.

Dynamically Continuous Innovation: A new product that changes behavior, but not necessarily consumption patterns. Examples include Palm Pilots, electric toothbrushes, and electric hair curlers.

Early Adopters: For new products, these are customers who, relying on their own intuition and vision, buy into new product concepts very early in the life cycle. For new processes, these are organizational entities that were willing to try out new processes rather than just maintaining the old.

Economic Value Added (EVA): The value added to or subtracted from shareholder value during the life of a project.

Empathic Design: A five-step method for uncovering customer needs and sparking ideas for new concepts. The method involves going to a customer's work site, watching as he or she performs functions associated with the customer needs your firm wants to solve, and then debriefing the customer about what they did, why they did those things, the problems they encountered as they were trying to perform the function, and what worked well. By spending time with customers, the team develops empathy for the problems customers encounter trying to perform their daily tasks. See also *Customer Site Visits*.

Engineering Design: A function in the product creation process where a good or service is configured and specific form is decided.

Engineering Model: The combination of hardware and software intended to demonstrate the simulated functioning of the intended product as currently designed.

Enhanced New Product: A form of derivative product. Enhanced products include additional features not previously found on the base platform, which provide increased value to consumers.

Entrance Requirement: The document(s) and reviews required before any phase of a stages and gates development process can be started. (See Chapter 7 of *The PDMA ToolBook 1*.)

Entrepreneur: A person who initiates, organizes, operates, assumes the risk, and reaps the potential reward for a new business venture.

Ethnography: A descriptive, qualitative market research methodology for studying the customer in relation to his or her environment. Researchers spend time in the field observing customers and their environment to acquire a deep understanding of the lifestyles or cultures as a basis for better understanding their needs and problems. (See customer site visits, Chapter 15 in *The PDMA HandBook 2nd ed.*, Chapter 8 in *The PDMA ToolBook 2 for New Product Development*, and Chapter 5 in *The PDMA ToolBook 3 for New Product Development*.)

Event: Marks the point in time when a task is completed.

Event Map: A chart showing important events in the future that is used to map out potential responses to probable or certain future events.

Excursion: An idea generation technique to force discontinuities into the idea set. Excursions consist of three generic steps: (1) Step away from the task; (2) Generate disconnected or irrelevant material; (3) Force a connection back to the task.

Exit Requirement: The document(s) and reviews required to complete a stage of a stages and gates development process. (See Chapter 7 of *The PDMA ToolBook 1* and Chapter 21 of *The PDMA HandBook 2nd ed.*)

Exit Strategy: A preplanned process for deleting a product or product line from the firm's portfolio. At a minimum, it includes plans for clearing inventory out of the supply chain pipeline at a minimum of losses, continuing to provide for after-sales parts supply and maintenance support, and converting customers of the deleted product line to a different one. (See Chapter 34 of *The PDMA HandBook 2nd ed.*)

Explicit Customer Requirement: What the customer asks for in a product.

Extrusion: A manufacturing process that utilizes a softened billet of material that is forced through a shape (or die) to allow for a continuous form, much like spaghetti.

Factory Cost: The cost of producing the product in the production location including materials, labor and overhead.

Failure Mode Effects Analysis (FMEA): A technique used at the development stage to determine the different ways in which a product may fail, and evaluate the consequences of each type of failure.

Failure Rate: The percentage of a firm's new products that make it to full market commercialization, but which fail to achieve the objectives set for them.

Feasibility Activity: The set of product development tasks in which major unknowns are examined to produce knowledge about how to resolve or overcome them or to clarify the nature of any limitations. Sometimes called *exploratory investigations*.

Feasibility Determination: The set of product development tasks in which major unknowns (technical or market) are examined to produce knowledge about how to resolve or overcome them or to clarify the nature of any limitations. Sometimes called *exploratory investigation*.

Feature: The solution to a consumer need or problem. Features provide benefits to consumers. The handle (feature) allows a laptop computer to be carried easily (benefit). Usually any one of several different features will be chosen to meet a customer need. For example, a carrying case with shoulder straps is another feature that allows a laptop computer to be carried easily.

Feature Creep: The tendency for designers or engineers to add more capability, functions and features to a product as development proceeds than were originally intended. These additions frequently cause schedule slip, development cost increases, and product cost increases.

Feature Roadmap: The evolution over time of the performance attributes associated with a product. Defines the specific features associated with each iteration/generation of a product over its lifetime, grouped into releases (sets of features that are commercialized). See also, *Product Life-Cycle Management* and *Cadence Plans*.

Field Testing: Product use testing with users from the target market in the actual context in which the product will be used.

Financial Success: The extent to which a new product meets its profit, margin, and return on investment goals.

Firefighting: An unplanned diversion of scarce resources, and the reassignment of some of them to fix problems discovered late in a product's development cycle. (See Repenning, *JPIM*, September 2001.)

Firm-Level Success: The aggregate impact of the firm's proficiency at developing and commercializing new products. Several different specific measures may be used to estimate performance. (See Chapter 36 in *The PDMA HandBook 2nd ed.*)

First-to-Market: The first product to create a new product category or a substantial subdivision of a category.

Flexible Gate: A permissive or permeable gate in a Stage-Gate™ process that is less rigid than the traditional go-stop-recycle gate. Flexible gates are useful in shortening time to market. A permissive gate is one where the next stage is authorized, although some work in the almost-completed stage has not yet been finished. A permeable gate is one where some work in a subsequent stage is authorized before a substantial amount of work in the prior stage is completed. (Robert G. Cooper, *JPIM*, 1994.)

Focus Groups: A qualitative market research technique where 8 to 12 market participants are gathered in one room for a discussion under the leadership of a trained moderator. Discussion focuses on a consumer problem, product, or potential solution to a problem. The results of these discussions are not projectable to the general market.

Forecast: A prediction, over some defined time, of the success or failure of implementing a business plan's decisions derived from an existing strategy. (See Chapter 23 of *The PDMA HandBook 2nd ed.* and Chapter 9 of *The PDMA ToolBook 3 for New Product Development*.)

Function: (1) An abstracted description of work that a product must perform to meet customer needs. A function is something the product or service must do. (2) Term describing an internal group within which resides a basic business capability such as engineering.

Functional Elements: The individual operations that a product performs. These elements are often used to describe a product schematically.

Functional Pipeline Management: Optimizing the flow of projects through all functional areas in the context of the company's priorities.

Functional Reviews: A technical evaluation of the product and the development process from a functional perspective (such as mechanical engineering or manufacturing), in which a group of experts and peers review the product design in detail to identify weaknesses, incorporate lessons learned from past products, and make decisions about the direction of the design going forward. The technical community may perform a single review that evaluates the design from all perspectives, or individual functional departments may conduct independent reviews.

Functional Schematic: A schematic drawing that is made up of all of the functional elements in a product. It shows the product's functions as well as how material, energy, and signal flow through the product.

Functional Testing: Testing either an element of or the complete product to determine whether it will function as planned and as actually used when sold.

Fuzzy Front End: The messy getting-started period of product development, when the product concept is still very fuzzy. Preceding the more formal product development process, it generally consists of three tasks: strategic planning, concept generation, and, especially, pre-technical evaluation. These activities are often chaotic, unpredictable, and unstructured. In comparison, the subsequent new product development process is typically structured, predictable, and formal, with prescribed sets of activities, questions to be answered, and decisions to be made. (See Chapter 6 of *The PDMA HandBook 2nd ed*. and Chapter 1 of *The PDMA ToolBook1 for New Product Development*.)

Fuzzy Gates: Fuzzy gates are conditional or situational, rather than full go decisions. Their purpose is to try to balance timely decisions and risk management. Conditional go decisions are go, subject to a task being successfully completed by a future, but specified, date. Situational gates have some criteria that must be met for all projects, and others that are only required for some projects. For example, a new-to-the world product may have distribution feasibility criteria that a line extension will not have. (R.G. Cooper, *JPIM*, 1994.) (See also *Flexible Gates*.)

Gamma Test: A product use test in which the developers measure the extent to which the item meets the needs of the target customers, solves the problems(s) targeted during development, and leaves the customer satisfied.

Gamma/In-Market Testing: Not to be confused with test marketing (which is an overall determination of marketability and financial viability), the in-market test is an evaluation of the product itself and its marketing plan through placement of the product in a field setting. Another way of thinking about this is to view it as an in-market test using a real distribution channel in a constrained geographic area or two, for a specific period of time, with advertising, promotion and all associated elements of the marketing plan working. In addition to an evaluation of the features and benefits of the product, the components of the marketing plan are tested in a real-world environment to make sure they deliver the desired results. The key element being evaluated is the synergy of the product and the marketing plan, not the individual components. the market test should deliver a more accurate forecast of dollar and unit sales volume, as opposed to the approximate range estimates produced earlier in the discovery phase. it should also produce diagnostic information on any facet of the proposed launch that may need adjustment, be it product, communications, packaging, positioning, or any other element of the launch plan.

Gantt Chart: A horizontal bar chart used in project scheduling and management that shows the start date, end date, and duration of tasks within the project.

Gap Analysis: The difference between projected outcomes and desired outcomes. In product development, the gap is frequently measured as the difference between expected and desired revenues or profits from currently planned new products if the corporation is to meet its objectives.

Garage Bill Scheduling: A scheduling tool that details every task, no matter how small, that must be completed to achieve a deliverable.

Gate: The point at which a management decision is made to allow the product development project to proceed to the next stage, to recycle back into the current stage to better complete some of the tasks, or to terminate. The number of gates varies by company. (See Chapter 21 in *The PDMA HandBook 2nd ed.*)

Gatekeepers: The group of managers who serve as advisors, decision-makers and investors in a Stage-Gate™ process. Using established business criteria, this multifunctional group reviews new product opportunities and project progress, and allocates resources accordingly at each gate. This group is also commonly called a *product approval committee* or *portfolio management team.*

Graceful Degradation: When a product, system or design slides into defective operation a little at a time, while providing ample opportunity to take corrective preventative action or protect against the worst consequences of failure before it happens. The opposite is catastrophic failure.

Gross Rating Points (GRPs): A measure of the overall media exposure of consumer households (reach times frequency).

Groupware: Software designed to facilitate group efforts such as communication, workflow coordination, and collaborative problem solving. The term generally refers to technologies relying on modern computer networks (external or internal).

Growth Stage: The second stage of the product life cycle. This stage is marked by a rapid surge in sales and market acceptance for the good or service. Products that reach the growth stage have successfully crossed the chasm.

Heavyweight Team: An empowered project team with adequate resources to complete the project. Personnel report to the team leader and are co-located as practical.

Hunting for Hunting Grounds: A structured methodology for completing the fuzzy front end of new product development. (See Chapter 2 of *The PDMA ToolBook 1.*)

Hunting Ground: A discontinuity in technology or the market that opens up a new product development opportunity.

Hurdle Rate: The minimum return on investment or internal rate of return percentage a new product must meet or exceed as it goes through development.

Idea: The most embryonic form of a new product or service. It often consists of a high-level view of the envisioned solution needed to solve the problem identified by a person, team, or firm.

Idea Generation (Ideation): All of those activities and processes that lead to creating broad sets of solutions to consumer problems. These techniques may be used in the early stages of product development to generate initial product concepts, in the intermediate stages for overcoming implementation issues, in the later stages for planning launch and in the postmortem stage to better understand success and failure in the marketplace. (See Chapter 17 in *The PDMA HandBook 2nd ed.*, Chapter 9 in the *PDMA ToolBook2 for New Product Development* and Chapters 1, 4, 5, and 6 in the *PDMA ToolBook3 for New Product Development.*)

Idea Exchange: A divergent thinking technique that provides a structure for building on different ideas in a quiet, nonjudgmental setting that encourages reflection.

Idea Merit Index: An internal metric used to impartially rank new product ideas.

Implementation Team: A team that converts the concepts and good intentions of the should-be process into practical reality.

Implicit Product Requirement: What the customer expects in a product, but does not ask for, and may not even be able to articulate.

Importance Surveys: A particular type of attribute testing in which respondents are asked to evaluate how important each of the product attributes are in their choice of products or services.

Incremental Improvement: A small change made to an existing product that serves to keep the product fresh in the eyes of customers.

Incremental Innovation: An innovation that improves the conveyance of a currently delivered benefit, but produces neither a behavior change nor a change in consumption.

Individual Depth Interviews (IDIs): A qualitative market research technique in which a skilled moderator conducts an open-ended, in-depth, guided conversation with an individual respondent, as opposed to in a (focus) group format. Such an interview can be used to better understand the respondent's thought processes, motivations, current behaviors, preferences, opinions, and desires. (See Chapter 7 of *The PDMA ToolBook 2 for New Product Development.*)

Industrial Design (ID): The professional service of creating and developing concepts and specifications that optimize the function, value, and appearance of products and systems for the mutual benefit of both user and manufacturer [Industrial Design Society of America]. (See Chapters 24 and 25 of *The PDMA HandBook 2nd ed.*)

Information: Knowledge and insight, often gained by examining data.

Information Acceleration: A concept testing method employing virtual reality. In it, a virtual buying environment is created that simulates the information available (product, societal, political, and technological) in a real purchase situation at some time several years or more into the future.

Informed Intuition: Using the gathered experiences and knowledge of the team in a structured manner.

Initial Screening: The first decision to spend resources (time or money) on a project. The project is born at this point. Sometimes called *idea screening*.

Injection Molding: A process that utilizes melted plastics injected into steel or aluminum molds, which ultimately result in finished production parts.

In-licensed: The acquisition from external sources of novel product concepts or technologies for inclusion in the aggregate NPD portfolio.

Innovation: A new idea, method, or device. The act of creating a new product or process. The act includes invention as well as the work required to bring an idea or concept into final form.

Innovation-Based Culture: A corporate culture where senior management teams and employees work habitually to reinforce best practices that *systematically* and *continuously* churn out valued new products to customers. (See Chapter 1 of *The PDMA ToolBook 2 for New Product Development.*)

Innovation Engine: The creative activities and people that actually think of new ideas. It represents the synthesis phase when someone first recognizes that customer and market opportunities can be translated into new product ideas.

Innovation Steering Committee: The senior management team or a subset of it responsible for gaining alignment on the strategic and financial goals for new product development, as well as setting expectations for *portfolio and development teams*.

Innovation Strategy: The firm's positioning for developing new technologies and products. One categorization divides firms into prospectors (those who lead in technology, product and market development, and commercialization, even though an individual product may not lead to profits), analyzers (fast followers, or imitators, who let the prospectors lead, but have a product development process organized to imitate and commercialize quickly any new product a prospector has put on the market), defenders (those who stake out a product turf and protect it by whatever means, not necessarily through developing new products), and reactors (those who have no coherent innovation strategy). (See Chapter 2 of *The PDMA HandBook 2nd ed.*)

Innovative Problem Solving: Methods that combine rigorous problem definition, pattern-breaking generation of ideas, and action planning that results in new, unique, and unexpected solutions. (See Chapter 1 of *The PDMA ToolBook 3 for New Product Development.*)

Integrated Architecture: A product architecture in which most or all of the functional elements map into a single or very small number of chunks. It is difficult to subdivide an integrally designed product into partially functioning components.

Integrated Product Development (IPD): A philosophy that systematically employs an integrated team effort from multiple functional disciplines to develop effectively and efficiently new products that satisfy customer needs.

Intellectual Property (IP): Information, including proprietary knowledge, technical competencies, and design information, which provides commercially exploitable competitive benefit to an organization. (See Chapter 10 of *The PDMA ToolBook3 for New Product Development.*)

Internal Rate of Return (IRR): The discount rate at which the present value of the future cash flows of an investment equals the cost of the investment. The discount rate with a net present value of 0.

Intrapreneur: The large-firm equivalent of an entrepreneur. Someone who develops new enterprises within the confines of a large corporation.

Introduction Stage: The first stage of a product's commercial launch and the product life cycle. This stage is generally seen as the point of market entry, user trial, and product adoption.

ISO-9000: A set of five auditable standards of the International Standards Organization that establishes the role of a quality system in a company and which is used to assess whether the company can be certified as compliant to the standards. ISO-9001 deals specifically with new products.

Issue: A certainty that will affect the outcome of a project, either negatively or positively. Issues require investigation as to their potential impacts, and decisions about how to deal with them. Open issues are those for which the appropriate actions have not been resolved, while closed issues are ones that the team has dealt with successfully.

Journal of Product Innovation Management: The premier academic journal in the field of innovation, new product development and management of technology. The *Journal,* which is owned by the PDMA, is dedicated to the advancement of management practice in all of the functions involved in the total process of product innovation. Its purpose is to bring to managers and students of product innovation the theoretical structures and the practical techniques that will enable them to operate at the cutting edge of effective management practice. Web site: *www.jpim.org.*

Kaizen: A Japanese term describing a process or philosophy of continuous, incremental improvement.

Launch: The process by which a new product is introduced into the market for initial sale. (See Chapter 30 of *The PDMA HandBook 2nd ed.*)

Lead Users: Users for whom finding a solution to one of their consumer needs is so important that they have modified a current product or invented a new product to solve the need themselves because they have not found a supplier who can solve it for them. When these consumers' needs are portents of needs that the center of the market will have in the future, their solutions are new product opportunities.

Learning Organization: An organization that continuously tests and updates the experience of those in the organization, and transforms that experience into improved work processes and knowledge that is accessible to the whole organization and relevant to its core purpose. (See Continuous Learning Activity.)

Life Cycle Cost: The total cost of acquiring, owning, and operating a product over its useful life. Associated costs may include: purchase price, training expenses, maintenance expenses, warrantee costs, support, disposal, and profit loss due to repair downtime.

Lightweight Team: New product team charged with successfully developing a product concept and delivering to the marketplace. Resources are, for the most part, not dedicated, and the team depends on the technical functions for resources necessary to get the work accomplished.

Line Extension: A form of derivative product that adds or modifies features without significantly changing the product functionality.

Long-term Success: The new product's performance in the long run or at some large fraction of the product's life cycle.

M Curve: An illustration of the volume of ideas generated over a given amount of time. The illustration often looks like two arches from the letter M.

Maintenance Activity: That set of product development tasks aimed at solving initial market and user problems with the new product or service. (See Chapter 33 of *The PDMA HandBook 2nd ed.*)

Manufacturability: The extent to which a new product can be easily and effectively manufactured at minimum cost and with maximum reliability.

Manufacturing Assembly Procedure: Procedural documents normally prepared by manufacturing personnel that describe how a component, subassembly, or system will be put together to create a final product.

Manufacturing Design: The process of determining the manufacturing process that will be used to make a new product. (See Chapter 23 of *The PDMA HandBook 1st ed.*)

Manufacturing Test Specification and Procedure: Documents prepared by development and manufacturing personnel that describe the performance specifications of a component, subassembly, or system that will be met during the manufacturing process, and that describe the procedure by which the specifications will be assessed.

Market Conditions: The characteristics of the market into which a new product will be placed, including the number of competing products, level of competitiveness, and growth rate.

Market Development: Taking current products to new consumers or users. This effort may involve making some product modifications.

Market-Driven: Allowing the marketplace to direct a firm's product innovation efforts.

Market Research: Information about the firm's customers, competitors, or markets. Information may be from secondary sources (already published and publicly available) or primary sources (from customers themselves). Market research may be qualitative in nature, or quantitative (see entries for these two types of market research).

Market Segmentation: Market segmentation is defined as a framework by which to subdivide a larger heterogeneous market into smaller, more homogeneous parts. These segments can be defined in many different ways: *demographic* (men versus women, young versus old, or richer versus poorer), *behavioral* (those who buy on the phone versusversus the Internet versusversus retail, or those who pay with cash versusversus credit cards), or *attitudinal* (those who believe that store brands are just as good as national brands versus those who don't). There are many analytical techniques used to identify segments such as cluster analysis, factor analysis, or discriminate analysis. But the most common method is simply to hypothesize a potential segmentation definition and then to test whether any differences that are observed are statistically significant. (See Chapter 13 of *The PDMA HandBook 2nd ed.* and Chapter 7 of *The PDMA ToolBook 3 for New Product Development.*)

Market Share: A company's sales in a product area as a percent of the total market sales in that area.

Market Testing: The product development stage when the new product and its marketing plan are tested together. A market test simulates the eventual marketing mix and takes many different forms, only one of which bears the name *test market*. (See Chapter 32 of *The PDMA HandBook 2nd ed.*)

Mating Part: A general reference to one of two parts that join together.

Matrix Converger: A convergent thinking tool that uses a matrix to help synthesize data into key concepts with numbered ratings.

Maturity Stage: The third stage of the product life cycle. This is the stage where sales begin to level off due to market saturation. It is a time when heavy competition, alternative product options, and (possibly) changing buyer or user preferences start to make it difficult to achieve profitability.

Metrics: A set of measurements to track product development and allow a firm to measure the impact of process improvements over time. These measures generally vary by firm but may include measures characterizing both aspects of the process, such as time to market, and duration of particular process stages, as well as outcomes from product development such as the number of products commercialized per year and percentage of sales due to new products. (See Chapter 29 of *The PDMA HandBook 2nd ed.* and Chapter 16 of *The PDMA ToolBook 3 for New Product Development.*)

Modular Architecture: A product architecture in which each functional element maps into its own physical chunk. Different chunks perform different functions, the interactions between the chunks are minimal, and they are generally well-defined.

Monitoring Frequency: The frequency with which performance indicators are measured.

Morphological Analysis: A matrix tool that breaks a product down by needs met and technology components, allowing for targeted analysis and idea creation.

Multifunctional Team: A group of individuals brought together from the different functional areas of a business to work on a problem or process that requires the knowledge, training and capabilities across the areas to successfully complete the work. (See Chapters 9 and 10 in *The PDMA HandBook 2nd* ed. and Chapter 6 in *The PDMA ToolBook 1*, Chapter 5 in *The PDMA ToolBook 2*, and Chapter 13 in the *PDMA ToolBook 3 for New Product Development*.) (See also *cross-functional team*.)

Needs Statement: Summary of consumer needs and wants, described in customer terms, to be addressed by a new product. (See Chapter 14 of *The PDMA HandBook 2nd ed*.)

Net Present Value (NPV): Method to evaluate comparable investments in very dissimilar projects by discounting the current and projected future cash inflows and outflows back to the present value based on the discount rate, or cost of capital, of the firm.

Network Diagram: A graphical diagram with boxes connected by lines that shows the sequence of development activities and the interrelationship of each task with another. Often used in conjunction with a *Gantt chart*.

New Concept Development Model: A theoretical construct that provides for a common terminology and vocabulary for the fuzzy front end. The model consists of three parts: the uncontrollable influencing factors, the controllable engine that drives the activities in the fuzzy front end and five activity elements: opportunity identification, opportunity analysis, idea generation and enrichment, idea selection, and concept definition. (See Chapter 1 of *The PDMA ToolBook.)*

New Product: A term of many opinions and practices, but most generally defined as a product (either a good or service) new to the firm marketing it. Excludes products that are only changed in promotion.

New Product Development (NPD): The overall process of strategy, organization, concept generation, product and marketing plan creation and evaluation, and commercialization of a new product. Also, frequently referred to just as product development.

New Product Introduction (NPI): The launch or commercialization of a new product into the marketplace. Takes place at the end of a successful product development project. (See Chapter 30 of *The PDMA HandBook 2nd ed*.)

New Product Development Process (NPD Process): A disciplined and defined set of tasks and steps that describe the normal means by which a company repetitively converts embryonic ideas into salable products or services. (See Chapters 4 and 5 of *The PDMA HandBook 2nd ed.*)

New Product Development Professional (NPDP): A New Product Development Professional is certified by the PDMA as having mastered the body of knowledge in new product development, as proven by performance on the certification test. To qualify for the NPDP certification examination, a candidate must hold a bachelor's or higher university degree (or an equivalent degree) from an accredited institution and have spent a minimum of two years working in the new product development field. For more information, see: *http://www.pdma.org/certification/*.

New Product Idea: A preliminary plan or purpose of action for formulating new products or services.

New-to-the-World Product: A good or service that has never before been available to either consumers or producers. The automobile was new-to-the-world when it was introduced, as were microwave ovens and pet rocks.

Nominal Group Process: A brainstorming process in which members of a group first write their ideas out individually, and then participate in group discussion about each idea.

Non-Destructive Test: A test of the product that retains the product's physical and operational integrity.

Non-Product Advantage: Elements of the marketing mix that create competitive advantage other than the product itself. These elements can include marketing communications, distribution, company reputation, technical support, and associated services.

Operational Strategy: An activity that determines the best way to develop a new product while minimizing costs, ensuring adherence to schedule, and delivering a quality product. For product development, the objective is to maximize the return on investment and deliver a high quality product in the optimal market window of opportunity.

Operations: A term that includes manufacturing but is much broader, usually including procurement, physical distribution, and, for services, management of the offices or other areas where the services are provided.

Operator's Manual: The written instructions to the users of a product or process. These may be intended for the ultimate customer or for the use of the manufacturing operation.

Opportunity: A business or technology gap that a company or individual realizes, by design or accident, that exists between the current situation and an envisioned future in order to capture competitive advantage, respond to a threat, solve a problem or ameliorate a difficulty.

Outsourcing: The process of procuring a good or service from someone else, rather than firms producing it themselves.

Outstanding Corporate Innovator Award: An annual PDMA award given to firms acknowledged through a formal vetting process as being outstanding innovators. The basic requirements for receiving this award, which is given yearly by the PDMA, are: (1) Sustained success in launching new products over a five-year time frame; (2) Significant company growth from new product success; (3) A defined new product development process, that can be described to others; (4) Distinctive innovative characteristics and intangibles. For more information, see: *http://www.pdma.org/innovators/*.

Pareto Chart: A bar graph with the bars sorted in descending order used to identify the largest opportunity for improvement. Pareto charts distinguish the vital few from the useful many.

Participatory Design: A democratic approach to design that does not simply make potential users the subjects of user testing, but empowers them to be a part of the design and decision-making process. (See Chapter 13 of *The PDMA ToolBook2 for New Product Development.*)

Payback: The time, usually in years, from some point in the development process until the commercialized product or service has recovered its costs of development and marketing. While some firms take the point of full-scale market introduction of a new product as the starting point, others begin the clock at the start of development expense.

Payout: The amount of profits and their timing expected from commercializing a new product.

Perceptual Mapping: A quantitative market research tool used to understand how customers think of current and future products. Perceptual maps are visual representations of the positions that fix the product in consumers' minds. (See Chapter 7 of *The PDMA ToolBook 3 for New Product Development.*)

Performance Indicators: Criteria on which the performance of a new product in the market are evaluated. (See Chapter 29 of *The PDMA HandBook 2nd ed.*)

Performance Measurement System: The system that enables the firm to monitor the relevant performance indicators of new products in the appropriate time frame. (See Chapter 16 of *The PDMA ToolBook 3 on New Product Development.*)

Performance/Satisfaction Surveys: A particular type of market research tool in which respondents are asked to evaluate how well a particular product or service is performing and/or how satisfied they are with that product or service on a specific list of attributes. It is often useful to ask respondents to evaluate more than one product or service on these attributes in order to be able to compare them and to better understand what they like and dislike about one versus the other. In this way, this information can become a key input to the development process for next-generation product modifications.

PERT (Program Evaluation and Review Technique): An event-oriented network analysis technique used to estimate project duration when there is a high degree of uncertainty in estimates of duration times for individual activities.

Phase Review Process: A staged product development process in which first one function completes a set of tasks, then passes the information it generated sequentially to another function, which in turn completes the next, set of tasks and then passes everything along to the next function. Multifunctional teamwork is largely absent in these types of product development processes, which may also be called *baton-passing processes*. Most firms have moved from these processes to Stage-Gate™ processes using multifunctional teams.

Physical Elements: The components that make up a product. These can be both components (or individual parts) in addition to minor subassemblies of components.

Pilot Gate Meeting: A trial, informal gate meeting usually held at the launch of a Stage-Gate™ process to test the design of the process and familiarize participants with the Stage-Gate™ process.

Pipeline (product pipeline): The scheduled stream of products in development for release to the market.

Pipeline Alignment: The balancing of project demand with resource supply. (See Chapter 5 in *The PDMA HandBook 1st ed.* and Chapter 3 in *The PDMA HandBook 2nd ed.*)

Pipeline Inventory: Production of a new product that has not yet been sold to end consumers, but that exists within the distribution chain.

Pipeline Loading: The volume and time phasing of new products in various stages of development within an organization.

Pipeline Management: A process that integrates product strategy, project management, and functional management to continually optimize the cross-project management of all development-related activities. (See Chapter 5 in *The PDMA HandBook 1st ed.* and Chapter 3 in *The PDMA HandBook 2nd ed.*)

Pipeline Management Enabling Tools: The decision-assistance and data-handling tools that aid managing the pipeline. The decision-assistance tools allow the pipeline team to systematically perform trade-offs without losing sight of priorities. The data-handling tools deal with the vast amount of information needed to analyze project priorities, understand resource and skill set loads, and perform pipeline analysis.

Pipeline Management Process: Consists of three elements: pipeline management teams, a structured methodology, and enabling tools.

Pipeline Management Teams: The teams of people at the strategic, project, and functional levels responsible for resolving pipeline issues.

Platform Product: The design and components that are shared by a set of products in a product family. From this platform, numerous derivative products can be designed. (See also *product platform*.)

Platform Roadmap: A graphical representation of the current and planned evolution of products developed by the organization, showing the relationship between the architecture and features of different generations of products.

Porter's Five Forces: Analysis framework developed by Michael Porter in which a company is evaluated based on its capabilities versus competitors, suppliers, customers, barriers to entry, and the threat of substitutes. (See Porter, Michael. 1998. *Competitive Strategy*. The Free Press.)

Portfolio: Commonly referred to as a set of projects or products that a company is investing in and making strategic trade-offs against. (See also *project portfolio* and *product portfolio*.)

Portfolio Criteria: The set of criteria against which the business judges both proposed and currently active product development projects to create a balanced and diverse mix of ongoing efforts.

Portfolio Management: A business process by which a business unit decides on the mix of active projects, staffing and dollar budget allocated to each project currently being undertaken. See also *pipeline management*. (See Chapter 13 of *The PDMA ToolBook 1* and Chapter 3 of *The PDMA HandBook 2nd ed.*)

Portfolio Map: A chart or graph that graphically displays the relative scalar strength and weakness of a portfolio of products, or competitors in two orthogonal dimensions of customer value or other parameters. Typical portfolio maps include price versus performance, newness to company versus newness to market; Risk versus return.

Portfolio Rollout Scenarios: Hypothetical illustrations of the number and magnitude of new products that would need to be launched over a certain time frame to reach the desired financial goals. They account for success/failure rates and consider company and competitive benchmarks.

Portfolio Team: A short-term, cross-functional, high-powered team focused on shaping the concepts and business cases for a portfolio of new product concepts within a market, category, brand or business to be launched over a two- to five-year time period, depending on the pace of the industry.

Preliminary Bill of Materials (PBOM): A forecasted listing of all the sub-assemblies, intermediate parts, raw materials, and engineering design, tool design, and customer inputs that are expected to go into a parent assembly showing the quantity of each required to make an assembly.

Pre-Production Unit: A product that looks like and acts like the intended final product, but is made either by hand or in pilot facilities rather than by the final production process.

Process Champion: The person responsible for the daily promotion of and encouragement to use a formal business process throughout the organization. They are also responsible for the ongoing training, innovation input and continuous improvement of the process.

Process Managers: The operational managers responsible for ensuring the orderly and timely flow of ideas and projects through the process.

Process Map: A workflow diagram that uses an x-axis for process time and a y-axis that shows participants and tasks.

Process Mapping: The act of identifying and defining all of the steps, participants, inputs, outputs, and decisions associated with completing any particular process.

Process Maturity Level: The amount of movement of a reengineered process from the *as-is map*, which describes how the process operated initially, to the *should-be map* of the desired future state of the operation.

Process Owner: The executive manager responsible for the strategic results of the NPD process. This includes process throughput, quality of output, and participation within the organization. (See Section 3 of *The PDMA ToolBook* for four tools that process owners might find useful, and see Chapter 5 of *The PDMA HandBook*.)

Process Reengineering: A discipline to measure and modify organizational effectiveness by documenting, analyzing, and comparing an existing process to best-in-class practice, and then implementing significant process improvements or installing a whole new process.

Product: Term used to describe all goods, services, and knowledge sold. Products are bundles of attributes (features, functions, benefits, and uses) and can be either tangible, as in the case of physical goods, or intangible, as in the case of those associated with service benefits, or can be a combination of the two.

Product and Process Performance Success: The extent to which a new product meets its technical performance and product development process performance criteria.

Product Approval Committee (PAC): The group of managers who serve as advisors, decision makers and investors in a Stage-Gate™ process: a company's NPD executive committee. Using established business criteria, this multifunctional group reviews new product opportunities and project progress, and allocates resources accordingly at each gate. (See Chapter 7 of *The PDMA ToolBook 1* and Chapters 21 and 22 of *The PDMA HandBook 2nd ed.*)

Product Architecture: The way in which the functional elements are assigned to the physical chunks of a product and the way in which those physical chunks interact to perform the overall function of the product. (See Chapter 16 of *The PDMA HandBook 1st ed.*)

Project Decision Making and Reviews: A series of go/no-go decisions about the viability of a project that ensure the completion of the project provides a product that meets the marketing and financial objectives of the company. This includes a systematic review of the viability of a project as it moves through the various phase stage gates in the development process. These periodic checks validate that the project is still close enough to the original plan to deliver against the business case. (See Chapters 21 and 22 of *The PDMA HandBook 2nd ed.*)

Product Definition: Defines the product, including the target market, product concept, benefits to be delivered, positioning strategy, price point, and even product requirements and design specifications.

Product Development: The overall process of strategy, organization, concept generation, product and marketing plan creation and evaluation, and commercialization of a new product. (See Chapters 19 to 22 of *The PDMA HandBook 1st ed.*)

Product Development and Management Association (PDMA): A not-for-profit professional organization whose purpose is to seek out, develop, organize, and disseminate leading-edge information on the theory and practice of product development and product development processes. The PDMA uses local, national, and international meetings and conferences, educational workshops, a quarterly newsletter (*Visions*), a bi-monthly scholarly journal (*Journal of Product Innovation Management*), research proposal and dissertation proposal competitions, *The PDMA HandBook of New Product Development 1st and 2nd eds.*, and *The PDMA ToolBook 1 for New Product Development* to achieve its purposes. The association also manages the certification process for New Product Development Professionals (*www.pdma.org*).

Product Development Check List: A predetermined list of activities and disciplines responsible for completing those activities used as a guideline to ensure that all the tasks of product development are considered prior to commercialization. (See Ray Riek, *JPIM*, 2001.)

Product Development Engine: The systematic set of corporate competencies, principles, processes, practices, tools, methods and skills that combine to define the how of an organization's ability to drive high-value products to the market in a competitive timely manner.

Product Development Portfolio: The collection of new product concepts and projects that are within the firm's ability to develop, are most attractive to the firm's customers and deliver short- and long-term corporate objectives, spreading risk and diversifying investments. (See Chapter 3 in *The PDMA ToolBook 1* and Chapter 3 of Chapters 21 and 22 of *The PDMA HandBook 2nd ed.*)

Product Development Process: A disciplined and defined set of tasks, steps, and phases that describe the normal means by which a company repetitively converts embryonic ideas into salable products or services. (See Chapters 4 and 5 of *The PDMA HandBook 2nd ed.*)

Product Development Strategy: The strategy that guides the product innovation program.

Product Development Team: A multifunctional group of individuals chartered to plan and execute a new product development project.

Product Discontinuation: A product or service that is withdrawn or removed from the market because it no longer provides an economic, strategic, or competitive advantage in the firm's portfolio of offerings. (See Chapter 28 of *The PDMA HandBook 1st ed.*)

Product Discontinuation Timeline: The process and time frame in which a product is carefully withdrawn from the marketplace. The product may be discontinued immediately after the decision is made, or it may take a year or more to implement the discontinuation timeline, depending on the nature and conditions of the market and product.

Product Failure: A product development project that does not meet the objective of its charter or marketplace.

Product Family: The set of products that have been derived from a common product platform. Members of a product family normally have many common parts and assemblies.

Product Innovation Charter: A critical strategic document, the Product Innovation Charter (PIC) is the heart of any organized effort to commercialize a new product. It contains the reasons the project has been started, as well as the goals, objectives, guidelines, and boundaries of the project. It is the who, what, where, when, and why of the product development project. In the discovery phase, the charter may contain assumptions about market preferences, customer needs, and sales and profit potential. As the project enters the development phase, these assumptions are challenged through prototype development and in-market testing. While business needs and market conditions can and will change as the project progresses, one must resist the strong tendency for projects to wander off as the development work takes place. The PIC must be constantly referenced during the development phase to make sure it is still valid, that the project is still within the defined arena, and that the opportunity envisioned in the discovery phase still exists.

Product Interfaces: Internal and external interfaces impacting the product development effort, including the nature of the interface, action required, and timing.

Product Life Cycle: The four stages that a new product is thought to go through from birth to death: introduction, growth, maturity, and decline. Controversy surrounds whether products go through this cycle in any predictable way.

Product Life-Cycle Management: Changing the features and benefits of the product, elements of the marketing mix, and manufacturing operations over time to maximize the profits obtainable from the product over its lifecycle (See Chapter 33 of *The PDMA HandBook 2nd ed.*)

Product Line: A group of products marketed by an organization to one general market. The products have some characteristics, customers, and uses in common and may also share technologies, distribution channels, prices, services, and other elements of the marketing mix.

Product Management: Ensuring over time that a product or service profitably meets the needs of customers by continually monitoring and modifying the elements of the marketing mix, including the product and its features, the communications strategy, distribution channels, and price.

Product Manager: The person assigned responsibility for overseeing all of the various activities that concern a particular product. Sometimes called a *brand manager* in consumer packaged goods firms.

Product Plan: Detailed summary of the key elements involved in a new product development effort, such as product description, schedule, resources, financial estimations, and interface management plan.

Product Platforms: Underlying structures or basic architectures that are common across a group of products or that will be the basis of a series of products commercialized over a number of years.

Product Portfolio: The set of products and product lines the firm has placed in the market. (See Chapter 13 of *The PDMA ToolBook 1*.)

Product Positioning: How a product will be marketed to customers. The product positioning refers to the set of features and value that is valued by (and therefore defined by) the target customer audience, relative to competing products.

Product Rejuvenation: The process by which a mature or declining product is altered, updated, repackaged, or redesigned to lengthen the product life cycle and, in turn, extend sales demand.

Product Requirements Document: The contract between, at a minimum, marketing and development, describing completely and unambiguously the necessary attributes (functional performance requirements) of the product to be developed, as well as information about how achievement of the attributes will be verified (i.e., through testing).

Product Superiority: Differentiation of a firm's products from those of competitors, achieved by providing consumers with greater benefits and value. This is one of the critical success factors in commercializing new products.

Program Manager: The organizational leader charged with responsibility of executing a portfolio of NPD projects. (See Section 4 of *The PDMA ToolBook 1* for 4 product development tools a program manager may find helpful.)

Project Leader: The person responsible for managing an individual new product development project through to completion. He or she is responsible for ensuring that milestones and deliverables are achieved and that resources are utilized effectively. See also *team leader*. (See Sections 1 and 2 of *The PDMA ToolBook 1* for 8 product development tools for project leaders.)

Project Management: The set of people, tools, techniques, and processes used to define the project's goal, plan all the work necessary to reach that goal, lead the project and support teams, monitor progress, and ensure that the project is completed in a satisfactory way.

Project Pipeline Management: Fine-tuning resource deployment smoothly for projects during ramp-up, ramp-down, and mid-course adjustments.

Project Plan: A formal, approved document used to guide both project execution and control. Documents planning assumptions and decisions, facilitates communication among stakeholders, and documents approved scope, cost, and schedule deadlines.

Project Portfolio: The set of projects in development at any point in time. These will vary in the extent of newness or innovativeness. (See Chapter 13 in *The PDMA ToolBook 1* and Chapter 3 of *The PDMA HandBook 2nd ed.*)

Project Resource Estimation: This activity provides one of the major contributions to the project cost calculation. Turning functional requirements into a realistic cost estimate is a key factor in the success of a product delivering against the business plan.

Project Sponsor: The authorization and funding source of the project. The person who defines the project goals and to whom the final results are presented. This is typically a senior manager.

Project Strategy: The goals and objectives for an individual product development project. It includes how that project fits into the firm's product portfolio, who the target market is, and what problems the product will solve for those customers. (See Chapter 2 in *The PDMA HandBook 2nd ed.*)

Project Team: A multifunctional group of individuals chartered to plan and execute a new product development project.

Prospectors: Firms that lead in technology, product and market development, and commercialization, even though an individual product may not lead to profits. Their general goal is to be first to market with any particular innovation.

Protocol: A statement of the attributes (mainly benefits; features only when required) that a new product is expected to have. A protocol is prepared prior to assigning the project to the technical development team. The benefits statement is agreed to by all parties involved in the project.

Prototype: A physical model of the new product concept. Depending upon the purpose, prototypes may be nonworking, functionally working, or both functionally and aesthetically complete.

Psychographics: Characteristics of consumers that, rather than being purely demographic, measure their attitudes, interests, opinions, and lifestyles.

Pull-Through: The revenue created when a new product or service positively impacts the sales of other, existing products or services (the obverse of cannibalization.)

Q-Sorts: A process for sorting and ranking complex issues.

Qualitative Cluster Analysis: An individual- or group-based process using informed intuition for clustering and connecting data points.

Qualitative Marketing Research: Research conducted with a very small number of respondents, either in groups or individually, to gain an impression of their beliefs, motivations, perceptions, and opinions. Frequently used to gather initial consumer needs and obtain initial reactions to ideas and concepts. Results are not representative of the market in general, and they are not projectable. Qualitative marketing research is used to show why people buy a particular product, whereas quantitative marketing research reveals how many people buy it. (See Chapters 14 to 16 of *The PDMA HandBook 2nd ed.*, Chapters 7 and 8 of *The PDMA ToolBook 2 for New*

Product Development, and Chapters 4 and 5 of *The PDMA TookBook 3 for New Product Development*.)

Quality: The collection of attributes, which, when present in a product, means a product has conformed to or exceeded customer expectations.

Quality Assurance/Compliance: Function responsible for monitoring and evaluating development policies and practices, to ensure they meet company and applicable regulatory standards.

Quality-by-Design: The process used to design quality into the product, service, or process from the inception of product development.

Quality Control Specification and Procedure: Documents that describe the specifications and the procedures by which they will be measured which a finished subassembly or system must meet before judged ready for shipment.

Quality Function Deployment (QFD): A structured method employing matrix analysis for linking what the market requires to how it will be accomplished in the development effort. This method is most frequently used during the stage of development when a multifunctional team agrees on how customer needs relate to product specifications and the features that deliver those needs. By explicitly linking these aspects of product design, QFD minimizes the possibility of omitting important design characteristics or interactions across design characteristics. QFD is also an important mechanism in promoting multifunctional teamwork. Developed and introduced by Japanese auto manufacturers, QFD is widely used in the automotive industry. (See Chapter 2 of *The PDMA ToolBook3 for New Product Development*.)

Quantitative Market Research: Consumer research, often surveys, conducted with a large enough sample of consumers to produce statistically reliable results that can be used to project outcomes to the general consumer population. Used to determine importance levels of different customer needs, performance ratings of and satisfaction with current products, probability of trial, repurchase rate, and product preferences. These techniques are used to reduce the uncertainty associated with many other aspects of product development. (See Chapter 18 of *The PDMA HandBook 2nd ed.* and Chapters 3 and 4 of *The PDMA ToolBook 3 for New Product Development*.)

Radical Innovation: A new product, generally containing new technologies, that significantly changes behaviors and consumption patterns in the marketplace. (See Chapter 2 of *The PDMA ToolBook2 for New Product Development*.)

Rapid Prototyping: Any of a variety of processes that avoid tooling time in producing prototypes or prototype parts and therefore allow (generally nonfunctioning) prototypes to be produced within hours or days, rather than weeks. These prototypes are frequently used to test quickly the product's technical feasibility or consumer interest.

Reactors: Firms that have no coherent innovation strategy. They only develop new products when absolutely forced to by the competitive situation.

Realization Gap: The time between first perception of a need and the launch of a product that fills that need.

Relay-Race Process: A staged product development process in which first one function completes a set of tasks, then passes the information they generate sequentially to another function, which in turn completes the next set of tasks and then passes everything along to the next function. Multifunctional teamwork is largely absent in these types of product development processes, which may also be called *phase review* or *baton-passing processes.*

Render: Process that industrial designers use to visualize their ideas by putting their thoughts on paper with any number of combinations of color markers, pencils and highlighters, or computer visualization software.

Reposition: To change the position of the product in the minds of customers, either on failure of the original positioning or to react to changes in the marketplace. Most frequently accomplished through changing the marketing mix rather than redeveloping the product.

Return on Ideas: Reflects the potential value of an idea.

Resource Matrix: An array that shows the percentage of each nonmanagerial person's time that is to be devoted to each of the current projects in the firm's portfolio.

Resource Plan: Detailed summary of all forms of resources required to complete a product development project, including personnel, equipment, time, and finances.

Responsibility Matrix: This matrix indicates the specific involvement of each functional department or individual in each task or activity in each stage.

Return on Investment (ROI): A standard measure of project profitability, this is the discounted profits over the life of the project expressed as a percentage of initial investment.

Rigid Gate: A review point in a Stage-Gate™ process at which all the prior stage's work and deliverables must be complete before work in the next stage can commence.

Risk: An event or condition that may or may not occur, but if it does occur will impact the ability to achieve a project's objectives. In new product development, risks may take the form of market, technical, or organizational issues. For more on managing product development risks. (See Chapters 8 and 15 in the *PDMA ToolBook 1* and Chapter 28 in *The PDMA HandBook 2nd ed.*)

Risk Acceptance: An uncertain event or condition for which the project team has decided not to change the project plan. A team may be forced to accept an identified risk when they are unable to identify any other suitable response to the risk.

Risk Avoidance: Changing the project plan to eliminate a risk or to protect the project objectives from any potential impact due to the risk.

Risk Management: The process of identifying, measuring, and mitigating the business risk in a product development project.

Risk Mitigation: Actions taken to reduce the probability and/or impact of a risk to below some threshold of acceptability.

Risk Tolerance: The level of risk that a project stakeholder is willing to accept. Tolerance levels are context specific. That is, stakeholders may be willing to accept different levels of risk for different types of risk, such as risks of project delay, price realization, and technical potential.

Risk Transference: Actions taken to shift the impact of a risk and the ownership of the risk response actions to a third party.

Roadmapping: A graphical multistep process to forecast future market and/or technology changes, and then plan the products to address these changes.

Robust Design: The design of products to be less sensitive to variations, including manufacturing variation and misuse, increasing the probability that they will perform as intended.

Rugby Process: A product development process in which stages are partially or heavily overlapped rather than sequential with crisp demarcations between one stage and its successor.

S-Curve (Technology S-Curve): Technology performance improvements tend to progress over time in the form of an S curve. When first invented, technology performance improves slowly and incrementally. Then, as experience with a new technology accrues, the rate of performance increase grows and technology performance increases by leaps and bounds. Finally, some of the performance limits of a new technology start to be reached and performance growth slows. At some point, the limits of the technology may be reached and further improvements are not made. Frequently, the technology then becomes vulnerable to a substitute technology that is capable of making additional performance improvements. The substitute technology is usually on the lower, slower portion of its own S curve and quickly overtakes the original technology when performance accelerates during the middle (vertical) portion of the S.

Scanner Test Markets: Special test markets that provide retail point-of-sale scanner data from panels of consumers to help assess the product's performance. First widely applied in the supermarket industry.

Scenario Analysis: A tool for envisioning alternate futures so that a strategy can be formulated to respond to future opportunities and challenges. (See Chapter 16 of the *PDMA ToolBook1 for New Product Development*.)

Screening: The process of evaluating and selecting new ideas or concepts to put into the project portfolio. Most firms now use a formal screening process with evaluation criteria that span customer, strategy, market, profitability, and feasibility dimensions.

Segmentation: The process of dividing a large and heterogeneous market into more homogeneous subgroups. Each subgroup, or segment, holds similar views about the product, and values, purchases, and uses the product in similar ways. (See Chapter 13 of *The PDMA HandBook 2nd Edition* and Chapter 7 of *The PDMA ToolBook3 for New Product Development*.)

Senior Management: That level of executive or operational management above the product development team that has approval authority or controls resources important to the development effort.

Sensitivity Analysis: A calculation of the impact that an uncertainty might have on the new product business case. It is conducted by setting upper and lower ranges on the assumptions involved and calculating the expected outcomes. (See Chapter 16 of *The PDMA ToolBook1 for New Product Development*.)

Services: Products, such as an airline flight or insurance policy, which are intangible or at least substantially so. If totally intangible, they are exchanged directly from producer to user, cannot be transported or stored and are instantly perishable. Service delivery usually involves customer participation in some important way. Services cannot be sold in the sense of ownership transfer, and they have no title of ownership.

Short-Term Success: The new product's performance shortly after launch, well within the first year of commercial sales.

Should-Be Map: A version of a process map depicting how a process will work in the future. A revised as-is process map. The result of the team's reengineering work.

Simulated Test Market: A form of quantitative market research and pretest marketing in which consumers are exposed to new products and to their claims in a staged advertising and purchase situation. Output of the test is an early forecast of expected sales or market share, based on mathematical forecasting models, management assumptions, and input of specific measurements from the simulation.

Six Sigma: A level of process performance that produces only 3.4 defects for every one million operations.

Slip Rate: Measures the accuracy of the planned project schedule according to the formula: Slip rate = ([Actual schedule/Planned schedule] − 1) * 100 percent.

Specification: A detailed description of the features and performance characteristics of a product. For example, a laptop computer's specification may read as a 90 megahertz Pentium, with 16 megabytes of RAM and 720 megabytes of hard disk space, 3.5 hours of battery life, weight of 4.5 pounds, with an active matrix 256 color screen.

Speed to Market: The length of time it takes to develop a new product from an early initial idea for a new product to initial market sales. Precise definitions of the start and end point vary from one company to another, and may vary from one project to another within a company. (See Chapter 12 of *The PDMA HandBook 2nd ed.*)

Sponsor: An informal role in a product development project, usually performed by a higher-ranking person in the firm who is not directly involved in the project, but who is ready to extend a helping hand if needed, or provide a barrier to interference by others.

Stage: One group of concurrently accomplished tasks, with specified outcomes and deliverables, of the overall product development process.

Stage-Gate™ Process: A widely employed product development process that divides the effort into distinct time-sequenced stages separated by management decision gates. Multifunctional teams must successfully complete a prescribed set of related cross-functional tasks in each stage prior to obtaining management approval to proceed to the next stage of product development. The framework of the Stage-Gate™ process includes work-flow and decision-flow paths and defines the supporting systems and practices necessary to ensure the process's ongoing smooth operation.

Staged Product Development Activity: The set of product development tasks commencing when it is believed that there are no major unknowns and that result in initial production of salable product, carried out in stages.

Standard Cost: See *factory cost*.

Stop-light Voting: A convergent thinking technique by which participants vote their idea preferences using colored adhesive dots. Also called *preference voting*.

Strategic Balance: Balancing the portfolio of development projects along one or more of many dimensions such as focus versus diversification, short versus long term, high versus low risk, extending platforms versus development of new platforms.

Strategic New Product Development (SNPD): The process that ties new product strategy to new product portfolio planning. (See Chapter 2 of both editions of *The PDMA HandBook*.)

Strategic Partnering: An alliance or partnership between two firms (frequently one large corporation and one smaller, entrepreneurial firm) to create a specialized new product. Typically, the large firm supplies capital, and the necessary product development, marketing, manufacturing, and distribution capabilities, while the small firm supplies specialized technical or creative expertise. (See Chapter 6 of *The PDMA ToolBook2 for New Product Development*.)

Strategic Pipeline Management: Strategic balancing, which entails setting priorities among the numerous opportunities and adjusting the organization's skill sets to deliver products.

Strategic Plan: Establishes the vision, mission, values, objectives, goals, and strategies of the organization's future state.

Strategy: The organization's vision, mission, and values. One subset of the firm's overall strategy is its innovation strategy.

Subassembly: A collection of components that can be put together as a single assembly to be inserted into a larger assembly or final product. Often, the subassembly is tested for its ability to meet some set of explicit specifications before inclusion in the larger product.

Success: A product that meet's its goals and performance expectations. Product development success has four dimensions. At the project level, there are three dimensions: financial, customer-based, and product technical performance. The fourth dimension is new product contribution to overall firm success. (See Chapters 1, 29, 31, 35 and 36 of *The PDMA HandBook 2nd ed.*)

Success Dimensions: Product development success has four dimensions. At the project level, there are three dimensions: financial, customer-based, and product and process performance. The fourth dimension of product development success is measured at the firm level.

Support Service: Any organizational function whose primary purpose is not product development but whose input is necessary to the successful completion of product development projects.

SWOT Analysis: Strengths, weaknesses, opportunities, and threats analysis. A SWOT analysis evaluates a company in terms of its advantages and disadvantages versus competitors, customer requirements, and market/economic environmental conditions.

System Hierarchy Diagram: The diagram used to represent product architectures. This diagram illustrates how the product is broken into its chunks.

Systems and Practices: Established methods, procedures, and activities that either drive or hinder product development. These may relate to the firm's day-to-day business or may be specific to product development.

Systems and Practices Team: Senior managers representing all functions who work together to identify and change those systems and practices hindering product development and who establish new tools, systems, and practices for improving product development.

Task: The smallest describable unit of accomplishment in completing a deliverable.

Target Cost: A cost objective established for a new product based on consideration of customer affordability. Target cost is treated as an independent variable that must be satisfied along with other customer requirements.

Target Market: The group of consumers or potential customers selected for marketing. This market segment is most likely to buy the products within a given category. These are sometimes called *prime prospects*.

Team: That group of persons who participate in the product development project. Frequently each team member represents a function, department, or specialty. Together they represent the full set of capabilities needed to complete the project. (See Chapter 9 in *The PDMA HandBook 2nd ed.* and Chapter 6 in *The PDMA ToolBook 1.*)

Team Leader: The person leading the new product team. Responsible for ensuring that milestones and deliverables are achieved, but may not have any authority over project participants. (See Sections 1 and 2 of *The PDMA ToolBook* for 8 product development tools for team leaders.)

Team Spotter's Guide: A questionnaire used by a team leader (or team members) to diagnose the quality of the team's functioning. (See Chapter 6 in the *PDMA ToolBook 1.*)

Technology-Driven: A new product or new product strategy based on the strength of a technical capability. Sometimes called solutions in search of problems.

Technology Road Map: A graphic representation of technology evolution or technology plans mapped against time. It is used to guide new technology development for or technology selection in developing new products.

Technology Stage Gate (TSG): A process for managing the technology development efforts when there is high uncertainty and risk. The process brings a structured methodology for managing new technology development without thwarting the creativity needed in this early stage of product development. It is specifically intended to manage high-risk technology development projects when there is uncertainty and risk that the technology discovery may never occur and therefore the ultimate desired product characteristics might never be achieved. (See Chapter 11 in *The PDMA ToolBook 1.*)

Technology Transfer: The process of converting scientific findings from research laboratories into useful products by the commercial sector. May also be referred to as the process of transferring technology between alliance partners.

Test Markets: The launching of a new product into one or more limited geographic regions in a very controlled manner, and measuring consumer response to the product and its launch. When multiple geographies are used in the test, different advertising or pricing policies may be tested and the results compared.

Think Links: Stimuli used in divergent thinking to help participants make new connections using seemingly unrelated concepts from a list of people, places, or things.

Think-Tank: Environments, frequently isolated from normal organizational activities, created by management to generate new ideas or approaches to solving organizational problems.

Thought Organizers: Tools that help categorize information associated with ideas such that the ideas can be placed into groups that can be more easily compared or evaluated.

Three Rs: The fundamental steps of *record, recall,* and *reconstruct* that most creative minds go through when generating new product ideas.

Threshold Criteria: The minimum acceptable performance targets for any proposed product development project.

Thumbnail: The most minimal form of sketching, usually using pencils, to represent a product idea.

Time to Market: The length of time it takes to develop a new product from an early initial idea for a new product to initial market sales. Precise definitions of the start and end point vary from one company to another, and may vary from one project to another within the company.

Tone: The feeling, emotion, or attitude most associated with using a product. The appropriate tone is important to include in consumer new product concepts and advertising.

Tornado: A mid- to late growth stage strategy that follows the bowling alley and that describes an often-frenzied period of rapid growth and acceptance for a product category. Activities of the tornado phase include commoditization of a product to become an industry standard, competitive pricing to maximize share and low-cost volume distribution channels. Success in the tornado is related to maintaining previously established product leadership and complementing it with operational excellence in a variety of strategic areas.

Total Quality Management (TQM): A business improvement philosophy that comprehensively and continuously involves all of an organization's functions in improvement activities.

Tracking Studies: Surveys of consumers (usually conducted by telephone) following the product's launch to measure consumer awareness, attitudes, trial, adoption, and repurchase rates.

TRIZ: The acronym for the *Theory of Inventive Problem Solving*, which is a Russian, systematic method of solving problems and creating multiple-alternative solutions. It is based on an analysis and codification of technology solutions from millions of patents. The method enhances creativity by getting individuals to think beyond their own experience and to reach across disciplines to solve problems using solutions from other areas of science. (See Chapter 1 of *The PDMA ToolBook 3 for New Product Development*.)

Uncertainty Range: The spread between the high (best case) and low (worst case) values in a business assumption. (See Chapter 9 of *The PDMA ToolBook 3 for New Product Development*.)

User: Any person who uses a product or service to solve a problem or obtain a benefit, whether or not they purchase it. Users may consume a product, as in the case of a person using shampoo to clean his or her hair or eating a potato chip to assuage hunger between meals. Users may not directly consume a product, but may interact with it over a longer period of time, like a family owning a car, with multiple family members using it for many purposes over a number of years. Products also are employed in the production of other products or services, where the users may be the manufacturing personnel who operate the equipment.

Utilities: The weights derived from conjoint analysis that measure how much a product feature contributes to purchase interest or preference.

Value: Any principle to which a person or company adheres with some degree of emotion. It is one of the elements that enter into formulating a strategy.

Value-added: The act or process by which tangible product features or intangible service attributes are bundled, combined or packaged with other features and attributes to create a competitive advantage, reposition a product or increase sales.

Value Analysis: A technique for analyzing systems and designs. Its purpose is to help develop a design that satisfies users by providing the needed user requirements in sufficient quality at an optimum (minimum) cost.

Value Chain: As a product moves from raw material to finished good delivered to the customer, value is added at each step in the manufacturing and delivery process. The value chain indicates the relative amount of value added at each of these steps.

Value Proposition: A short, clear, and simple statement of how and on what dimensions a product concept will deliver value to prospective customers. The essence of value is embedded in the trade-off between the benefits a customer receives from a new product and the price a customer pays for it. (See Chapter 3 of the *PDMA ToolBook 1 for New Product Development*.)

Vertical Integration: A firm's operation across multiple levels of the value chain. In the early 1900s, Ford Motor Company was extremely vertically integrated, as it owned forests and operated logging and wood finishing and glass-making businesses. They made all of the components that went into automobiles, as well as most of the raw materials used in those components.

Virtual Customer: A set of Web-based market research methods for gathering Voice of the Customer data in all phases of product development. (See Dahan and Hauser, *JPIM*, July 2002.)

Virtual Product Development: Paperless product development. All design and analysis is computer-based.

Virtual Reality: Technology that enables a designer or user to enter and navigate a computer-generated 3-D environment. Users can change their viewpoint and interact with the objects in the scene in a way that simulates real-world experiences.

Virtual Team: Dispersed teams that communicate and work primarily electronically may be called virtual teams. (See Chapter 5 of *The PDMA ToolBook 2 for New Product Development*.)

Vision: An act of imagining, guided by both foresight and informed discernment, that reveals the possibilities as well as the practical limits in new product development. It depicts the most desirable, future state of a product or organization.

Visionary Companies: Leading innovators in their industries, they rank first or second in market share, profitability, growth, and shareholder performance. A substantial portion (e.g., 30 percent or more) of their sales are from products introduced in the last three years. Many firms want to benchmark these firms.

Visions: The new product development practitioner-oriented magazine of the PDMA.

Voice of the Customer (VOC): A process for eliciting needs from consumers that uses structured in-depth interviews to lead interviewees through a series of situations in which they have experienced and found solutions to the set of problems being investigated. Needs are obtained through indirect questioning by coming to understand how the consumers found ways to meet their needs, and, more important, why they chose the particular solutions they found. (See Chapter 4 of *The PDMA ToolBook 2 for New Product Development*.)

Waste: Any activity that utilizes equipment, materials, parts, space, employee time, or other corporate resource beyond minimum amount required for value-added operations to ensure manufacturability. These activities could include waiting, accumulating semi-processed parts, reloading, passing materials from one hand to the other, and other nonproductive processes. The seven basic categories of waste that a business should strive to eliminate: overproduction, waiting for machines, transportation time, process time, excess inventory, excess motion, and defects.

Whole Product: A product definition concept that emphasizes delivering all aspects of a product that are required for it to deliver its full value. This would include training materials, support systems, cables, how to recipes, additional hardware/software, standards and procedures, implementation, applications consulting—any constitutive elements necessary to assure the customer will have a successful experience and achieve at least minimum required value from the product. Often elements of the whole product are provided via alliances with others. This term is most often used in the context of planning high technology products.

Workflow Design Team: Functional contributors who work together to create and execute the work-flow component of a Stage-Gate™ system. They decide how the firm's Stage-Gate™ process will be structured, what tasks it will include, what decision points will be included and who is involved at all points.

Workplan: Detailed plan for executing the project, identifying each phase of the project, the major steps associated with them, and the specific tasks to be performed along the way. Best practice workplans identify the specific functional resources assigned to each task, the planned task duration, and the dependencies between tasks. See also *Gantt chart*.

Worth What Paid For (WWPF): The quantitative evaluation by a person in your customer segment of the question: Considering the products and services that your vendor offers, are they worth what you paid for them?

Acknowledgment: Some of the definitions for terms in this glossary have been adapted from the glossary in *New Products Management*, by C. Merle Crawford and C. Anthony Di Benedetto. Terms, phrases, and definitions have generously been contributed to this list by the PDMA Board of Directors, the design teams for the PDMA Body of Knowledge, the editors and authors of *The PDMA ToolBooks 1, 2, and 3 for New Product Development* (John Wiley & Sons, 2002, 2004, 2007), the editors and authors of *The PDMA Handbook of New Product Development,* both 1st and 2nd editions (John Wiley & Sons, 1996, 2005) and several other individuals knowledgeable in the science, skills, and art of new product development. We thank all of these volunteer contributors for their continuing support.

Index